COMMUNICATIONS SATELLITE SYSTEMS

A James Martin BOOK

Prentice-Hall
Series in Automatic Computation

AHO, ed., *Currents in the Theory of Computing*
AHO and ULLMAN, *The Theory of Parsing, Translation, and Compiling,*
Volume I: *Parsing;* Volume II: *Compiling*
ANDREE, *Computer Programming: Techniques, Analysis, and Mathematics*
ANSELONE, *Collectively Compact Operator Approximation Theory and Applications to Integral Equations*
AVRIEL, *Nonlinear Programming: Analysis and Methods*
BENNETT, JR., *Scientific and Engineering Problem-Solving with the Computer*
BLAAUW, *Digital System Implementation*
BLUMENTHAL, *Management Information Systems*
BRENT, *Algorithms for Minimization without Derivatives*
BRINCH HANSEN, *The Architecture of Concurrent Programs*
BRINCH HANSEN, *Operating System Principles*
BRZOZOWSKI and YOELL, *Digital Networks*
COFFMAN and DENNING, *Operating Systems Theory*
CRESS, et al., *FORTRAN IV with WATFOR and WATFIV*
DAHLQUIST, BJÖRCK, and ANDERSON, *Numerical Methods*
DANIEL, *The Approximate Minimization of Functionals*
DEO, *Graph Theory with Applications to Engineering and Computer Science*
DESMONDE, *Computers and Their Uses,* 2nd ed.
DIJKSTRA, *A Discipline of Programming*
DRUMMOND, *Evaluation and Measurement Techniques for Digital Computer Systems*
ECKHOUSE, *Minicomputer Systems: Organization and Programming (PDP-11)*
FIKE, *Computer Evaluation of Mathematical Functions*
FIKE, *PL/1 for Scientific Programmers*
FORSYTHE, MALCOLM, and MOLER, *Computer Methods for Mathematical Computations*
FORSYTHE and MOLER, *Computer Solution of Linear Algebraic Systems*
GEAR, *Numerical Initial Value Problems in Ordinary Differential Equations*
GILL, *Applied Algebra for the Computer Sciences*
GORDON, *System Simulation*
GRISWOLD, *String and List Processing in SNOBOL4: Techniques and Applications*
HANSEN, *A Table of Series and Products*
HARTMANIS and STEARNS, *Algebraic Structure Theory of Sequential Machines*
HILBURN and JULICH, *Microcomputers/Microprocessor; Hardware, Software, and Applications*
HUGHES and MICHTOM, *A Structured Approach to Programming*
JACOBY, et al., *Iterative Methods for Nonlinear Optimization Problems*
JOHNSON, *System Structure in Data, Programs, and Computers*
KIVIAT, et al., *The SIMSCRIPT II Programming Language*
LAWSON and HANSON, *Solving Least Squares Problems*
LORIN, *Parallelism in Hardware and Software: Real and Apparent Concurrency*
LOUDEN and LEDIN, *Programming the IBM 1130,* 2nd ed.
MARTIN, *Communications Satellite Systems*
MARTIN, *Computer Data-Base Organization,* 2nd ed.
MARTIN, *Design of Man-Computer Dialogues*
MARTIN, *Design of Real-Time Computer Systems*

MARTIN, *Future Developments in Telecommunications*, 2nd ed.
MARTIN, *Introduction to Teleprocessing*
MARTIN, *Principles of Data-Base Management*
MARTIN, *Programming Real-Time Computing Systems*
MARTIN, *Security, Accuracy, and Privacy in Computer Systems*
MARTIN, *Systems Analysis for Data Transmission*
MARTIN, *Telecommunications and the Computer*, 2nd ed.
MARTIN, *Teleprocessing Network Organization*
MARTIN and NORMAN, *The Computerized Society*
MCKEEMAN, et al., *A Compiler Generator*
MEYERS, *Time-Sharing Computation in the Social Sciences*
MINSKY, *Computation: Finite and Infinite Machines*
NIEVERGELT, et al., *Computer Approaches to Mathematical Problems*
PLANE and MCMILLAN, *Discrete Optimization: Integer Programming and Network Analysis for Management Decisions*
POLIVKA and PAKIN, *APL: The Language and Its Usage*
PRITSKER and KIVIAT, *Simulation with GASP II: A FORTRAN-based Simulation Language*
PYLYSHYN, ed., *Perspectives on the Computer Revolution*
RICH, *Internal Sorting Methods Illustrated with PL/1 Programs*
RUDD, *Assembly Language Programming and the IBM 360 and 370 Computers*
SACKMAN and CITRENBAUM, eds., *On-Line Planning: Towards Creative Problem-Solving*
SALTON, ed., *The SMART Retrieval System: Experiments in Automatic Document Processing*
SAMMET, *Programming Languages: History and Fundamentals*
SCHAEFER, *A Mathematical Theory of Global Program Optimization*
SCHULTZ, *Spline Analysis*
SCHWARZ, et al., *Numerical Analysis of Symmetric Matrices*
SHAH, *Engineering Simulation Using Small Scientific Computers*
SHAW, *The Logical Design of Operating Systems*
SHERMAN, *Techniques in Computer Programming*
SIMON and SIKLOSSY, eds., *Representation and Meaning: Experiments with Information Processing Systems*
STERBENZ, *Floating-Point Computation*
STOUTEMYER, *PL/1 Programming for Engineering and Science*
STRANG and FIX, *An Analysis of the Finite Element Method*
STROUD, *Approximate Calculation of Multiple Integrals*
TANENBAUM, *Structured Computer Organization*
TAVISS, ed., *The Computer Impact*
UHR, *Pattern Recognition, Learning, and Thought: Computer-Programmed Models of Higher Mental Processes*
VAN TASSEL, *Computer Security Management*
VARGA, *Matrix Iterative Analysis*
WAITE, *Implementing Software for Non-Numeric Application*
WILKINSON, *Rounding Errors in Algebraic Processes*
WIRTH, *Algorithms + Data Structures = Programs*
WIRTH, *Systematic Programming: An Introduction*
YEH, ed., *Applied Computation Theory: Analysis, Design, Modeling*

COMMUNICATIONS

The views expressed in this book are those of the author and do not necessarily reflect those of the IBM Corporation.

SATELLITE SYSTEMS

JAMES MARTIN

PRENTICE-HALL, INC., Englewood Cliffs, New Jersey 07632

Library of Congress Cataloging in Publication Data

Martin, James (date)
Communications satellite systems.

Includes bibliographies and index.
1. Artificial satellites in telecommunication.
I. Title.
TK5104.M37 621.38'0422 78-5247
ISBN 0-13-153163-8

Communications Satellite Systems
James Martin

Englewood Cliffs, N.J. 07632

Printed in the United States of America

10 9 8 7 6 5 4 3

PRENTICE-HALL INTERNATIONAL, INC., *London*
PRENTICE-HALL OF AUSTRALIA PTY. LIMITED, *Sydney*
PRENTICE-HALL OF CANADA, LTD., *Toronto*
PRENTICE-HALL OF INDIA PRIVATE LIMITED, *New Delhi*
PRENTICE-HALL OF JAPAN, INC., *Tokyo*
PRENTICE-HALL OF SOUTHEAST ASIA PTE. LTD., *Singapore*
WHITEHALL BOOKS LIMITED, *Wellington, New Zealand*

TO CHARITY

CONTENTS

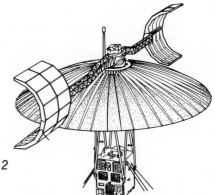

PART II

THE GROUND SEGMENT

History of early communications satellites

1958: SCORE (NASA)

A 150-pound satellite rebroadcast an on-board tape recording of President Eisenhower's Christmas message. Orbit height, 110 to 920 miles.

1960: ECHO (NASA)

A 100-foot-diameter plastic balloon with an aluminum coating which passively reflected radio signals transmitted from a huge earth antenna. Orbit height, 1000 miles; ECHO II, 600 to 800 miles.

1960: COURIER (Department of Defense)

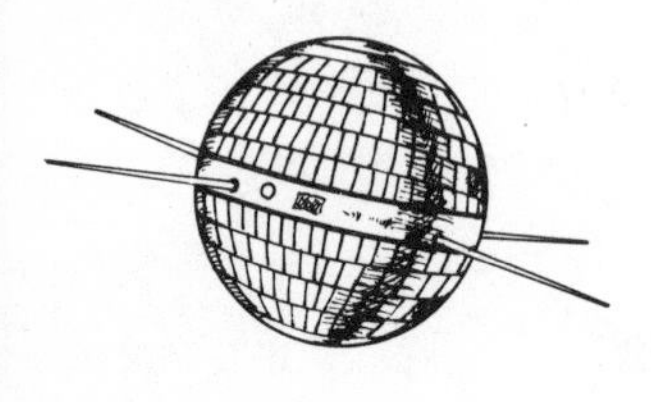

The first active radio repeater satellite. It accepted and stored up to 360,000 teletype words as it passed overhead and rebroadcast them to ground stations further along its orbit. It operated for 17 days with 3 watts of output power. 600- to 700-mile orbit.

1962: TELSTAR (AT&T)

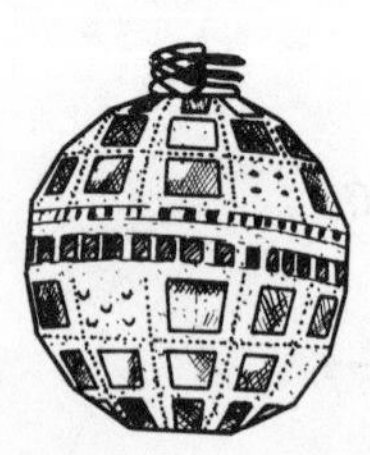

The first satellite to receive and transmit simultaneously. 4/6 GHz. Used for telephone, television, facsimile, and data. 3 watts of output power. Orbit height, 682 to 4030 miles.

1962: RELAY (RCA and NASA)

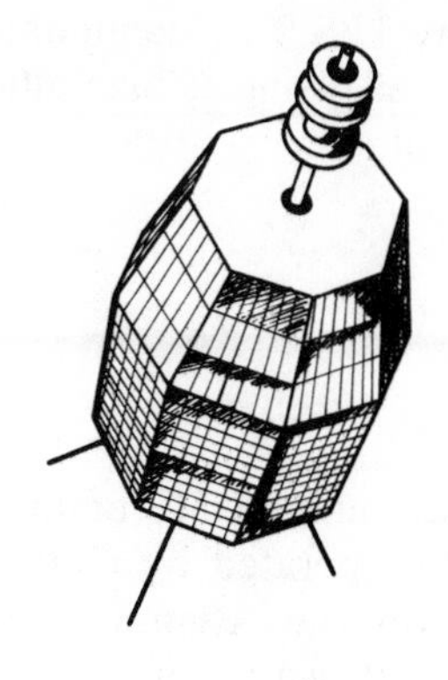

4.2/1.7-GHz satellite of 10 watts output power. Orbit height, 942 to 5303 miles.

(U.S. Air Force)
1963: PROJECT WEST FORD

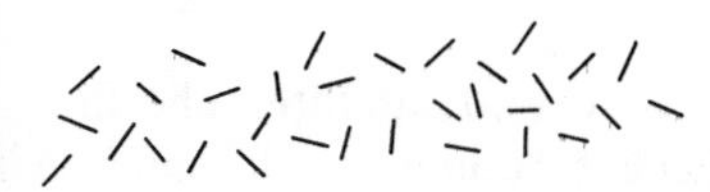

An orbital belt of small needles was launched (to the fury of some radio astronomers), 2300 miles high, to act as passive radio reflectors. Speech in digitized form was transmitted intelligibly.

1963: SYNCOM (NASA)

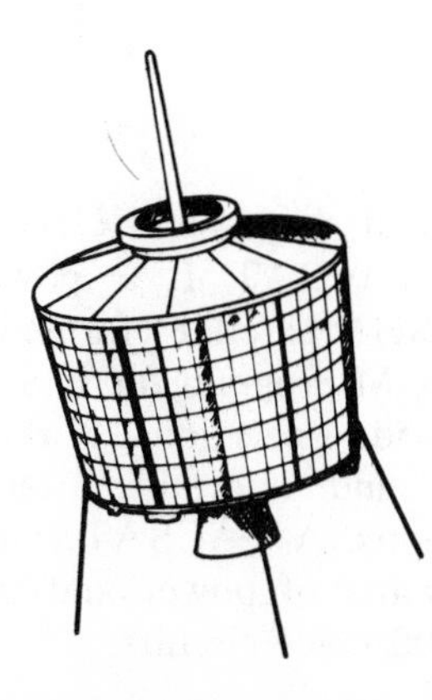

The first communications satellite in geosynchronous orbit. Used for many experiments. Transmitted television of the Tokyo Olympic Games in 1964.

1965: MOLNIYA

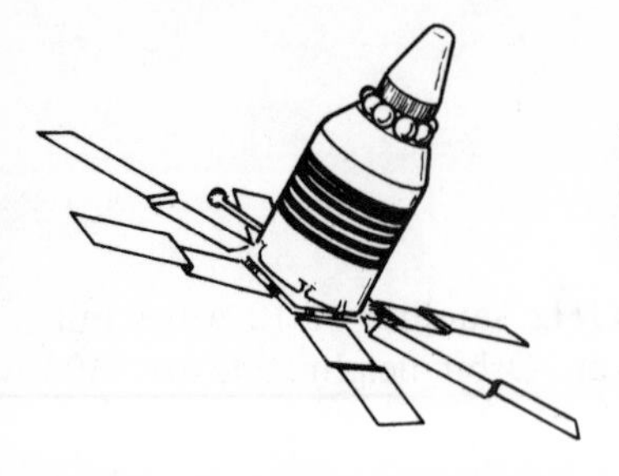

The first of many U.S.S.R. communications satellites all using a high-altitude elliptical orbit.

1965: EARLY BIRD (INTELSAT)

The world's first commercial communications satellite, operated by COMSAT, in orbit over the Atlantic. 240 voice channels. 40 watts of output power.

1966: INTELSAT II

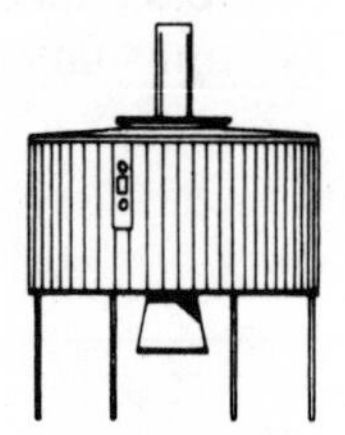

COMSAT's second satellite. The first multiple-access commercial satellite with multidestination capability. 240 voice circuits. 75 watts of output power.

1968: LES-6
1969: TACSAT I
(U.S. Military)

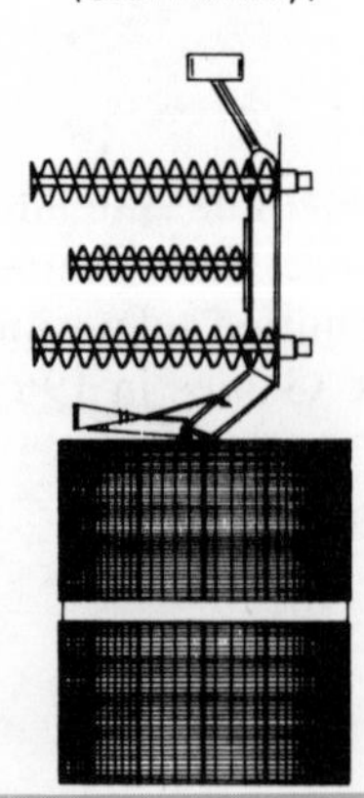

LES-6, a small single-band UHF satellite, and TACSAT I, a powerful UHF and SHF satellite, formed the TACSATCOM program for U.S. military worldwide operations. Airborne, shipborne, and portable tactical ground stations. A TACSAT satellite had 1000 watts of power and transmitted 10,000 voice circuits.

1968: INTELSAT III

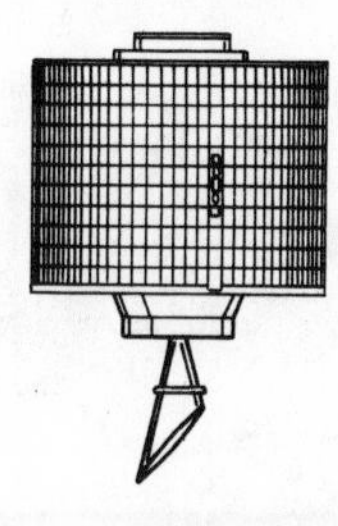

COMSAT's third generation. 1200 voice circuits. A directional earth-coverage antenna. 120 watts of output power.

1971: INTELSAT IV

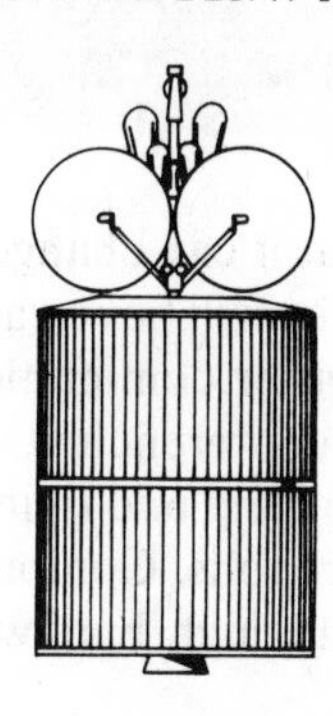

COMSAT's fourth generation. 6000 voice circuits. An earth-coverage and two spot-beam antennas. 400 watts of output power.

1972: ANIK (Telesat Canada)

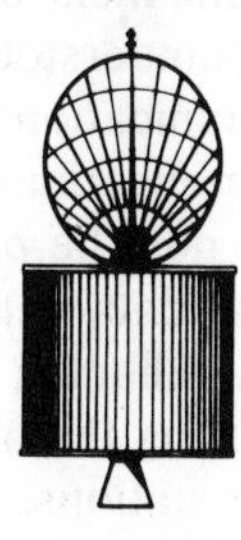

The world's first domestic satellite, designed for Canada. 5000 voice circuits. 300 watts of power.

1974: WESTAR (Western Union)

The first domestic satellite in the United States. The beginning of a new era in U.S. communications.

PREFACE

The potential of satellite communications is so great that it can change the entire fabric of society. At some future time, advanced countries will have vast numbers of small antennas receiving mail, television, and computer communications via satellites. As petroleum costs rise, large-screen teleconferencing, interactive education, and new electronic networks will change working and living patterns. The office of the future will not need to be in a big metropolis. Computer system architectures and uses will be different because of the high bandwidth transmission links.

We are rushing headlong towards a world of immensely powerful microprocessors, giant data banks, and very high capacity channels spanning the world. Communications satellite technology is often misunderstood. In some ways it is fundamentally different to terrestrial telecommunications. This book attempts to explain the subject, the tradeoffs, and the implications for system design. Corporations in countries which permit it should now be planning how the new satellite facilities will change their telephone and data networks, their word processing and mail, their travel budgets, training, and human communications. There will be many new business opportunities in using the new satellites which are accessed by small rooftop and parking lot earth stations.

Systems analysts should understand the implications of satellites because they change what is needed in data communications protocols, terminal selection, design of end-user dialogues, and distributed processing. The redesign of corporate networks to take advantage of satellites affects DP strategy.

Satellite communications is plunging downwards in cost. However, it competes with much that is traditional. Old, established telecommunications organizations see the new technology as a threat. It is being resisted by telephone administrations in many countries. In some countries, new legislative barriers have been erected against satellite communications technology. It is desirable that there should be widespread comprehension of what large satellites launched with the space shuttle could do, so that corporations, the public, and the economy are not robbed of the immense benefits.

JAMES MARTIN

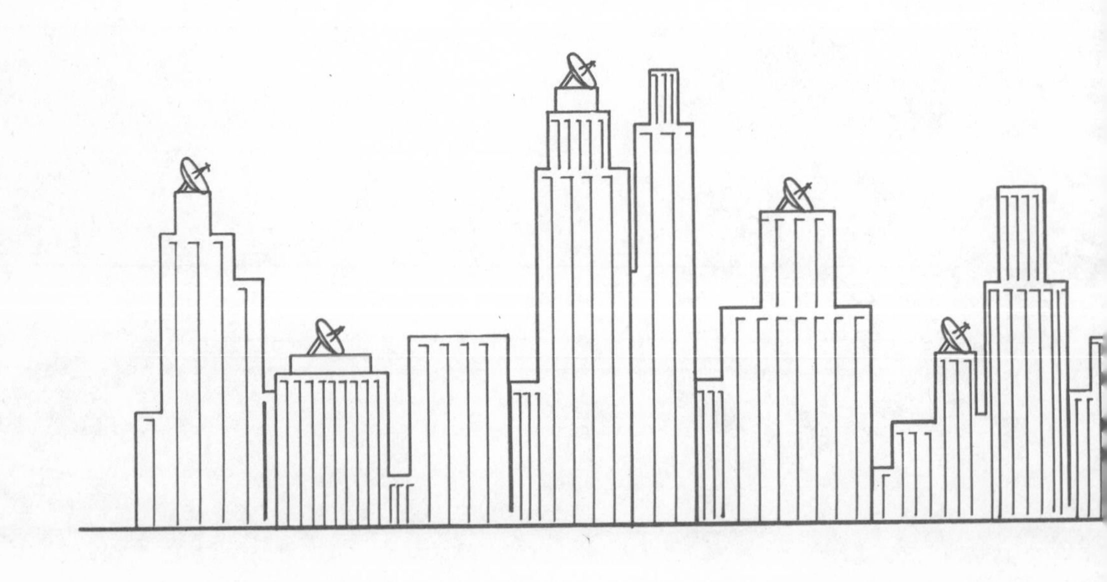

"For a long time business communications was regarded as the weather is: something you have to live with and can't do much about. We hope to change that perception, not only with better communications but also with the chance for the user to *manage* it for the first time."

Philip Whittaker
President, Satellite Business Systems

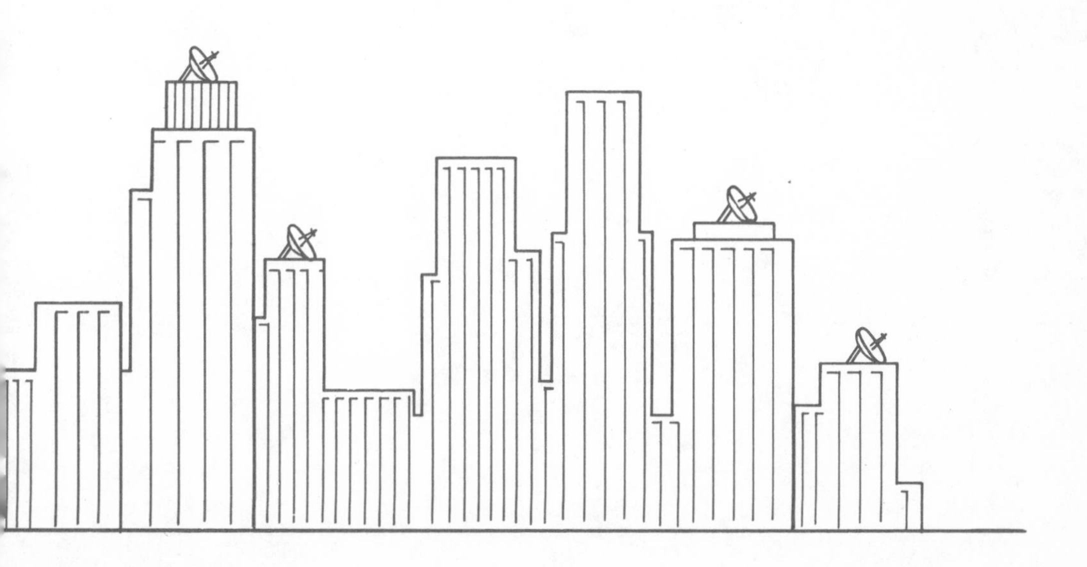

PART I THE SPACE SEGMENT

1 THE PROMISE OF SPACE

A communications satellite is a very simple concept, but simple concepts sometimes change the world. The potential of satellites is dawning on us slowly.

A communications satellite, in essence, is a radio relay in the sky. Signals are sent to it from antennas on earth; it amplifies the signals and sends them back. The power of satellites lies in the fact that they can handle a large amount of traffic and send it over most of the earth. Three satellites can cover almost all of the inhabited regions of the earth. The cost of satellite channels is dropping remarkably fast, and new directions are emerging in the technology. A major thrust seeks to have satellite antennas at corporate and government locations where the channels can be put to good use.

In the last quarter of the twentieth century, different societies will be profoundly changed (for good and ill) by communications media—television, person-to-person communications, communication between people and computers, computer networks, educational facilities, facilities for running industry, vast data banks, electronic funds transfer, military command and control systems, and so on. Satellites, both nationally and internationally, promise to be one of the most powerful and cost-effective communications media.

The perception of the value of communications satellites has changed since man's first satellites were launched. At first satellites were perceived largely as a means to reach isolated places. Most of the world's population is not served by the telephone and television networks that so greatly influence Western society. The cost of lacing Africa and South America with Bell System engineering would be unthinkable. Satellites were perceived as a counter technology, and earth stations began to appear in the remotest parts of the world. Countries with only the most primitive telecommunications put satellites on their postage stamps.

In the developing world, communications facilities are built in a sequence different from that in the West. The West had railroads, and then, later, roads,

Figure 1.1 Postage stamps from developing nations who perceived the satellite as a link to the rest of the world. (*Courtesy Hughes Aircraft Company.*)

and later the airplane. In Brazil many towns can be reached only by air; later, roads are built to them, and finally, if at all, railroads. Similarly, satellites will bring the links of culture and commerce to towns without telephone trunks. By the late 1980s Brazil might have far better satellite facilities than Europe.

As satellites dropped from their initial exorbitant cost it was realized that they could compete with the world's suboceanic cables; satellites then had a part to play in the industrial nations, linking the continental landmasses. The owners of the suboceanic cables took political steps to protect their investment at the expense of satellites, but soon more transoceanic telephone calls were made by satellite than by cable. Television relayed across the ocean by satellite became common, because cables of the 1960s did not have the capacity to send live television.

Comsat (the Communications Satellite Corporation) launched four generations of satellites in six years. EARLY BIRD, the first commercial satellite to retransmit signals from a fixed position in space, was followed by INTELSAT II in 1967, INTELSAT III in 1968, and INTELSAT IV in 1971. The second half of the 1970's brought INTELSAT IVA and INTELSAT V. Figure 1.2 shows the INTELSAT network.

When the first INTELSAT birds brought competition to suboceanic telephone cables, the domestic telephone networks seemed immune from the threat. The cost per telephone channel of the early satellites was high, and the U.S. Communication Satellite Act said that only Comsat could operate satellites and that they could be used only for international transmission.

As often, technology changed more rapidly than the law, and the first North American domestic satellite was launched in 1972 by Canada. It was originally perceived as a means to communicate with Canadians in the frozen north and was called ANIK, which means "brother" in Eskimo language. However, it was soon realized that the ANIK satellites would provide cheaper long-distance telephone or television circuits than those of the established common carriers. Antennas were set up in the United States to use the ANIK satellites, and for their first two years in orbit these satellites earned a return on capital investment that was virtually unprecedented in the telecommunications industry.

A flurry of legislation in 1972 resulted in the U.S. Federal Communications Commission's *Open Skies Policy,* which encouraged private industry to submit proposals for launching and operating communications satellites. The first U.S. common carrier to take advantage of the Open Skies Policy was Western Union, which launched two WESTAR satellites in 1974—the first U.S. domestic satellites.

A price war ensued for long-distance leased communications channels. A leased telephone circuit from coast to coast via WESTAR was a fraction of the cost of a similar channel on the ground leased from the terrestrial common carriers. It seemed clear that the price could drop further with more advanced equipment.

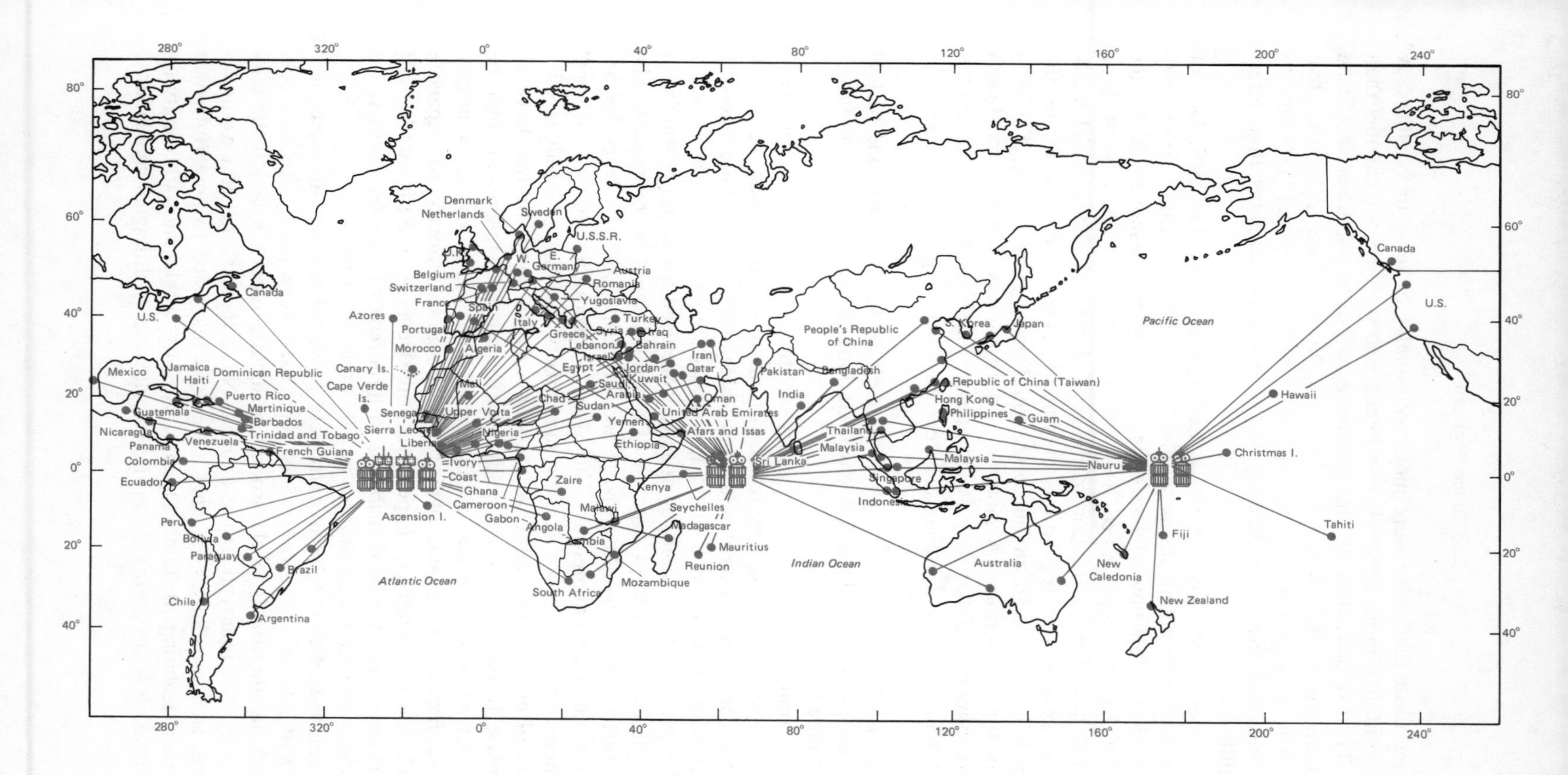

Figure 1.2 The global network of Intelsat serving over 150 earth stations in 80 countries. More than $1 billion invested in the network generates over $100 million revenue per year. Intelsat traffic is doubling every three years. Many new earth stations of lower cost will be added during the next 10 years.

It became clear that there were major *economies of scale* in satellites. A big satellite could give more channels and hence a lower cost per channel than a small one. To take advantage of the economies of scale, satellites should be employed where the traffic volume was heaviest. Nowhere was it heavier than in U.S. domestic telecommunications, and so it began to appear, contrary to the earlier view, that there was more profit in domestic satellites than international satellites.

This was perceived most clearly at Bell Laboratories, the birthplace of the first commercial communications satellite, TELSTAR. A Bell Laboratories study showed that a few powerful satellites of advanced design could handle far more traffic than the entire AT&T long-distance network. The cost of these satellites would have been a fraction of the cost of equivalent terrestrial facilities. However, government regulations prevented AT&T from developing the satellites which it, more than anyone else, could make good use of. The field was left open for competition. A number of corporations, which were dwarfs beside AT&T, announced that they would operate satellites, and AT&T continued to spend many billions of dollars per year on expanding its earth-bound facilities.

For WESTAR users the perception of satellites had now become that of communications pipelines linking five earth stations in one country. A further perceptual change was to follow.

While corporations and computer users saw the satellite as providing two-way channels between the relatively few earth stations, broadcasters or would-be broadcasters perceived it as a potentially ideal way to distribute one-way signals. Television or music sent up to the satellite could be received over a vast area. If a portion of the satellite capacity were used for sound channels for education or news, a very large number of channels could be broadcast. The transmitting earth stations would be large and expensive, but the receiving antennas could be small and numerous. The Musak Corporation envisioned small receiving antennas on the roofs of their subscribers' buildings. Satellites offer the possibility of broadcasting television to vast areas of the world that have no television today. If more powerful satellites were launched, television could be broadcast directly to the hundreds of millions of homes in industrial countries. The Japanese broadcast satellite will beam programs directly to Japanese homes, which can use relatively expensive home receivers. With satellites of lesser power, television can be distributed to hundreds of regional stations for rebroadcasting over today's transmitter or cable television links.

Television used well can be an extremely powerful medium for education. The majority of television, however, has been used very poorly for this purpose. America has thousands of classrooms or lecture rooms in which the television sets have been removed or are unused. Nevertheless, the best examples are very good. In some classes teachers have used television to powerful effect to augment their teaching. Britain's Open University, a television university on a massive scale, has programs deserving worldwide multilingual availability.

And what better way for most people to learn history than from programs such as Alistair Cooke's "America"? The dream of superb educational facilities via satellite will probably be achieved one day, but much more than advanced technology is needed to bring it about.

Broadcasting is usually thought of as having one transmitter and many receivers. However, when a satellite is used for two-way signals, a form of broadcasting is taking place in which there are many transmitters. Each earth station is, in effect, a broadcasting transmitter because its signal reaches all other earth stations, whether they want it or not. Each earth station, like a radio set, tunes in only to that signal it wants to receive.

Because of this broadcasting nature of satellites, it is limiting to think of a satellite as a "cable in the sky." It is much more than that. A signal sent up to the satellite comes down everywhere over a very wide area. To maximize the usefulness of the satellite for telecommunications any user in that area should be able to request a small portion of the vast satellite capacity at any time and have it allocated to him if at that moment there is any capacity free. Just as with a telephone network, to make it really useful any user should be able to call any other user when he wants; however, it is not desirable to put a telephone exchange in the satellite, at least not yet. The equipment on the satellite needs to be simple and reliable, because equipment failures cannot be repaired, and needs to be light and consume as little power as possible. To achieve the desirable *multiple-access* capability, ingenious ways have been devised of allocating satellite capacity to geographically scattered users, permitting them to intercommunicate.

For the first decade of communications satellite operation most of the capacity of the satellites was used for telephone traffic and television. The technology has evolved, however, so that, in a sense, satellites are much more powerful for the transmission of *data* between computers and computer users or between telegraph machines. It now appears desirable technically to carry telephone traffic, and possibly television, in a *digital* form, as we shall discuss later. When the telephone voice is digitized in a simple manner it becomes 64,000 bits per second. When digitized in a more compressed fashion, fewer bits can be used. Digitized television requires a bit rate between 40 million and 92.5 million bits per second depending on the technique used. A substantial quantity of data can be sent using these bit rates. Such bit rates make telephone voice appear expensive by comparison with data transmission as a means of transmitting information. A voice message requiring millions of bits of transmission is equivalent in verbal content to a telegram requiring only a thousand bits.

The potential power of satellites for the computer industry can be illustrated by means of a simple calculation. When a person uses a computer terminal he does not transmit continuously at the full speed of the line to which the terminal is connected. He transmits a number of characters, reads the response, thinks about it, and keys in more characters. The dialogue between the terminal user and the computer results in bursts of data, often short bursts, passing across the line with pauses between them. Much more data could be sent in the pauses than are actually transmitted in most cases. As we shall illustrate later, most man-terminal dialogues result in not more than 10 bits per second passing backwards and forwards *on average,* although in *certain* seconds a much larger number of bits is sent.

Today's domestic satellites carry a number of separate units, called transponders, each of which can relay one television program or equivalent. RCA SATCOM users can transmit 60 million bits per second via one transponder. The satellite has 24 such transponders. Such a satellite could thus have a total data throughput of 24×60 million bits per second. If this capacity were employed entirely for interactive terminal users, it would not be possible to achieve 100% efficiency. A conservative assumption is that 15% efficiency could be achieved (much lower than today's equivalent on well-organized terrestrial lines). 15% efficiency would give a usable capacity of *216 million bits per second.*

The combined population of the United States and Canada is about 240 million. Let us suppose, for the sake of this illustration, that *every* person makes substantial use of computer terminals. The *average* working person uses them one hour per day, and the average nonworking person half an hour per day. This gives a total of about 160 million hours of terminal usage per day. Let us suppose that in the peak hour of the day the usage is three times the daily average. The total data rate in the peak hour is then

$$\frac{160 \text{ million} \times 3 \times 10}{24} = \textit{200 million bits per second}$$

In other words, one satellite using today's state of the art could have enough transmission capacity to provide every man, woman, and child in the United States and Canada with a computer terminal. In addition, because we have done the calculation for the peak hour, twice as much data such as mail which can be deferred to non-peak hours could be sent over the same satellite.

This calculation assumes, as does other such discussion of the power of satellites, that it is possible to organize terrestrial facilities for the satellite channels in an appropriate manner. In the above case, how does one enable an extremely large number of users to share the same channel without interfering with one another excessively? As we shall see later, there *are* types of organization that could work well. However, when a large number of users share the large capacity available, the design becomes dominated not by the satellite relay itself as on a point-to-point link but by the architecture of the ground facilities that permit multiple access to the satellite.

In spite of the satellite's power for data transmission it would not be a sound business operation to launch a satellite solely for use with computers. Of all the traffic that might be sent by satellite, a relatively small proportion of it is computer traffic. Whatever the mix in the future, most of *today's* traffic is "Plain Old Telephone Service." To maximize its potential profit a satellite should be capable of carrying many different types of signals—real-time and non-real-time, voice, data, facsimile, and video. For all these signals it should be regarded as a broadcasting medium accessible from anywhere beneath it, not as a set of cables in the sky.

To summarize, the perception of what a communications satellite is has changed, and different perceptions have been

1. A means to reach isolated places on earth.
2. An alternative to suboceanic cables.
3. Long-distance domestic telephone and television links.
4. Television and music broadcasting facilities.
5. A data facility capable of interlinking computer terminals everywhere.
6. A multiple-access facility capable of carrying all types of signals on a demand basis.

The changing perception of satellite potential has been related to the change in satellite cost. This has been dramatic.

The first four generations of INTELSAT satellite carried increasing numbers of channels and have progressively longer design lives, as shown in Fig. 1.3. Consequently the cost per voice channel per year dropped dramatically. The process will continue with INTELSAT V.

The bottom line of Fig. 1.3 shows the drop in cost per satellite voice channel per year. Figure 1.4 plots the trend. The figure shows the *investment* cost of the satellite and its launch. The cost to a subscriber will be much higher because it must include the earth station and links to it and must take into consideration the fact that the average channel utilization may be low.

The extraordinary cost reduction shown in Fig. 1.4 will continue, if sufficient traffic is sent over satellite to permit the possible economies of scale. Massive reductions in the cost per voice channel could result if satellites with a much larger capacity than today's were launched. In the 1980's the space shuttle and associated equipment will lower the launch costs substantially.

The satellites and their launch costs are referred to as the *space segment* of satellite communications. The comment is sometimes made among systems planners that the space segment costs are dropping to such a low level that overall system costs will be dominated by the organization of the ground facilities.

The cost of an earth station, however, has dropped more spectacularly than that of a satellite. The first COMSAT earth stations cost more than $10 million. (The first Bell System earth stations for TELSTAR cost several times that.) Earth stations have dropped in cost until now a powerful transmit/receive facility such as that in Fig. 2.3 can be purchased for about $100,000. Receive-only facilities are a fraction of this cost. At the same time the traffic that can be

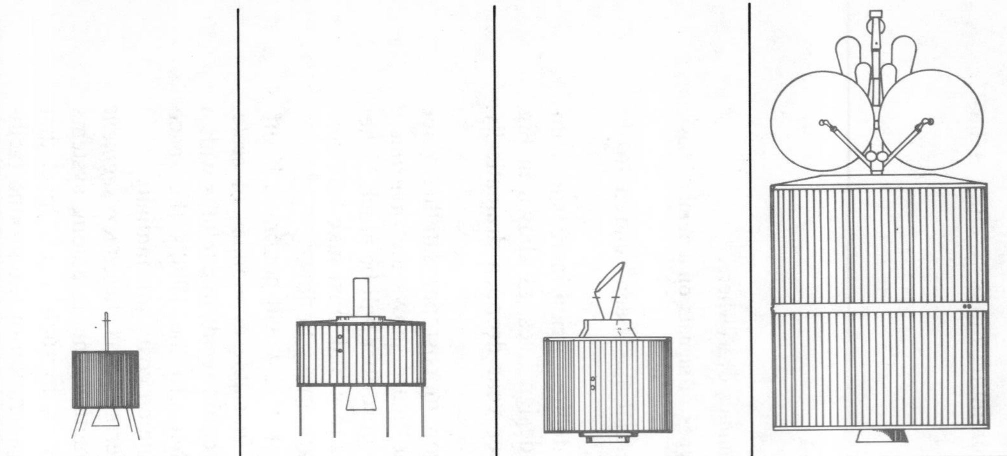

					Being designed
Name	Intelsat I (Early Bird)	Intelsat II	Intelsat III	Intelsat IV	Intelsat V
Year of launch	1965	1967	1968	1971	≈ 1979
Diameter	28 inches	56 inches	56 inches	93 inches	600 inch sails
Height	23 inches	26 inches	78 inches	111 inches	264 inches
Weight in orbit	85 lbs	192 lbs	322 lbs	1547 lbs	3200 lbs
Number of antennas	1	1	1	3	6
Primary power (watts)	40	75	120	400	1000
No. of transponders	2	1	2	12	27
Bandwidth of transponder	25 MHz	130 MHz	225 MHz	36 MHz	
Cost of satellite	$3.6 million	$3.5 million	$4.5 million	$14 million	≈ $ 25 million
Cost of launch	$4.6 million	$4.6 million	$6 million	$20 million	≈ $ 23 million
Design lifetime	1.5 years	3 years	5 years	7 years	10 years
Total cost per year	$5.47 million	$2.70 million	$1.90 million	$4.85 million	≈ $ 4.8 million
Maximum No. of voice circuits	240	240	1200	6000	≈ 24,000
Cost/voice circuit/year	$23,000	$11,000	$1600	$810	≈ $200

Figure 1.3 **The INTELSAT birds—four generations of satellites in six years.**

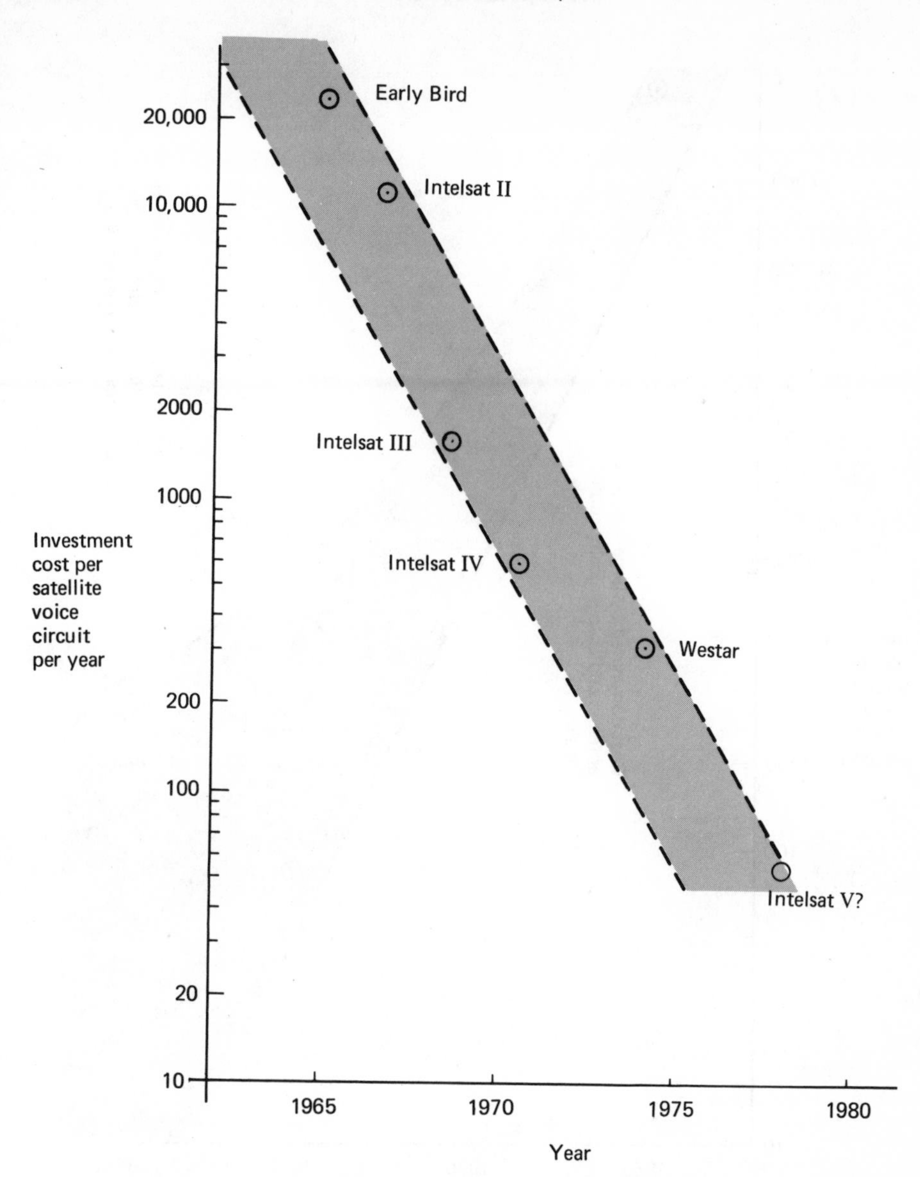

Figure 1.4 The falling cost of satellite voice circuits. The trend could continue if economies of scale are permitted.

handled by an earth station is increasing as satellite capacity increases. Combining these two trends we find that the investment cost per channel per earth station is dropping, as shown in Fig. 1.4.

The *total* earth segment costs are *not* dropping so fast as in Fig. 1.5 because to provide increased accessibility to the satellites many earth stations are being built. Prior to 1973 the United States had only a handful of earth sta-

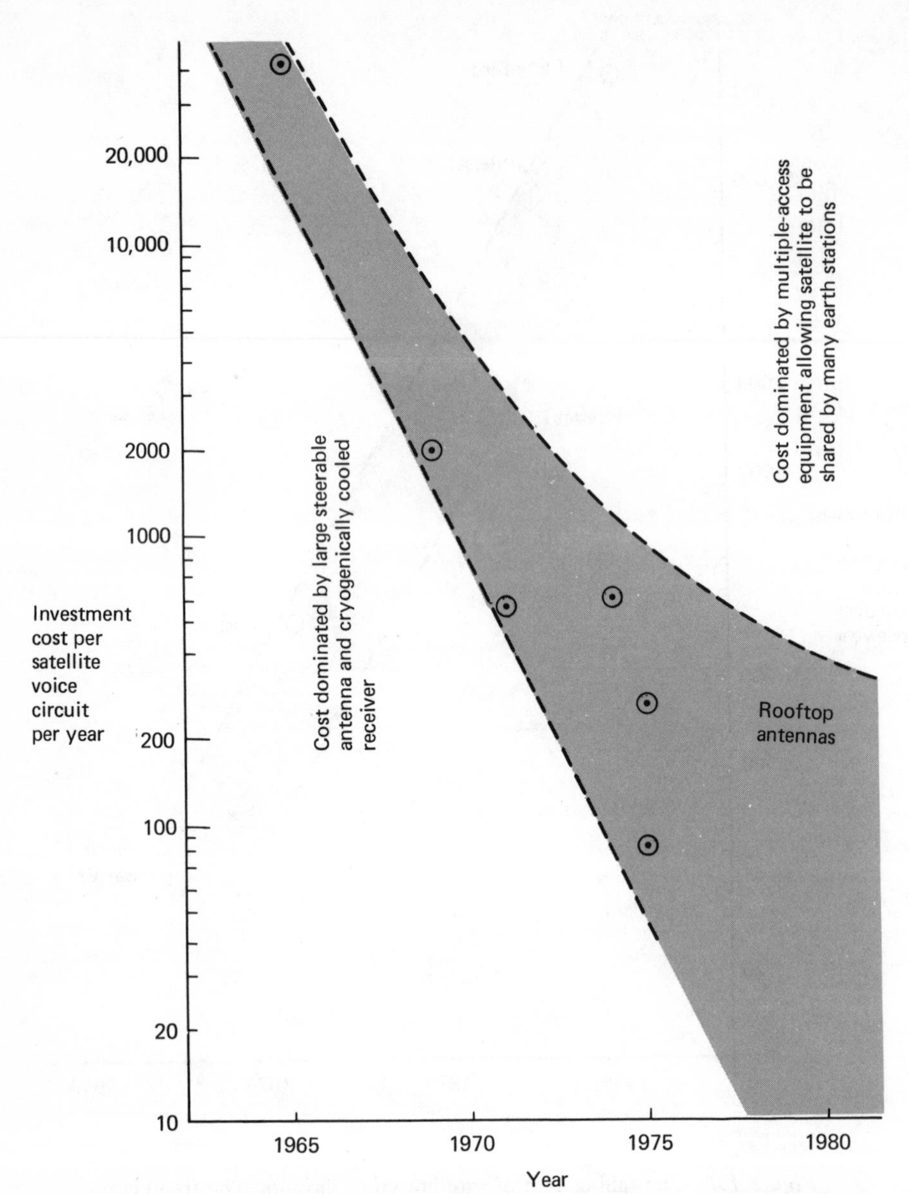

Figure 1.5 The falling cost of satellite earth stations. The cost per channel with private satellite antennas will vary greatly with the number of channels handled.

tions. Now many are being installed. Some corporations are setting up their own satellite antennas, and we can look forward to an era of many small private earth stations. The earth-station cost in the 1960's was dominated by its large steerable antenna, 30 meters in diameter, with automatic tracking facilities and hypersensitive cryogenically cooled receivers. Small corporate earth stations will use relatively low-cost nontracking antennas and uncooled cheaper receivers; their cost will be dominated by equipment which enables them to share the satellite with many other earth stations. A low earth-station cost does not necessarily mean a low cost per channel because it may be used at a location which employs relatively few channels.

There is a trade-off between the cost of the satellite and the cost of its earth station. If the satellite has a large antenna and considerable power, smaller stations can be used. If the satellite makes more efficient use of its frequency allocation, the cost per channel will be lower. There is a limit to satellite efficiency, and so the main effect of increasing satellite cost will be to reduce earth antenna size and cost.

As the earth facilities drop in cost, more antennas will be constructed and more traffic will be sent, making it economical to use more powerful satellites, which will make the earth facilities drop further in cost. If satellites use large numbers of small earth stations, however, the overall system architecture which permits the earth stations to share the satellite is extremely important and eventually will dominate the cost of satellite systems.

Box 1.1 lists the major categories of communications satellite. They differ in purpose, and differ enormously in earth station cost.

BOX 1.1 Major categories of communications satellite

Categories of Satellite	Examples	Earth Stations	Functions
Intercontinental Satellites	• EARLY BIRD INTELSAT II INTELSAT III INTELSAT IV INTELSAT IVa INTELSAT V • The Russian MOLNIYA & STATSIONAR Satellites • Europe's SYMPHONIE	• Large, expensive, highly reliable, designed for interconnection to circuits of national telephone administrations	• To provide world-wide common-carrier telephone and data circuits • Point-to-point television relays
Traditional Common-Carrier Domestic Satellites	• Canada's ANIK • Western Union's WESTAR • RCA's SATCOM • AT&T and GT&E COMSTAR	• Large common-carrier earth stations • Medium-size private earth stations • 10 and 4.5 meter • receive-only television earth stations	• To enhance common-carrier telephone networks • To provide leased long-distance circuits at lower cost • Point-to-point television relays • Television broadcasting

	• Indonesia's PALAPA • ARABSAT		to local transmitting stations or CATV systems • Music broadcasting (Musak Corporation)
Domestic Multiple-Access Satellites	• SBS (Satellite Business Systems)	• 5 or 7 meter 12/14 GHz installable on corporate premises and on city rooftops	• To provide private networks for corporations and government, relaying telephone, data, and image traffic
Television Broadcast Satellites	• Japanese broadcast satellite (JBS) • Canada's experimental CTS • NASA's experimental ATS-6	• Large transmitting station • Inexpensive receive-only antennas small enough for use in homes and schools	• Television broadcasting direct to homes or schools
Mobile Terminal Satellites	• MARISAT • MAROTS (both for ships) • Military satellites	• Very small shipboard or portable. Use UHF frequencies to give low-cost, low-capacity earth stations	• To provide telephone or data circuits to ships, vehicles, remote sites, or mobile military units

BOX 1.1—*Cont.*

Special Purpose Satellites	• Aeronautical	• Aircraft antennas	• Aircraft communications and navigation.
	• Amateur satellites	• Home antennas	• For amateur radio communications and experiments
Research Satellites	• NASA's ATS satellites • Canada's CTS	• All types often small	• Research into higher frequency propagation • Research into new satellite construction • Research into new applications of satellites

2 SATELLITE LINKS

Like the microwave relay stations with antennas on towers across the countryside, a satellite receives radio signals in a given frequency band, changes their frequency, and retransmits them. The radio signals are of high bandwidth (i.e., a wide spread of frequencies), and this is important because the information-carrying capacity of a signal is proportional to its bandwidth.

Many satellites use the same frequencies that are used by terrestrial microwave relays. Some have antennas of a similar size, though many satellite antennas are smaller. The main difference is that whereas the distance between microwave relays on earth is typically 30 miles, the distance of a communications satellite is about 25,000 miles. The strength of a radio signal diminishes proportionately to the square of the distance it travels, so the signal is very weak by the time it reaches earth from the satellite. Much of the concern in space communications is with overcoming the effects of this great loss in signal power.

DELAY

A disadvantage of satellite transmission is that a delay occurs because the signal has to travel far into space and back. The signal propagation time is about 270 milliseconds and varies slightly with the earth-station locations. A telephone user waits for the reply of the person he is talking to for an extra 540 milliseconds if the call goes via satellite in both directions.

The bad effects of this delay have been much exaggerated by organizations which operate long-distance terrestrial links. The claim is frequently heard that the delay is psychologically harmful in telephone conversations and renders satellite links useless for interactive data transmission. Some common carriers (without prospects of owning any satellites) have claimed that the delay is inacceptable to telephone users. Arthur C. Clarke, normally the most optimistic of writers about technology, once suggested that satellite users would end each stretch of conversation with the word "Over" [1]. In practice a telephone user certainly notices the delay but very quickly becomes used to it if he makes many satellite calls. It is much less annoying than having a noisy local loop. Assessment of psychological effect should not be based on the first call a person makes (an error made in some of the published studies). In a report to the Federal Communications Commission on the experimental use of a satellite link in its corporate telephone network, IBM commented on the delay as follows [2]: "For most cases with one talker unaware of the interposition of the satellite path (the originating talker is aware and primed in his expectations) the conversation suffered only initial awkwardness with a rapid adjustment made in terms of accommodating the delay effect by more care in interruption. It has been observed by several participants that a widespread use of satellite connectivity would gradually increase the politeness of telephone conversations."

On many of today's transatlantic telephone calls, the satellite is used for one direction of transmission and suboceanic cable is used for the other direction. The result is a total delay slightly greater than a quarter of a second. It also sometimes results in the disturbing effect of person A being able to hear person B very distinctly, as though he were in the next room, but person B being able to hear person A only poorly and hence feeling that there is a need to shout. The more clearly heard person shouts; the more poorly heard person reacts to the shouting by talking softly; the shouter reacts to the soft voice by shouting louder; and so on.

On transoceanic circuits, two-way delay seems less harmful to a person who is used to it than the effects of TASI. TASI (time-assigned speech interpolation) snatches the channel away from a person when he pauses in his speech and may allocate it to another speaker. This procedure increases the overall utilization of the circuit. When the circuit is heavily loaded, a speaker may not be reassigned a channel quickly enough when he speaks again, so his first spoken syllables are sometimes deleted. Intercontinental callers sometimes confuse the effects of TASI with the effects of satellite delay.

For telephone users it is particularly important to remove the *echo* on a satellite channel. If a talker hears his own voice echoed back to him with a delay of 540 milliseconds, this proves very disturbing. The echo is removed by means of an echo suppressor, which inserts an impedance into the reverse path when a person talks and removes the impedance when he stops talking.

While a telephone user can learn to ignore one or two 270-millisecond delays in a conversational response, four such delays (1080 milliseconds) may strain his tolerance. It is therefore desirable that the switching of calls should be organized so that no connection contains two or more round trips by satellite. Where satellites supplement the terrestrial toll telephone network the switching can usually be organized to limit the delay to 270 milliseconds.

THE EFFECTS OF DELAY ON DATA TRANSMISSION

In interactive data transmission via satellite a terminal user will experience a constant increase in response time of about 540 milliseconds. A systems designer has to take this into consideration in designing the overall system response time. In many interactive systems it is desirable that the mean response time should not be greater than 2 seconds. This is achieved satisfactorily on many interactive systems using satellites today.

Satellite delay can have a serious effect on data transmission when protocols and mechanisms are used which are designed for terrestrial links having little propagation delay. Systems designers often substitute a satellite circuit for a terrestrial voice circuit because it is less costly. They sometimes discover that this substitution severely lowers the throughput. If polling is used, it can make the response times inacceptably long. In some cases, a remote device, such as a printer, ceases to function correctly.

We discuss this important topic in Chapter 18. The conclusion can be summarized succinctly. Satellites are extremely efficient for data transmission if the protocols and control mechanisms in the user devices are appropriately designed. If certain terrestrial protocols are used without modification on a satellite link, the propagation delay may cause a severe degradation in performance. The change needed in the software or hardware of the using devices is often fairly simple. Sometimes, however, it has not been made when those devices are connected to a circuit going via satellite.

To gain full advantage of broadcasting capability satellites, the organization of the links will become fundamentally different from the organization of traditional data links on earth. This is discussed in Part II of the book.

TRANSPONDERS

Like a terrestrial microwave relay, the satellite must use different frequencies for receiving and transmitting; otherwise the powerful transmitted signal would interfere with the weak incoming signal. The equipment which receives a signal, amplifies it, changes its frequency, and retransmits it is called a *transponder.*

The frequencies used in a satellite link are referred to with phrases such as 4/6 GHz, 12/14 GHz, and 20/30 GHz†. The first number in each case refers to the frequency of the down-link, and the second number refers to the frequency of the up-link.

Box 2.1 shows the names of the radio-frequency bands. Most satellites use the UHF and SHF (microwave) frequencies, mainly SHF. Commercial satellites have mainly used the 4/6-GHz band. These frequencies are some-

† 1 Hz (hertz) means 1 cycle per second of frequency.
1 KHz (kilohertz) means 1000 cycles per second.
1 MHz (megahertz) means 1 million cycles per second.
1 GHz (gigahertz) means 1000 million cycles per second.

BOX 2.1 Designation of radio-frequency bands

Band Number†	Band Name	Frequency Range (including lower figure, excluding higher figure)	Metric Subdivision
4	VLF, very low frequency	3–30 kHz	Myriametric waves
5	LF, low frequency	30–300 kHz	Kilometric waves
6	MF, medium frequency	300–3000 kHz	Hectometric waves
7	HF, high frequency	3–30 MHz	Decametric waves
8	VHF, very high frequency	30–300 MHz	Metric waves
9	UHF, ultra high frequency	300–3000 MHz	Decimetric waves
10	SHF, super high frequency	3–30 GHz	Centimetric waves
11	EHF, extra high frequency	30–300 GHz	Millimetric waves
12		300–3000 GHz	Decimillimetric waves

† Band number N extends from 0.3×10^N to 3×10^N Hz.

Frequency bands are also given letter designations as follows:

Band		Frequency Range (GHz)	
P		0.225–0.39	
	J		0.35–0.53
L		0.39–1.55	
S		1.55–5.2	
	C		3.9–6.2
X		5.2–10.9	
K		10.9–36.0	
	Ku		15.35–17.25
Q		36–46	
V		46–56	
W		56–100	

times referred to as the C-band and are the main frequencies used for microwave transmission on earth. This has the advantage that there is much experience in designing for these frequencies but has the severe disadvantage that satellite and terrestrial microwave links can interfere with one another, as we shall discuss later. To avoid this problem 12/14 GHz transmission is coming into use with a new generation of satellite technology. UHF frequencies give smaller cheaper earth stations with a much low capacity than SHF. They are used for mobile stations such as on board ship, portable stations, and military field stations.

Figure 2.1 INTELSAT IV, used for international transmission of television, voice, and data.

Most satellites have more than one transponder. The bandwidth handled by a transponder has differed from one satellite design to another, but most contemporary satellites (e.g., INTELSAT IVA, ANIK, Western Union's WESTAR, and RCA's SATCOM) have transponders with a bandwidth of 36 MHz. How this bandwidth is utilized depends on the earth-station equipment. The WESTAR satellites, which are typical, can relay any of the following with one transponder:

1. One color television channel with program sound.
2. 1200 voice channels.
3. A data rate of 50 Mbps.

4. The center 24 MHz of each band may relay either
   a. 16 channels of 1.544 Mbps, or
   b. 400 channels of 64,000 bps, or
   c. 600 channels of 40,000 bps.

The WESTAR satellites each have 12 such transponders, 2 of which are spares used to back up the other 10 in case of failure. The RCA SATCOM satellites have 24 transponders. Some future satellites may have a larger number of transponders. The SBS satellites scheduled for 1980 launch will each have 10 transponders of 43-MHz bandwidth and 6 spares. SBS (Satellite Business Systems), a corporation formed by IBM, Comsat General, and the Aetna insurance company, intends to provide private switched voice, data, and image transmission for corporations and government.

EARTH STATIONS

A satellite earth station consists of a large dish-shaped antenna, such as that in Fig. 2.2, which points at the satellite in basically the same way that an earthbound microwave relay dish points at the next tower in the chain. The earth-station antenna is usually larger, transmitting a narrower beam angle.

The earth stations of the 1960s were massive, like that in Fig. 2.2. The earth station at Andover, Maine, originally built for AT&T's TELSTAR satellite and then used with EARLY BIRD and its successors, has a dome 18 stories high housing a huge steerable horn-shaped antenna weighing 380 tons, with electronics cooled by liquid helium. Many of today's earth stations, such as some in Fig. 2.3, are small enough to be erected quickly in a parking lot behind a factory or office building. The SBS satellite system will use 5-meter and 7-meter dishes, whereas the large Comsat earth stations use dishes 30 meters in diameter.† Initially, earth stations were owned only by the common carriers (and the military). Now, in countries where the regulations permit, the small earth stations are owned or leased by private organizations, and employ common carrier satellites.

While most earth stations simply transmit and receive the telecommunications signal with a fixed antenna, at least one must carry out the additional function of *controlling* the satellite. Western Union's earth stations, for example, are unmanned with the exception of one at Glenwood, N.J., which monitors both the satellites and the other earth stations.

During a satellite launch the Glenwood station has the critical function of maneuvering the satellite into position once it has separated from its launch vehicle. For 7 years or more after the launch it must maintain the satellite in its correct position by occasionally firing small gas jets on board the satellite. It can send commands to the satellite to turn transponders on and off (to save

†Although metric units are not used for all measurements in this book, it is conventional to quote satellite and earth-station antenna diameters in meters.

Figure 2.2 The earth stations of the 1960s were huge and very expensive. (*Courtesy British Post Office.*)

power), to switch redundant and backup equipment, and to control the charging of batteries and positioning of antennas.

ANTENNAS IN SPACE Except on the very earliest satellites, satellite antennas, like all microwave antennas, are directional. Some point toward the earth as a whole. The earth subtends an angle of 17.34° at the satellite and an earth-coverage antenna distributes the transmitted energy over this angle. Many satellite antennas cover a smaller angle, aiming the signal at a portion of the earth. INTELSAT IV carries two earth-coverage antennas and two narrower-angle antennas subtending 4.5° (Figs. 2.1 and 2.4). A domes-

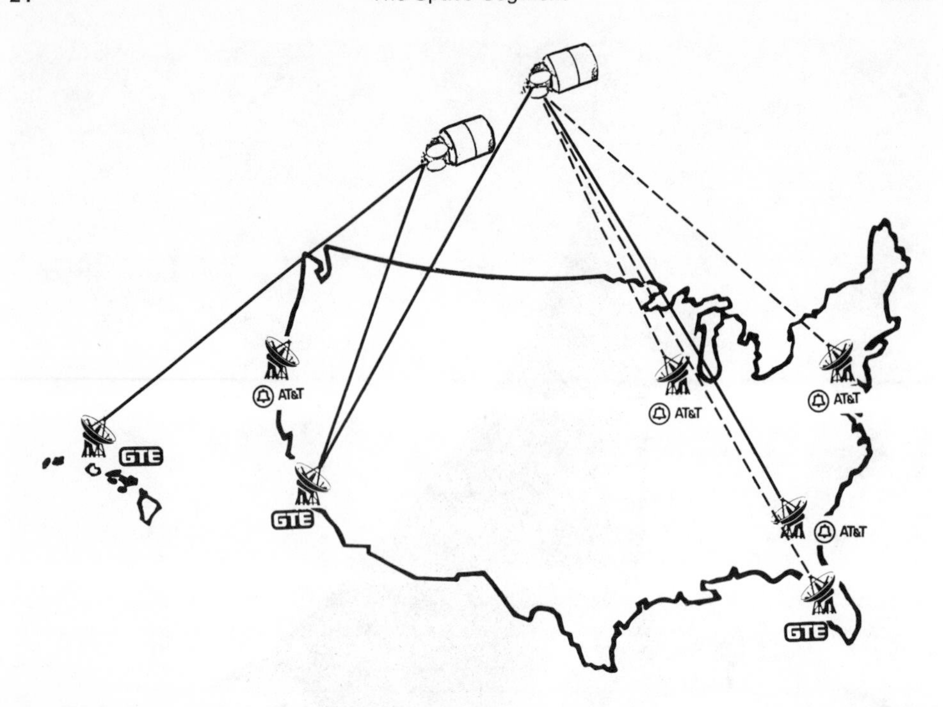

Figure 2.3 A satellite system to add long-distance circuits to the telephone networks of AT&T and GT&E. The earth antennas are 105 feet in diameter.

Canada's CTS (Communications Technology Satellite) makes possible excellent quality television reception with this 32-inch earth station. (*Photo courtesy Canadian Department of Communications.*)

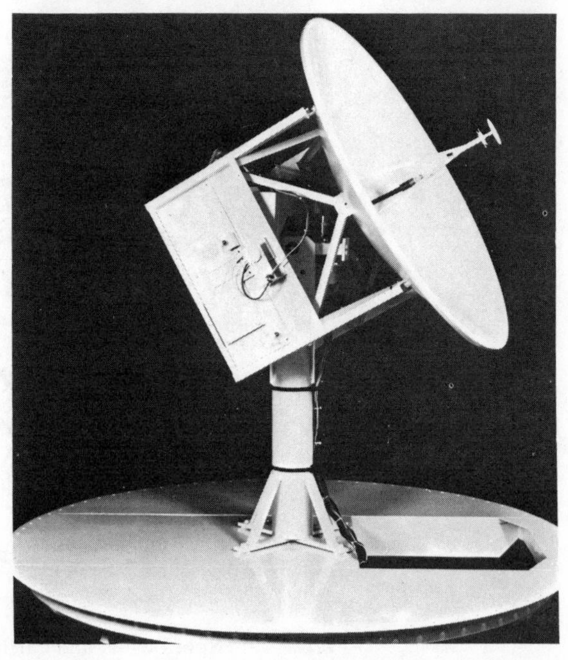

This earth station is 4 feet in diameter. It is used on board ships for worldwide voice and data communications using the Marisat satellite. (Its protective dome is removed in this photograph.) (*Photo courtesy Comsat General.*)

Figure 2.3 (continued).

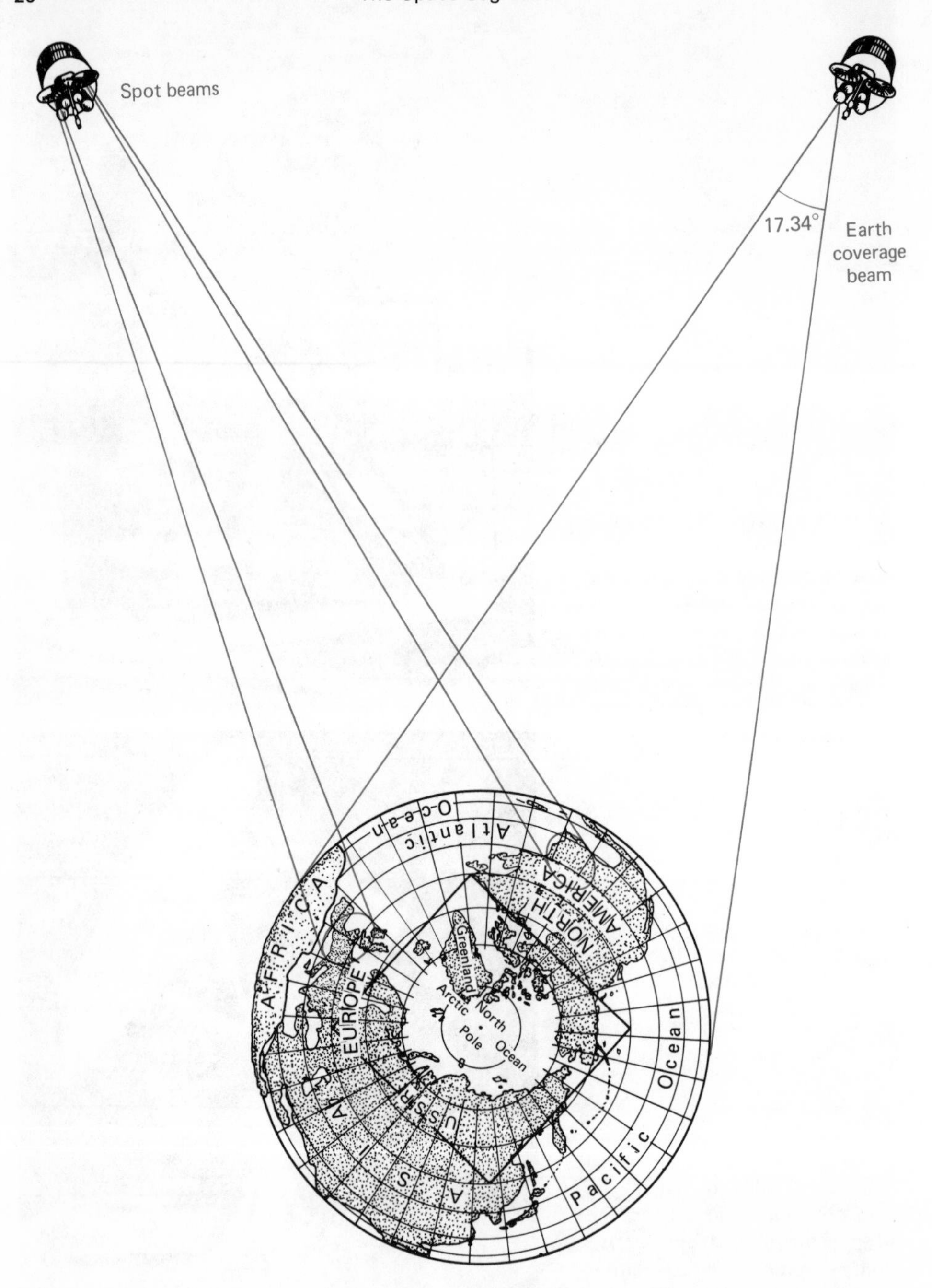

Figure 2.4 Spot beams and earth coverage beams.

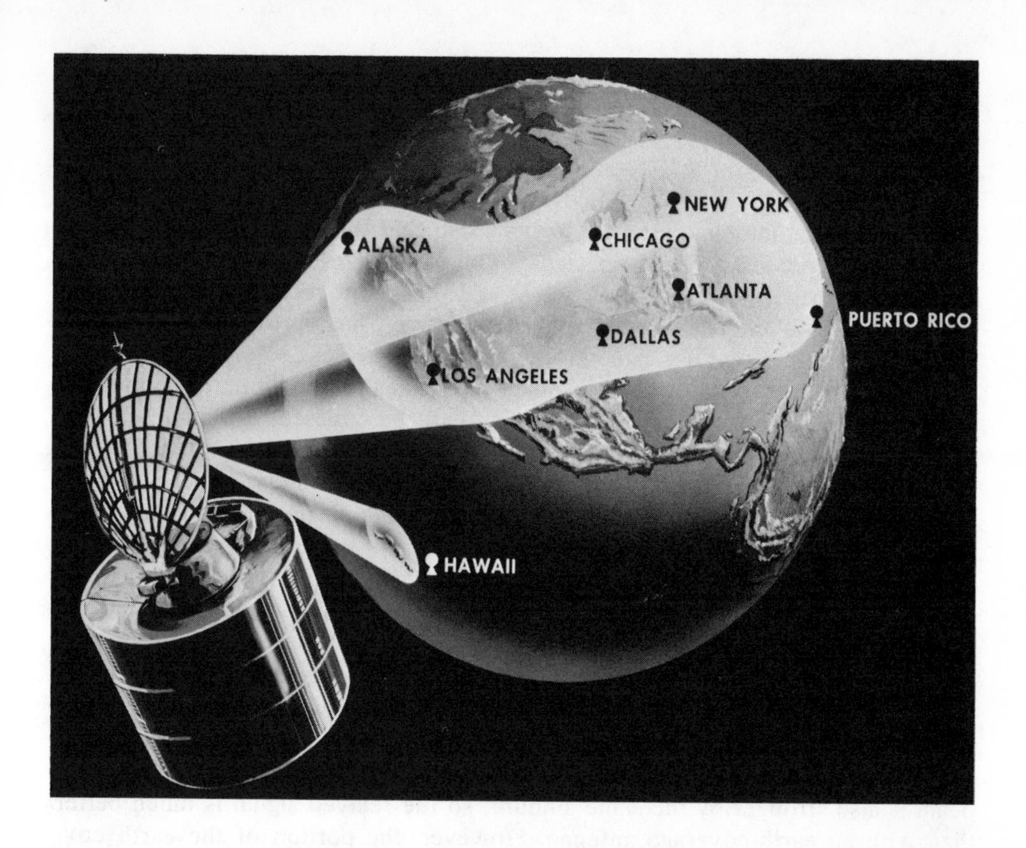

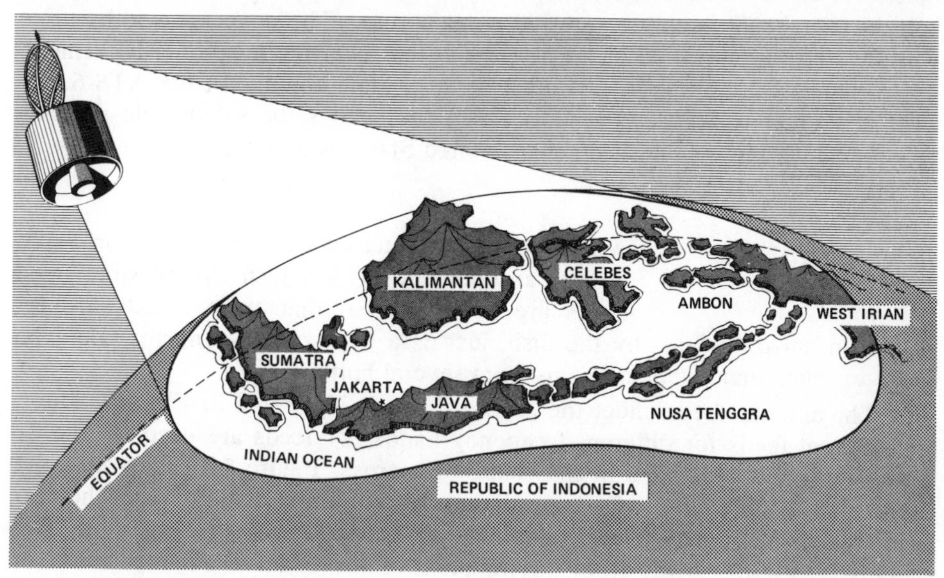

Figure 2.5 Domestic satellite systems. (*Courtesy Hughes Aircraft.*)

tic satellite designed for a country such as Canada or Brazil requires an antenna which focuses on that country (Fig. 2.5).

Figure 2.6 shows the planned contours of signal strength from an SBS satellite [3]. The antenna pattern is designed so that the East Coast area of the United States receives a stronger signal. The majority of the customers will be in the area labeled Region 1 in Fig. 2.6, and these will receive a signal strong enough to permit the use of relatively inexpensive 5-meter, antennas. The customers in Region 2 will use 7-meter antennas.

The reflector of an antenna is shaped so that the signal contours on earth follow the shape of the country or area that is to be covered. A Japanese satellite for example shapes its signal to the outline of the Japanese islands. Figure 2.7 shows antenna coverage patterns for the Intelsat V international satellites.

Highly directional antennas which cover small portions of the earth are also used. These are larger in diameter than wide-coverage antennas. Relatively narrow beams are called *spot beams.* They receive over a similarly narrow angle. The NASA satellite ATS-6 has an antenna with an unusually narrow beamwidth of 1°.

The smaller the coverage of the antenna, the greater the power received from it in the area it covers. A signal from an antenna with a beamwidth of 1° is stronger than that from an earth-coverage antenna by a factor of $17.34^2 = 301$. The signal received at the satellite from an earth station in its beam is also stronger by the same amount, so the relayed signal is much better than with an earth-coverage antenna. However, the portion of the earth covered by a 1° beam is only about 500 miles across. (The coverage width varies with the angle of elevation of the satellite at the receiving location.) To achieve a narrow-angle beam an antenna must be large. The antenna of the ATS-6 satellite is 10 meters across (Fig. 2.8) and so dominates the satellite design. An antenna which covers the continental United States is about 2 meters across.

MULTIPLE SPOT BEAMS

A single large antenna dish can produce many spot beams, as does the ATS-6 dish. Many small feed horns are positioned so that their signals are reflected in narrow beams by the dish, just as a huge concave mirror could be made to reflect narrow light beams from several bulbs.

The antenna feeds under the ATS-6 dish are illustrated in Fig. 2.9. There are different feeds for different frequency bands; 21 feeds are used for the S-band frequencies, arranged in the shape of a cross. Figure 2.10 illustrates the beams that result from the feeds on the north-south arms of the cross. The two shaded feeds in Fig. 2.9 correspond to the two shaded beams in Fig. 2.10. The signal strength 1/2° away from the midpoint of the beam is slightly less than half the power at the center. *Beamwidth* usually refers to the width of the beam

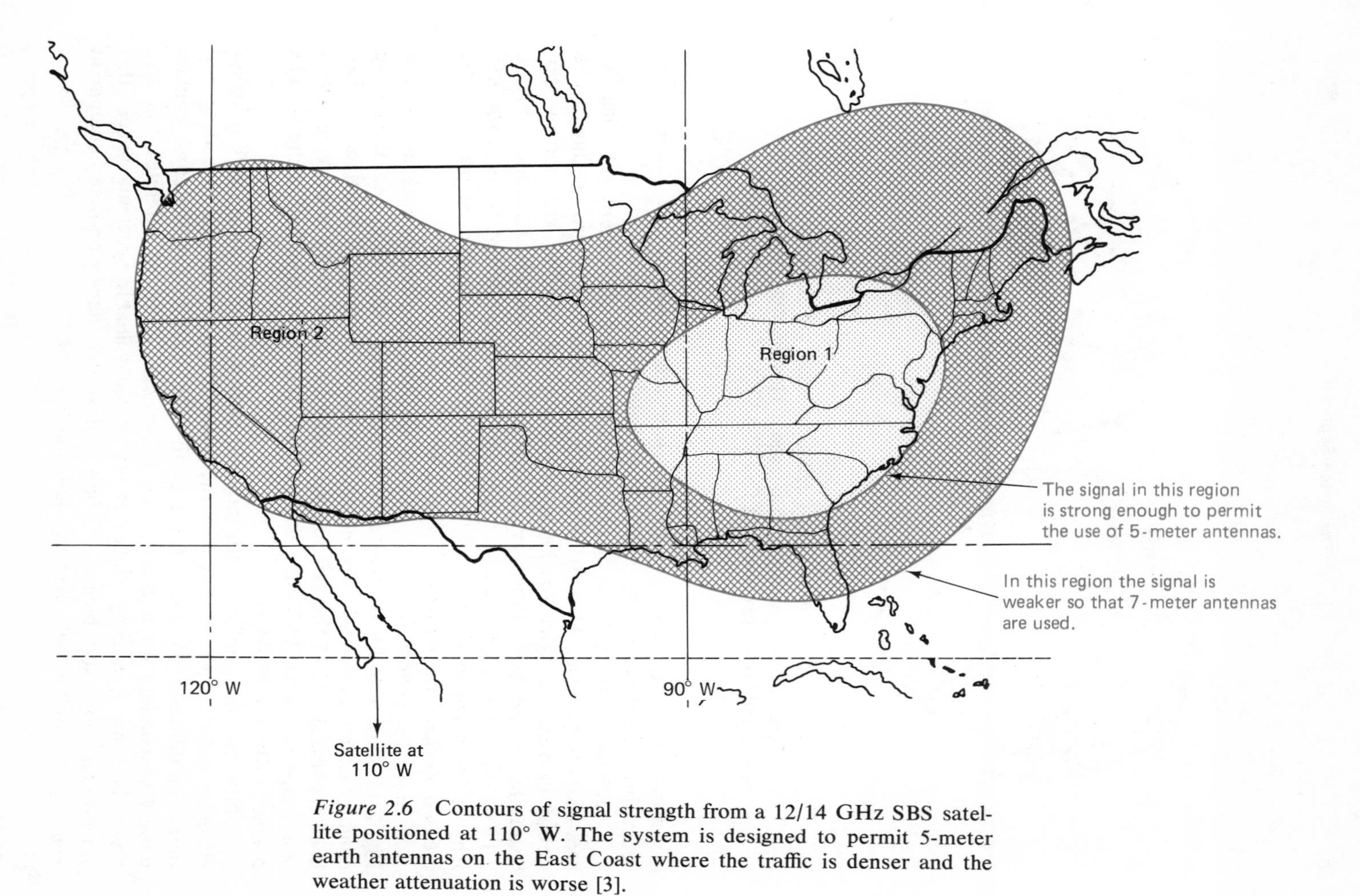

Region 2
Region 1
The signal in this region
is strong enough to permit
the use of 5-meter antennas.
In this region the signal is
weaker so that 7-meter antennas
are used.
120° W
90° W
Satellite at
110° W

Figure 2.6 Contours of signal strength from a 12/14 GHz SBS satellite positioned at 110° W. The system is designed to permit 5-meter earth antennas on the East Coast where the traffic is denser and the weather attenuation is worse [3].

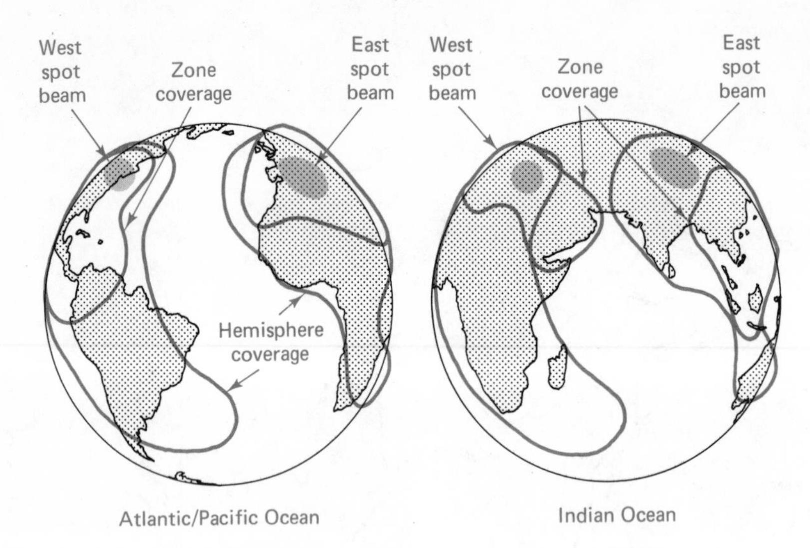

Figure 2.7 The reflectors of satellite antennas are shaped to contour the signal to the geography in question. This diagram shows the spot beams and large area beams of INTELSAT V.

measured to those points which are half the power at the center. The beamwidth illustrated in Fig. 2.10 is thus slightly less than 1°.

Different antenna feeds can be switched on and off, thereby selecting the spot beam to be used. Also, for positioning the beams the entire satellite can be slewed in orbit, swinging the beam as it moves. A satellite with narrow spot beams has to be held stationary in orbit with considerable accuracy. Figure 2.11 shows the area covered by the beams on earth, referred to as *footprints*. Any of the north-south pairs of footprints in Fig. 2.11 could be produced by the two shaded antenna feeds in Fig. 2.9.

In its first year of life of the ATS-6 satellite, spot beams such as those in Fig. 2.11 were used by the Department of Health, Education, and Welfare to conduct experiments with educational television and medical assistance in isolated regions. In 1975 the satellite was moved over to India to provide television transmission to 5000 villages previously (like more than half of the world's people) without television.

Not only do satellites with multiple narrow-beam antennas give a higher effective radiated power, but also the same frequency can be reused several times for different portions of the earth. The satellite can therefore contain more transponders for a given fixed bandwidth. Different satellites can also reuse the same frequency, thus conserving the valuable spectrum space. If a satellite has many spot beams, it might be desirable to have on-board equipment which can switch signals from one antenna to another.

Figure 2.8 The 10-meter antenna of the NASA ATS-6 satellite. This can give narrow spot beams, as shown in Figs. 2.10 and 2.11, which give a strong received signal on earth and hence permit low earth station costs.

SATELLITE ATTITUDE CONTROL

To keep satellite antennas pointing at the earth, or at a selected portion of the earth, the satellite must be stabilized so that it does not swing about. A satellite can spin about its three axes without moving from its orbital position. The three motions are referred to as *yaw, pitch,* and *roll.* They are illustrated in Fig. 2.12.

Stablilization of a satellite can be three-dimensional so that there is negligible yaw, pitch, or roll, or it can be about one axis. A cylindrical satellite may be spun about the axis of the cylinder, and the axis of the cylinder may be aligned with the earth. An antenna covering the earth may be used, which is

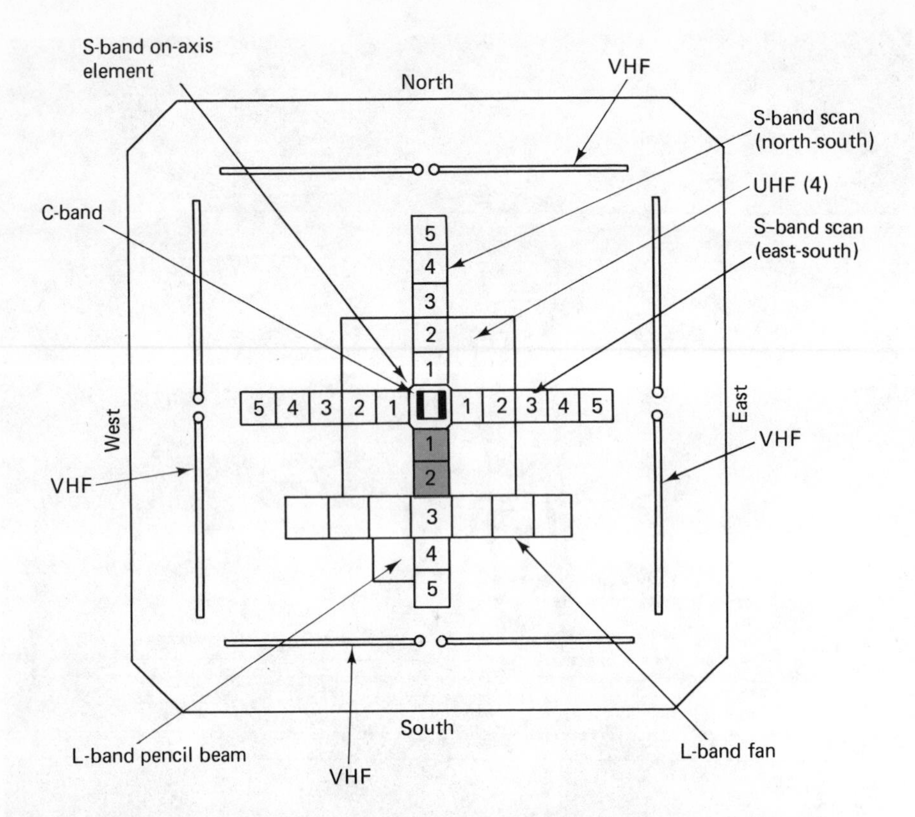

Figure 2.9 Multiple antenna feeds underneath the large reflector of the ATS-6 satellite. These feeds are on top of the large electronics compartment seen in Fig. 2.8. In addition to selecting the required feeds to give appropriate spot beams, the satellite can be slowed to point to the correct part of the earth.

symmetrical about the axis of the cylinder. In this way the satellite's momentum stablizes it.

A satellite can be stabilized in space either by its own spinning or by some form of gyroscope or flywheel on board. A system employing the motion of fluid such as mercury in an enclosed circuit has been proposed and has the advantage of being relatively friction-free. If a satellite is controlled by its own spinning and yet uses spot beams, the antenna subsystem must be despun; i.e., it must be held stationary while the body of the satellite spins.

All the satellites in Fig. 2.13 have one-axis spin stabilization. INTELSAT IV, TACSAT, the Canadian ANIK satellites, and the Western Union WESTAR satellites all have despun antenna subsystems. The INTELSAT IV despin mechanism has sensors which detect the edge of the earth and make adjustments to the despinning, as necessary. The ANIK satellites have the despinning controlled automatically from earth with control signals sent to the satel-

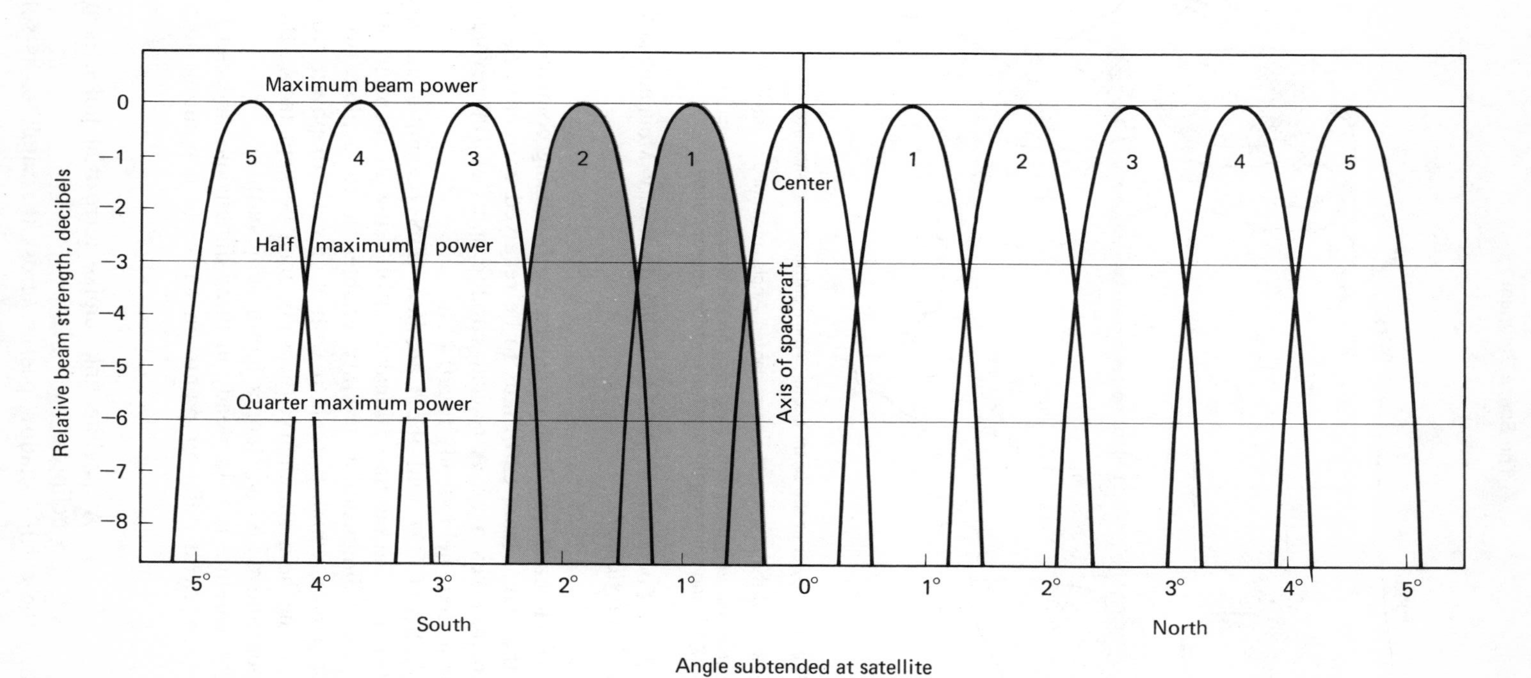

Figure 2.10 Antenna beams from the north-south row of antenna feeds shown in Fig. 2.9 (The red shaded beams correspond to the red shaded feeds of Fig. 2.9)

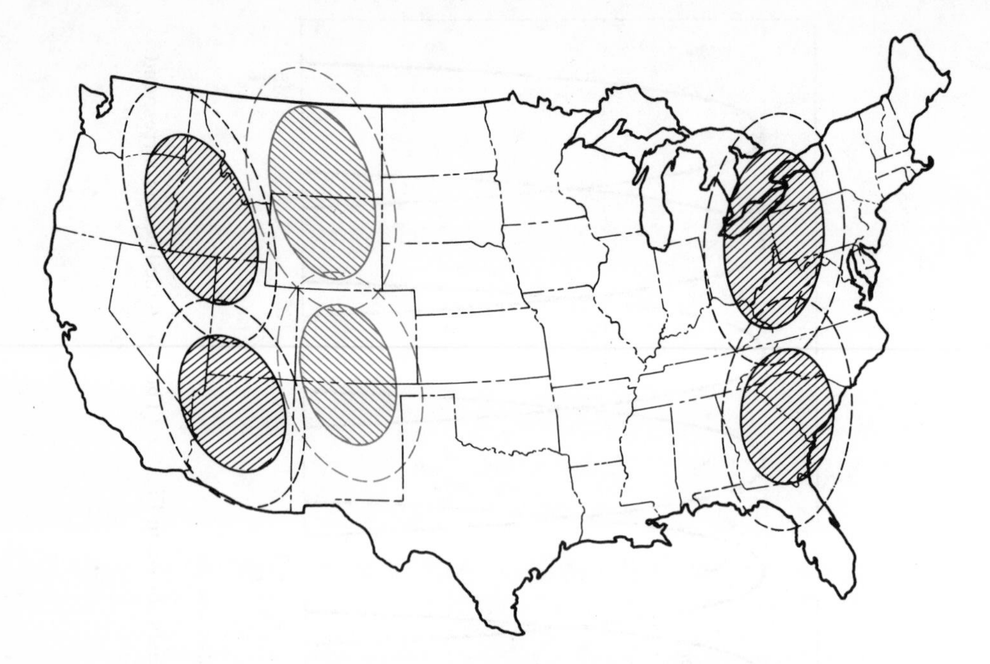

——— Half of the power at the beam center. Angle subtended at satellite = 0.9°

- - - - A quarter of the power at the beam center. Angle subtended at satellite = 1.2°

One north-south pair may be used by switching on the antenna feeds colored in Figs. 2.9 and 2.10, and adjusting the satellite position appropriately.

Figure 2.11 Footprints of ATS-6 satellite in Rocky Mountain and Appalachian regions.

lites. The WESTAR birds use both on-board earth-sensing control and control from earth. If the signal from earth fails to be received correctly, the on-board control mechanism takes over. A typical satellite spins at 100 revolutions per minute and has its axis held steady to ±0.1°.

The RCA SATCOM and the Canadian CTS (Communications Technology Satellite) have three-dimensional control using momentum wheels on board. CTS's three-dimensional stability enables it to deploy two 28-foot "sails" of solar panels, which provide it with 1 kilowatt of power. In addition to the satellite being held stationary, the sails must be kept facing the sun. A sun-tracking mechanism and dc stepper motor achieve this.

The ATS-6 satellite is held steady in three dimensions and can be swung to align its antenna with different points on earth, to a pointing accuracy of ±0.1°.

ECLIPSES

A satellite link, unlike terrestrial links, is prone to eclipses.

First, when the earth's shadow passes across the satellite, its solar cells

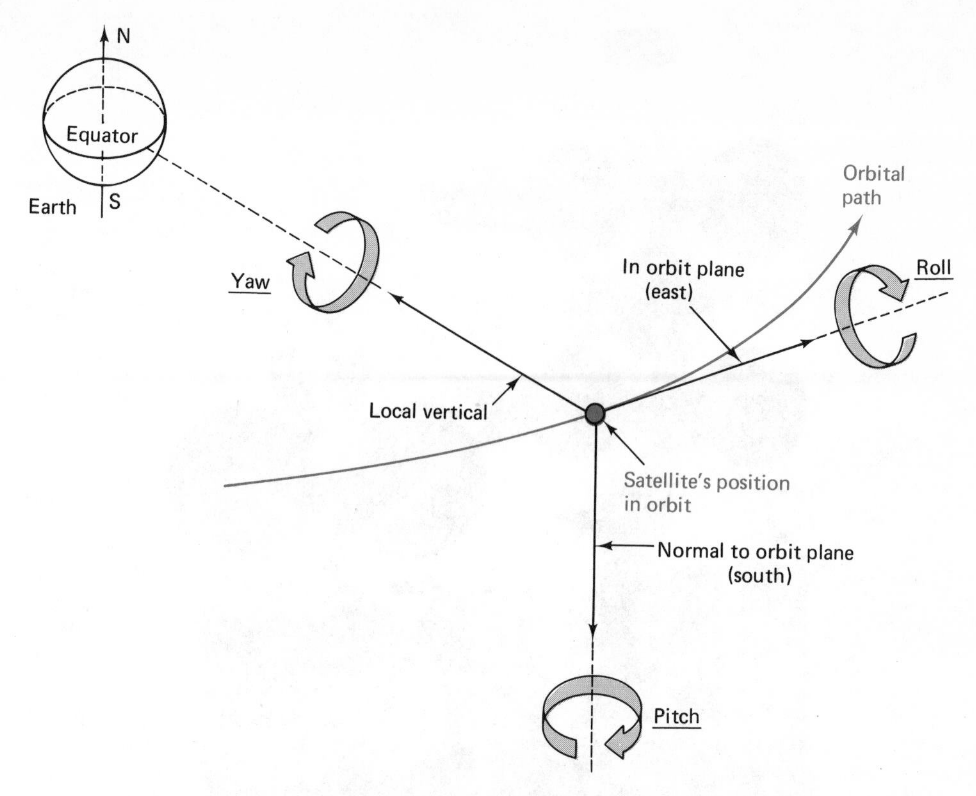

Figure 2.12 Yaw, pitch, and roll, the three spinning motions of a satellite.

stop operating. This occurs on 44 nights in the spring and 44 nights in the fall; 277 days of the year are free of eclipses. The maximum eclipses occur on the days of the equinoxes and last for 65 minutes. Eclipses on other days are shorter, as shown in Fig. 2.14. Less commonly, the moon's shadow passes across the satellite, like a solar eclipse on earth.

When a solar eclipse occurs, the time on earth beneath the satellite is close to midnight. Fortunately this is a period when the telephone or data traffic is low. By positioning a domestic satellite slightly to the west of the country it serves, its solar eclipses can be made to occur in the hours after midnight.

Most satellites carry storage batteries which are charged continuously from the solar cells. These batteries can keep the transponders operating during an eclipse. A satellite may not carry enough batteries because of weight to keep *all* the transponders in operation. This may not justify the weight requirement. It might, for example, be designed to keep half of its transponders operating during the eclipses. This may be thought to be sufficient for the marketplace served. The SBS satellites are designed with enough on-board batteries to continue full operation during eclipses.

A more serious form of outage occurs when the satellite passes directly in

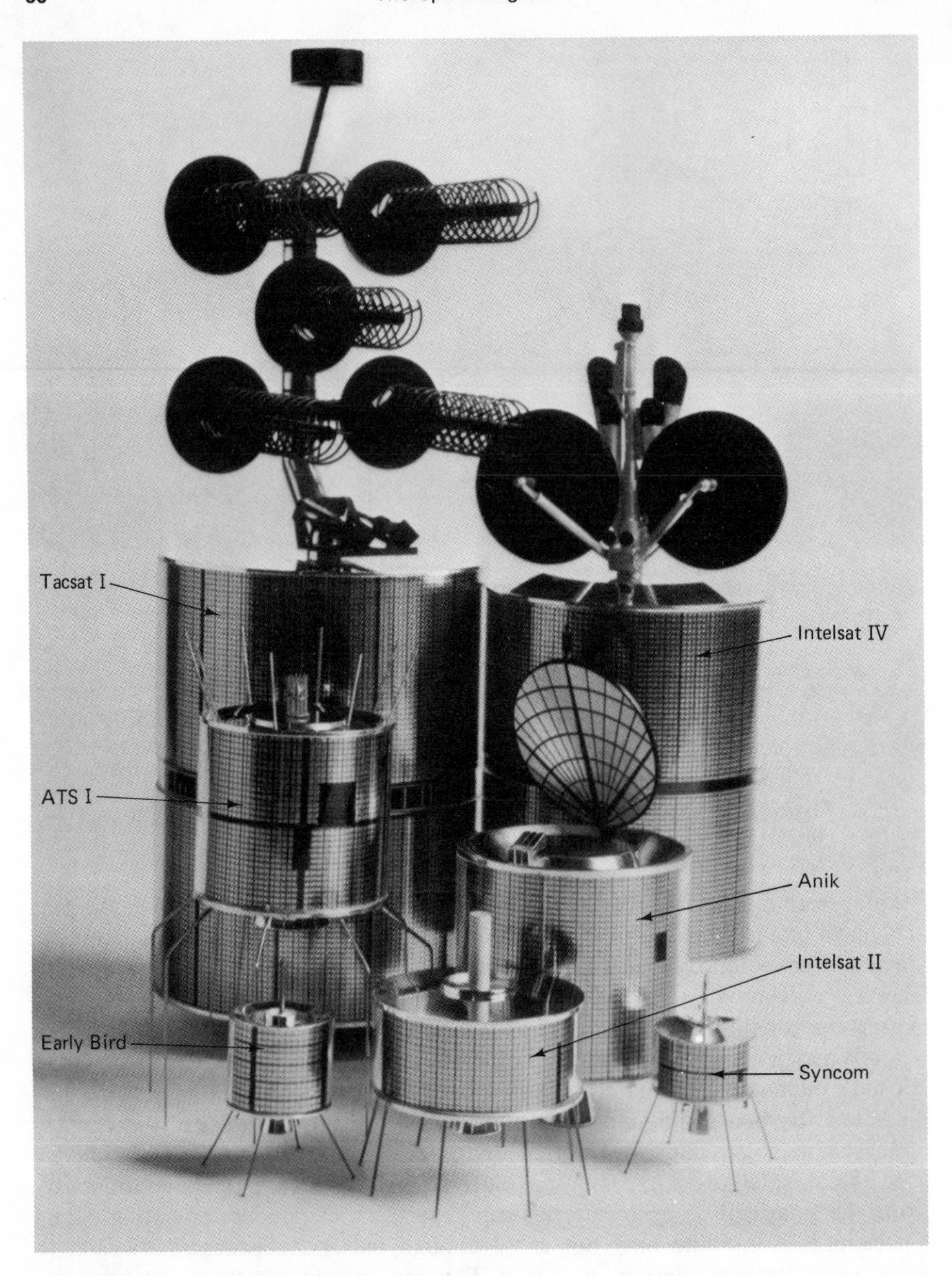

Figure 2.13 Satellites launched during the first decade of geosynchronous activity. All of these satellites were spin stabilized. TACSAT, INTELSAT IV, and ANIK had their antenna subsystems despun to point to their targets.

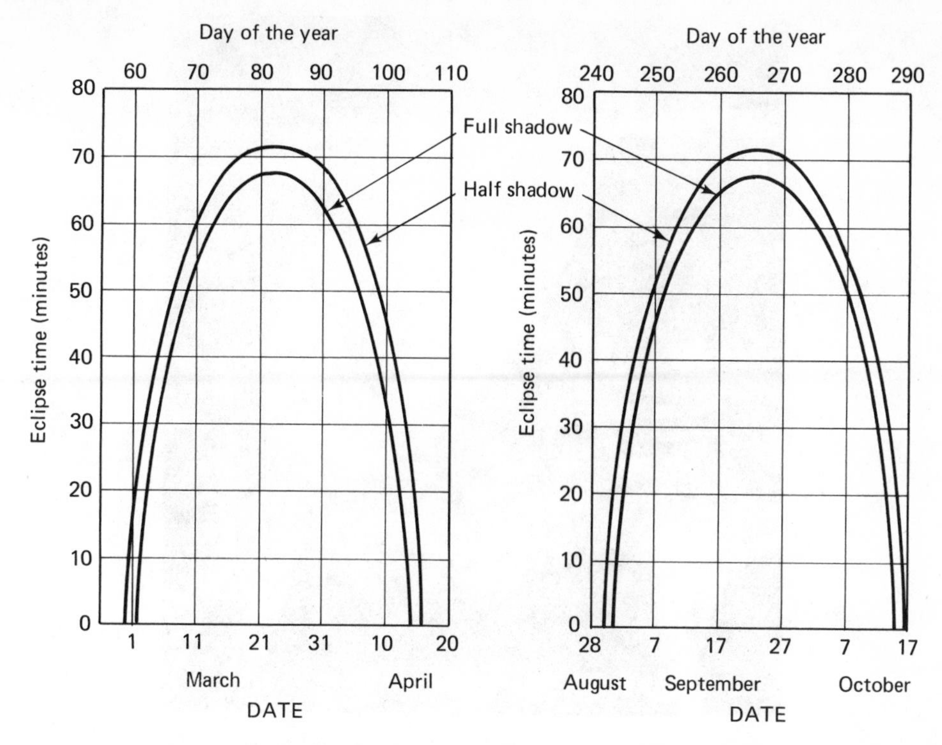

Figure 2.14 The dates and durations of satellite eclipses.

front of the sun. The sun, being of such a high temperature, is an extremely powerful noise source and so blots out transmission from the satellite. This outage lasts about 10 minutes on 5 consecutive days twice a year. The only way to achieve continuous transmission is to have two satellites and switch channels to the unaffected one before the sun outage begins. Most satellites *are* duplicated in orbit, not only to provide protection from sun outages but also protection if one satellite should fail. SBS will provide its customers with dual antenna systems if they require to be uninterrupted by sun outages. Sun-outage service is an optional extra.

The moon also passes directly behind the satellite on occasions. The moon shining directly into the earth antenna does not blot out the transmission, as does the sun, but it does increase the noise received and so degrades the transmission (as discussed in Chapter 7).

REFERENCES

1. Arthur C. Clarke, *Voices from the Sky,* Harper & Row, New York, 1965, p. 147.

Figure 2.15 The 8-foot antenna, above, at the Texaco executive offices in Harrison, N.Y., was in use for business conferences with their regional office in Bellaire, Texas. This was part of the Project Prelude experiment of Satellite Business Systems Inc., using the Canadian CTS satellite. Full-motion and freeze-frame color video is used, along with high-speed data and document (facsimile) transmission. All are transmitted digitally.

Figure 2.15 (continued) Industrial training courses have been conducted via satellite by SBS, using the Canadian CTS satellite, transportable earth stations, and the 7-foot Advent television screen. Companies participating were Rockwell International, Texaco, and Montgomery Ward – a representative cross-section of the potential user community. Both full-motion video and freeze-frame video are being used, the latter requiring a fraction of the bandwidth of full motion. Each color frame is transmitted in half a second. They show the teacher, audience, charts, and visuals typical of a business conference. A digital transmission rate of 1.544 Mbps is used. The technique is proving to be highly effective. Nearly three quarters of the participants in Project Prelude experiments said that teleconferencing is better than travel as a way of getting meetings organized [4] *(Pictures courtesy Satellite Business Systems.)*

2. "Report to the Federal Communications Commission on a Satellite Link Test Conducted by IBM Corporation at Poughkeepsie, New York," IBM System Development Division, Poughkeepsie, 1974.

3. *FCC Application of Satellite Business Systems for a Domestic Communications Satellite System,* Federal Communications Commission, Washington, D.C., 1976.

4. A copy of the Project Prelude evaluation report may be obtained from the SBA Public Affairs Department, 8003 Westpark Drive, McLean, Va. 22102.

3 ORBITS AND INCLINATION

The popular explanation of why a satellite remains in the sky rather than falling to earth is that the centrifugal force caused by its rotation around the earth exactly balances the earth's gravitational pull. A better-worded explanation is that the satellite's velocity would carry it away from the earth if gravity did not exist, but gravity constantly pulls it toward the earth and the acceleration due to gravity exactly balances the effect of its own velocity—like a stone being whirled around on a piece of elastic.

The closer the satellite is to the earth, the stronger is the earth's gravitational pull, and so the faster the satellite must travel to avoid falling into the earth. Low-earth-orbit satellites travel at about 17,500 miles per hour, and this speed carries them around the earth in 1½ hours. Communications satellites travel at 6,879 miles per hour and pass around the earth in 24 hours—the earth's own rotation time.

Figure 3.1 plots the time a satellite takes to travel around the earth against its height. The orbit at a height of 22,282 miles (35,860 kilometers) is special in that a satellite travels around the earth in exactly the earth's rotation time. If its orbit is over the equator and it travels in the same direction as the earth's surface, then it appears to be stationary over one point on earth. This orbit is called a *geosynchronous* orbit. The apparently stationary satellite is called a geosynchronous satellite. The INTELSAT satellites hang staionary in the sky over the Atlantic and Pacific. The U.S. domestic satellites hang over South America or the Pacific, above the equator. Such satellites have orbits very different from their experimental predecessors such as AT&T's TELSTAR satellites and RCA's RELAY satellites. The latter traveled rapidly around the earth at a relatively low height. The TELSTAR satellites had highly elliptical orbits, TELSTAR I from about 600 to 3800 miles and TELSTAR II from 600 to 6200 miles. The apogee of the elipse was positioned so that the satellite was within line-of-sight of certain stations for as long as possible. As with early manned orbital flights and most other satellites launched in the first decade of

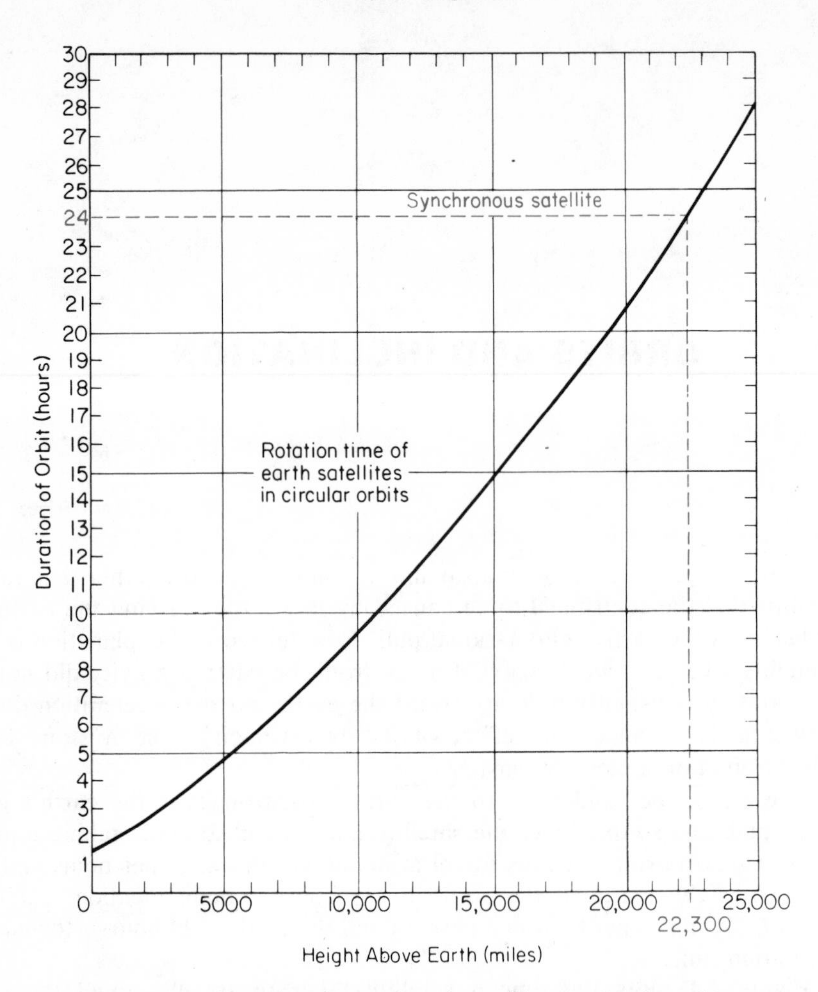

Figure 3.1 Rotation times of earth satellites in circular orbits.

space flight, they traveled around the earth in a few hours: TELSTAR I, 2 hours and 38 minutes; and TELSTAR II, 3 hours and 35 minutes. Herein lay their disadvantage for telecommunications; they were within line-of-sight of the tracking station for only a brief period of time, often less than half an hour. The Russians also use elliptical orbits for their MOLNIYA communications satellites, but their orbits are larger so that the satellites are within sight for longer periods.

Figure 3.2 shows three types of satellite orbits. As shown at the bottom of Fig. 3.2, three geosynchronous satellites can cover the entire earth with the exception of most unpopulated regions close to the poles. An advantage of using satellites in such a high orbit is that they cover a large portion of the earth. Figure 3.3 shows the maximum spacing between earth stations for different satellite heights, assuming that 5° is the minimum angle of elevation of the ground-

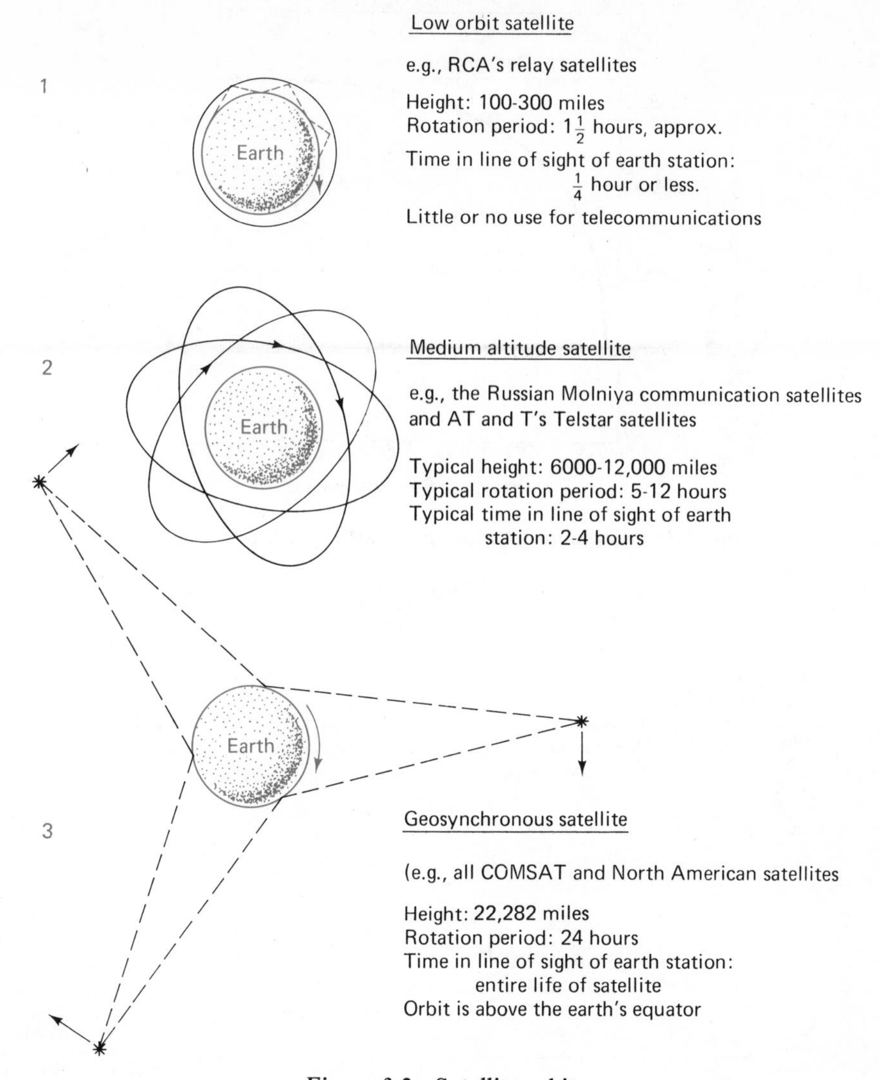

Low orbit satellite
e.g., RCA's relay satellites
Height: 100-300 miles
Rotation period: 1½ hours, approx.
Time in line of sight of earth station: ¼ hour or less.
Little or no use for telecommunications
1
Earth
Medium altitude satellite
e.g., the Russian Molniya communication satellites and AT and T's Telstar satellites
Typical height: 6000-12,000 miles
Typical rotation period: 5-12 hours
Typical time in line of sight of earth station: 2-4 hours
2
Earth
Geosynchronous satellite
(e.g., all COMSAT and North American satellites
Height: 22,282 miles
Rotation period: 24 hours
Time in line of sight of earth station: entire life of satellite
Orbit is above the earth's equator
3
Earth

Figure 3.2 Satellite orbits.

station antennas. Figure 3.4 shows the distances and propagation times with a geosynchronous satellite, drawn to scale.

The placement of a satellite in a synchronous orbit needs high-precision spacemanship. The launch vehicle first places it into a lengthy elliptical orbit with the highest part of the ellipse about 22,282 miles from earth. The craft is spin-stabilized in this orbit so that earth stations can communicate with its telemetry system. This orbit is then measured as exactly as possible and the satellite orientation adjusted so that it will be in precisely the right altitude for the next step. When the satellite is at the farthest end of its ellipse, traveling approximately at right angles to the earth's radius, a motor is fired at precisely

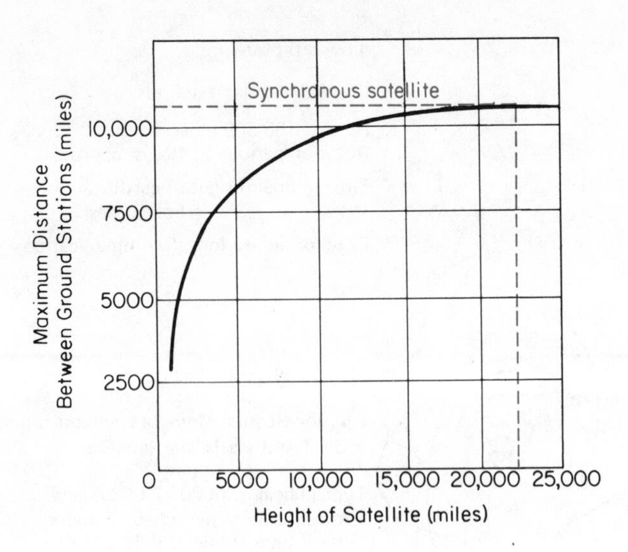

Figure 3.3 **Maximum separation of earth satellite stations.**

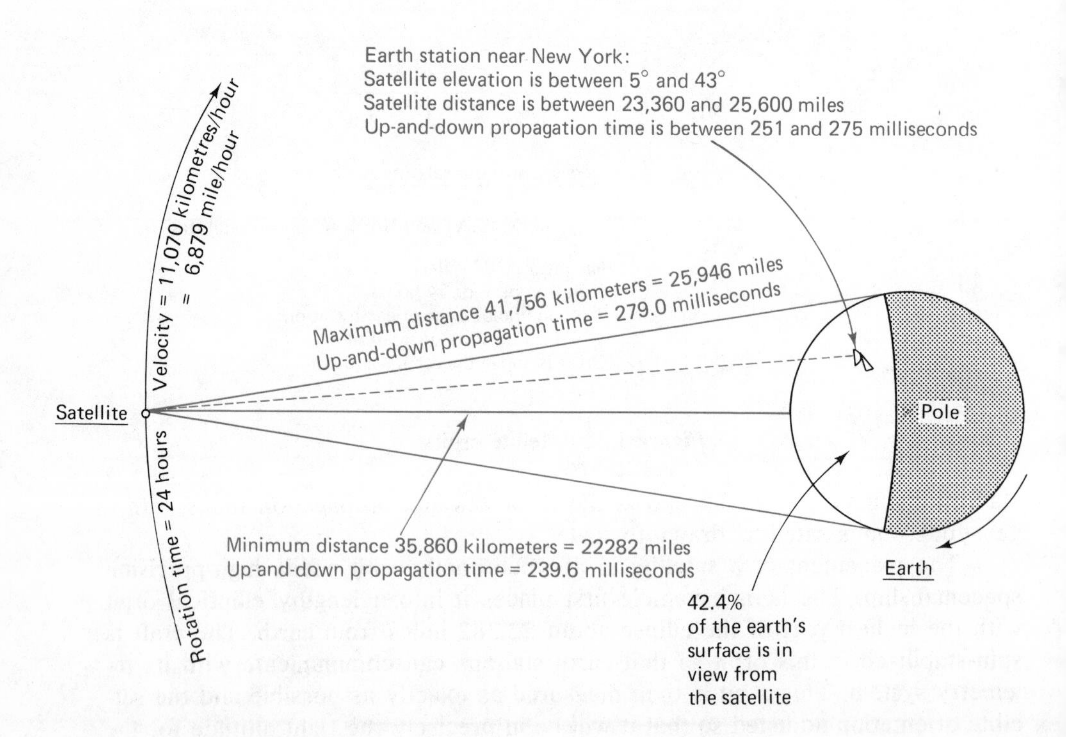

Figure 3.4 **Distances and propagation times with a geosynchronous satellite.**

the right instant to put the satellite in a circular orbit around the earth. The satellite's velocity is then adjusted to synchronize with the earth's rotation, and its attitude is swung so that its antenna points in the right direction. Figure 3.5 illustrates such a launching.

During the launching of the first INTELSAT II satellite, the "apogee" motor, which should change the elliptical orbit into the circular one, terminated its 16-second thrust prematurely and left the satellite plunging through space on a large elliptical nonsynchronous orbit. Comsat, however, managed to use it. Following its unplanned journey through space with their big antennas, they succeeded in transmitting the first live color TV between Hawaii and the American mainland. It was also used for commercial telephone circuits during those periods when its wanderings brought it within line-of-sight of suitable earth stations.

The geosynchronous orbit has great advantages for the systems designer, as follows:

1. The satellite remains almost stationary relative to the earth antennas, so the cost of computer-controlled tracking of the satellite is avoided. A fixed antenna is satisfactory (with provision for manual adjustments).
2. There is no necessity to switch from one satellite to another as one disappears over the horizon.
3. There are no breaks in transmission. A geosynchronous satellite is permanently in view.
4. Because of its distance, a geosynchronous satellite is in line-of-sight from 42.4% of the earth's surface (38% if angles of elevation below 5% are not used). A large number of earth stations may thus intercommunicate.
5. Three satellites give global coverage with the exception of the polar regions.
6. There is almost no Doppler shift, i.e., change in the apparent frequency of the radiation to and from the satellite caused by motion of the satellite to and from the earth station. Satellites in elliptical orbits have different Doppler shifts for different earth stations, and these increase the complexity of the receivers, especially when large numbers of earth stations intercommunicate.

The disadvantages of geosynchronous satellites are

1. Latitudes greater than 81.25° north and south (or 77° if angles of elevation below 5° are excluded) are not covered. There is little other than polar ice at these latitudes.
2. Because of the distance of the satellite, the received signal power, which is inversely proportional to the square of the distance, is weak, and the signal propagation delay is 270 milliseconds.

ORBITAL ADJUSTMENTS

Geosynchronous satellites must have their orbits adjusted occasionally to keep the satellite in position. Even if the satellite were launched into a perfect orbit, natural forces would introduce a slight drift of the orbit. However, with a

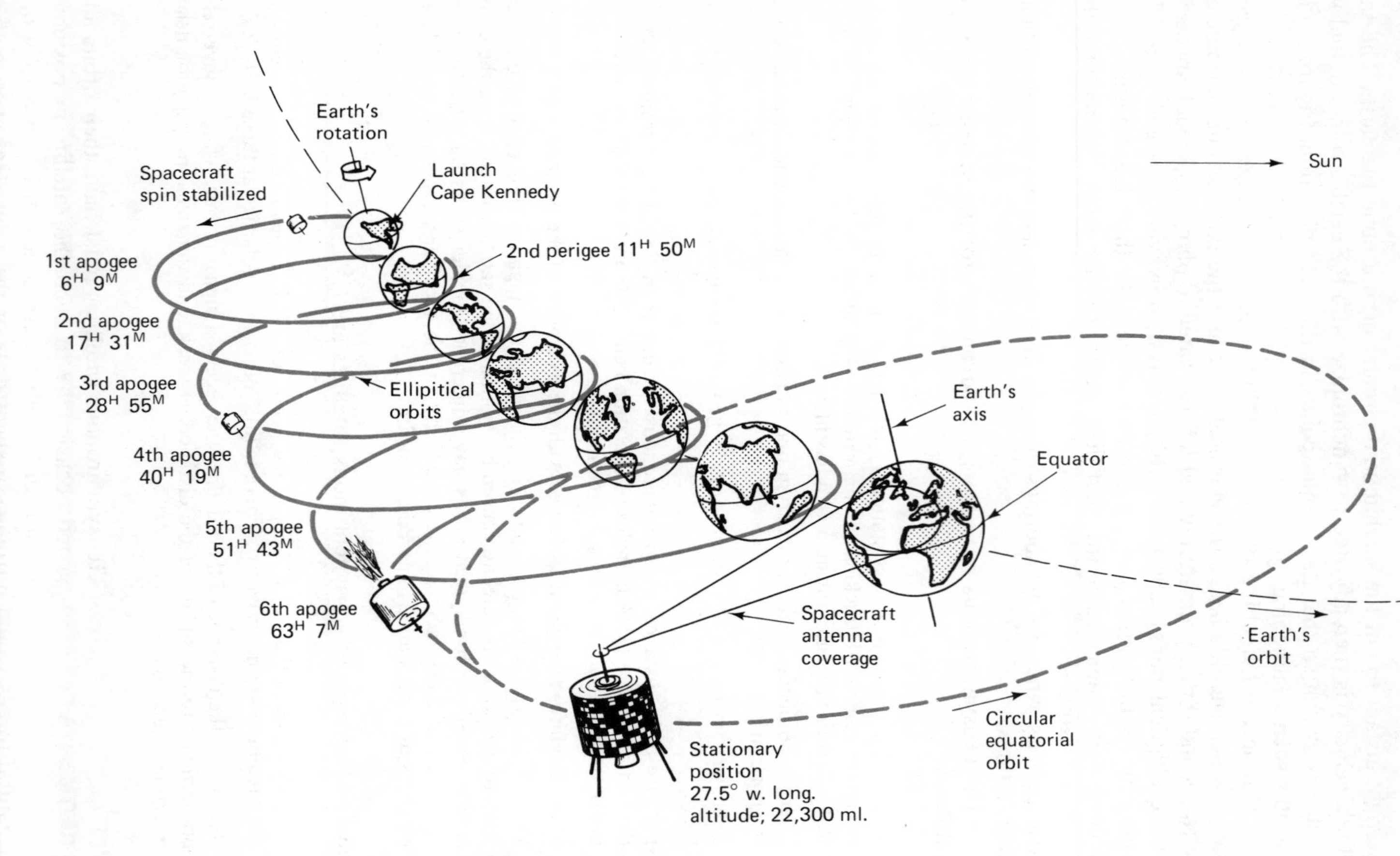
Earth's rotation
Spacecraft spin stabilized
Launch Cape Kennedy
1st apogee $6^H\ 9^M$
2nd perigee $11^H\ 50^M$
2nd apogee $17^H\ 31^M$
3rd apogee $28^H\ 55^M$
Ellipitical orbits
4th apogee $40^H\ 19^M$
5th apogee $51^H\ 43^M$
6th apogee $63^H\ 7^M$
Sun
Earth's axis
Equator
Spacecraft antenna coverage
Earth's orbit
Circular equatorial orbit
Stationary position
27.5° w. long.
altitude; 22,300 ml.

Figure 3.5 The launch and positioning of Early Bird, the first commercial communication satellite. (*Courtesy Hughes Aircraft Company.*)

The primary advantage of a synchronous orbit communications satellite is that ground installations are greatly simplified. By virtue of the satellite's "fixed" position, complex and expensive tracking antennas are not required. However, in order to hold its position in relation to the earth's axis, the satellite must be placed directly above the equator. Since Early Bird was launched from Cape Kennedy, which is north of the equator, certain maneuvers were necessary to properly position the satellite. These were performed on command from the earth station located at Andover, Maine. The launch vehicle for Early Bird was the Thrust Augmented Delta (TAD), a three-stage rocket built by Douglas Aircraft Company. Launched with its apogee motor in a forward position, Early Bird was aligned by the Delta's third stage at an angle of 16.7 degrees to the equator and spin stabilized. It was separated from the third stage at 26 minutes, 32 seconds after lift-off and coasted to its first apogee of 23,081 miles. During the elliptical orbits, Early Bird was precisely oriented by ground control.

With the firing of the apogee motor on the sixth apogee, Early Bird was thrust into an almost circular equatorial orbit near 32° W longitude with an eastward drift rate of 1.5° per day. Final synchronization to slow the drift rate and more nearly match the earth's rotation rate was accomplished on April 14, 1965. This maneuver placed Early Bird at its planned position, 28° W longitude. The attitude of the satellite was then changed to concentrate the antenna beam on the Andover and European earth stations. The earth station at Andover, Maine was equipped with Hughes developed telemetry and command equipment. Launching services and facilities were provided by NASA in accordance with an agreement with COMSAT.

good launch, the movement of the satellite relative to an earth station will be very slow.

The drift from a stationary position is caused by minor gravitational perturbations of the orbit due to the sun, the moon, and the oblateness of the earth and by the pressure of solar radiation.

The sun and moon act in a north-south direction, tending to incline the orbit away from the equatorial plane. A satellite at geosynchronous altitude in an orbit slightly inclined to the equatorial plane appears from earth to be describing a figure eight around its original stationary position. The orbit is pushed into the inclined position at the rate of about 0.86° per year. A velocity change of about 155 feet per second per year (106 miles per hour per year) is needed to correct this orbital change. To save power, a satellite is nudged into a slightly inclined orbit so that it drifts into an equatorial orbit and then into an orbit slightly inclined in the other direction before it receives another nudge.

The earth's gravitational field is not perfectly spherical because of the oblateness of the earth. There is therefore a very slight gravity gradient in the circular geosynchronous orbit, and the "bottom of the hill" is at longitudes of 79°E and 101°W. A geosynchronous satellite at these longitudes is therefore more stable than one elsewhere. 101°W is a good longitude for a North American domestic satellite, but most satellites will be far from the bottom of this

Figure 3.6 The sun shines on the solar cylinder of a spin-stabilized satellite at all times, except during eclipses.

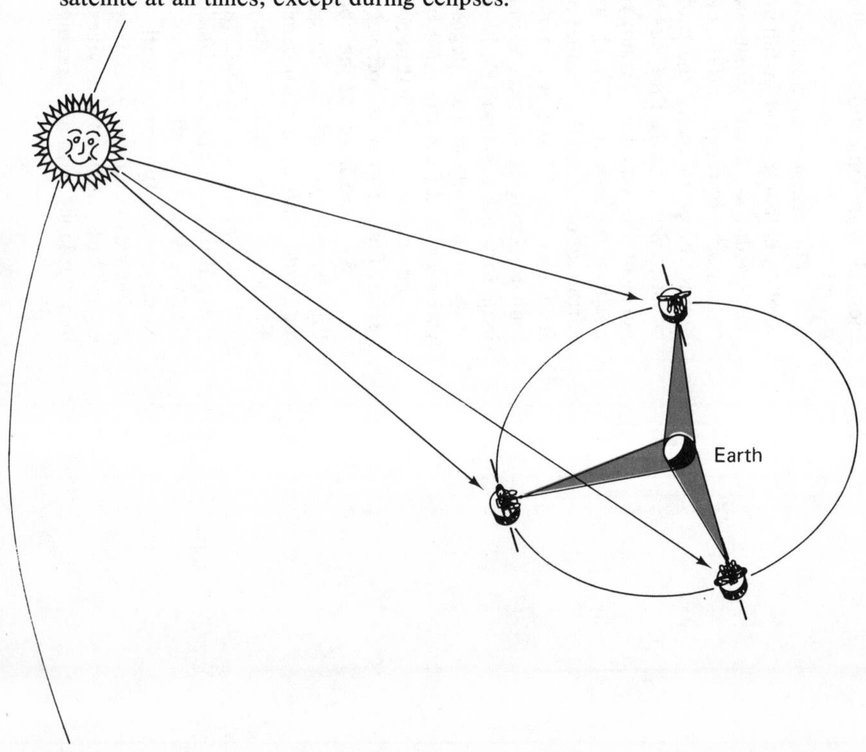

gravity gradient. A velocity increment of about 7 feet per second per year is enough to cancel the effect of the gradient.

Drift caused by the pressure of solar radiation will vary with the size of the satellite but is small compared with gravitational drift. Satellites are being designed with increasingly large solar panels, and their drift will be greater.

The orbit of the satellite needs adjusting periodically. This can be done either by the release of gas under pressure or, in a larger satellite, by small rockets. The release of hydrazine gas under pressure through a catalyst which causes explosive decomposition is normally used.

How often the gas jets are fired depends on how accurately the satellite position needs to be maintained. When EARLY BIRD was the only satellite over the Atlantic its gas jets were fired about once a year. The orbit over North America threatens to become crowded with domestic satellites, and these must be prevented from drifting too close together so that radio beams to adjacent satellites interfere. Minor adjustments every few weeks are necessary for accurate station keeping.

ANGLE OF ELEVATION

The angle of elevation of the satellite is the angle subtended at the antenna between the satellite and the earth's horizon.

If the angle of elevation is too small, the radio beams have to pass through much of the earth's atmosphere and are severely affected by noise and absorbtion. Usually 5° is regarded as the minimum practical angle of elevation. Figure 3.7 shows an earth station in the far north of Canada with an elevation

Figure 3.7 An earth station with an angle of elevation near to 5° in the frozen north of Canada.

of 5°. It is advantageous to keep the angle of elevation greater when possible. With satellites every decibel counts.

Figure 3.8 shows how the attentuation of satellite signals, by heavy rain and fog, varies with the angle of elevation. It will be seen that the percentage of the signal lost rises rapidly for angles of elevation below 10°. Figure 3.8 is plotted for the 4/6-GHz frequency bands. At 12/14 GHz the loss is much greater, as shown in Fig. 3.9. Most commercial satellites today use the 4/6-GHz band; however, the 12/14-GHz frequencies will make it possible (as we shall discuss later) to have satellite antennas in major cities where they are most needed. With the 12/14 frequencies it is desirable to use satellites and earth stations positioned to give an angle of elevation greater than 30° whenever possible.

The angle of elevation depends on the latitude of the earth station and the

Figure 3.8 Attenuation of 4-6 GHz signals in the earth's atmosphere. 5° is about the minimum useful angle of attenuation.

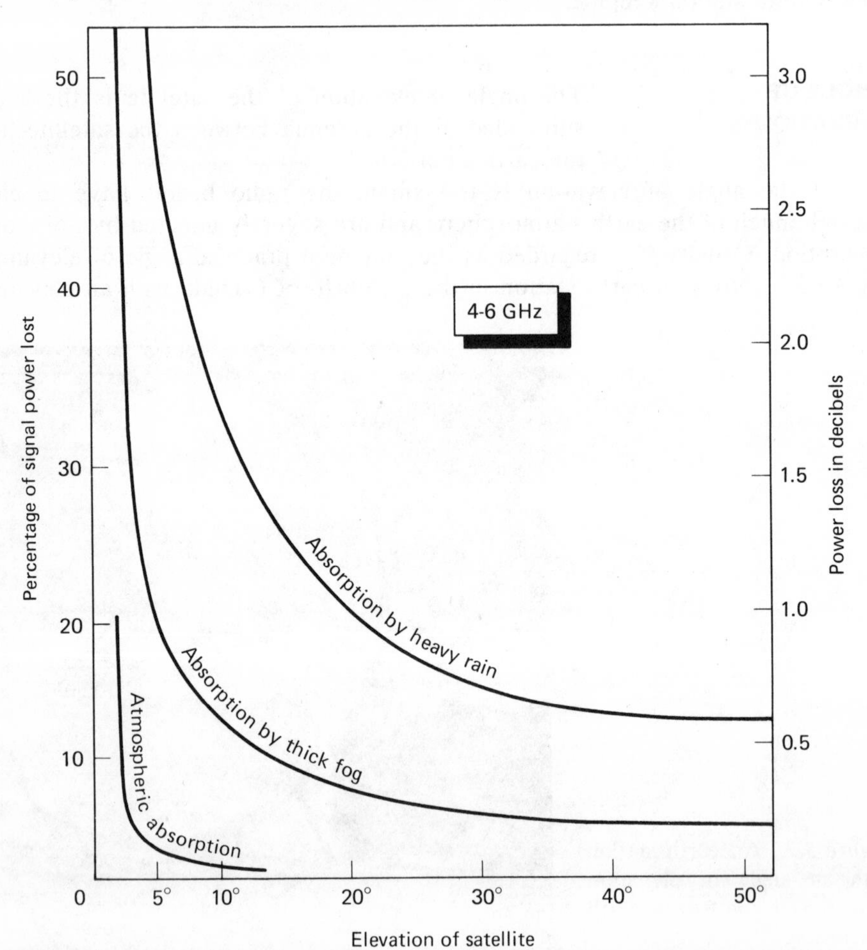

difference in longitude between the earth station and the satellite, ΔL. Figure 3.10 is a chart which may be used to estimate angles of elevation.

Let us suppose that a satellite is intended to link Boston and Los Angeles and points in between. The latitudes and longitudes are as follows:

	Latitude	Longitude
Boston	42.20°N	71.05°W
Los Angeles	34.00°N	118.15°W

A promising position for the satellite would be at a longitude halfway, i.e., 94.60°W. ΔL (the difference in longitude between the earth station and the satellite) would then be 23.55° for both Los Angeles and Boston. From Fig. 3.10 the angle of elevation at Los Angeles would be 43° and at Boston, 36°. It would be better to have the satellite closer to the location with the largest lati-

Figure 3.9 Attenuation of 12-14 GHz signals in the atmosphere.

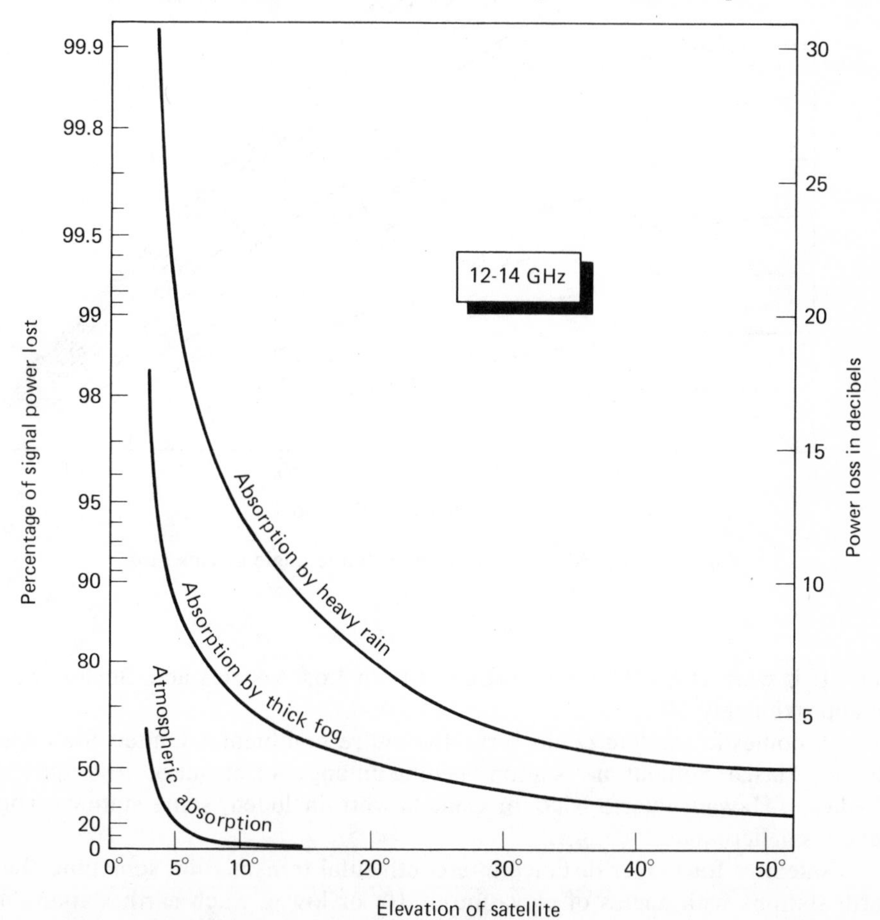

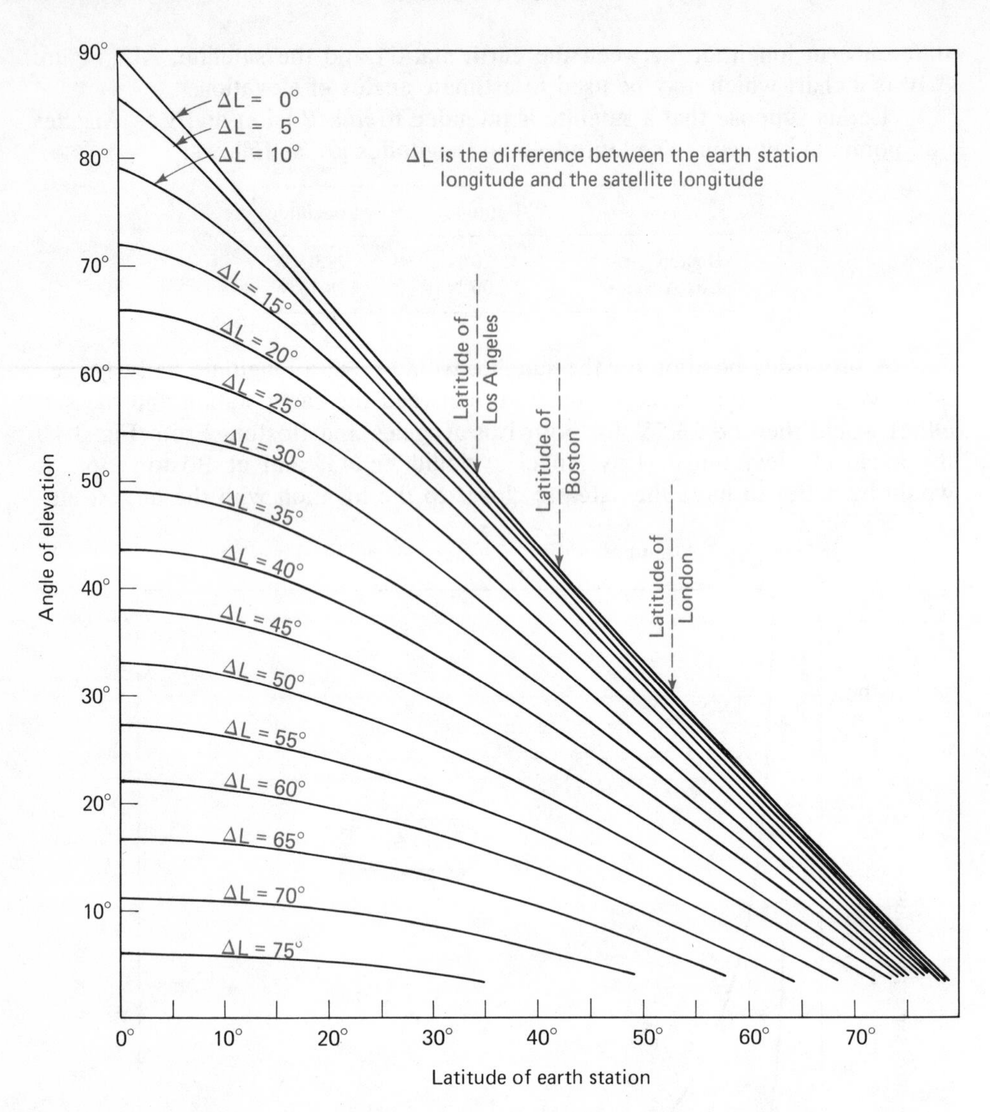

Figure 3.10 A chart giving the satellite angle of elevation.

tude. If it were at 88°W, the elevation at both Los Angeles and Boston would be approximately 38°.

A domestic satellite could serve the entire continental United States, excluding Alaska, without any station needing an angle of elevation less than 35°. If Alaska, Hawaii, Puerto Rico, or Canada were included, some stations would have a smaller angle.

Satellites for longer-distance intercontinental transmission sometimes have earth stations with angles of elevation of 10° or lower. Such earth stations use larger antennas and the 4/6-GHz frequencies. Figure 3.11 shows the earth coverage that corresponds to elevations of 10°, 15°, and 20°.

PROPAGATION DELAY

The distance between satellite and earth station varies somewhat with the angle of elevation, and consequently the propagation delay varies also. A satellite vertically overhead is 22,283 miles away. Its maximum distance from an earth station is 25,946 miles (assuming the earth to be spherical and the angle of elevation to be zero).

The time the signal takes to reach the satellite is the distance divided by the velocity of light, 186,000 miles per second. A round trip, up and down, thus varies from 239.6 to 279.0 milliseconds, as shown in Fig. 3.4.

Figure 3.12 shows how the distance and propagation delay varies with the angle of elevation. The minimum one-way propagation delay at New York City is 125 milliseconds—assuming the satellite has the same longitude as New York. For a satellite with 5° elevation at New York, or elsewhere, the one-way trip is 137 milliseconds.

The calculations later in the book assume a satellite distance of 25,000 miles, or a propagation delay of 270 milliseconds for an up-and-down trip. The reader should note that these are variables, as shown in Fig. 3.12.

ORBITAL SPACING

Satellites using the same frequencies must not be placed too close together, or there will be interference. At the time of writing, 4/6 GHz satellites are being kept at least 4° apart.

How close they can be without interference is a complex question because it depends on many factors in the design both of the space segment and the ground segment. The spacing is affected by the bandwidth of the transmitting earth stations. The beamwidth varies with the antenna size and frequency band used. The width of the beam in space is inversely proportional to the width of the transmitting antenna. A larger antenna can focus the beam more, like a searchlight. There is a trend toward smaller earth antennas, suggesting fewer satellite positions. On the other hand, the use of higher-frequency bands permits more satellites to share the orbit, first because different frequencies do not interfere, and second because the beamwidth is inversely proportional to the frequency.

The up-link normally operates at a higher frequency than the down-link because this gives a narrower beam shining into space. However, the number of satellites could be approximately doubled if the up-link and down-link frequencies were reversed for half the satellites, e.g., 4/6 GHz for some and 6/4 GHz for others. The 12/14 up-link frequency, 14.25 GHz, is 2.31 times the 4/6-GHz frequency, 6.175 GHz. If the same sized earth antennas are used, it is theoretically possible to have 2.31 times as many 12/14-GHz satellites as 4/6-GHz satellites. There could be five times as many 20/30-GHz satellites.

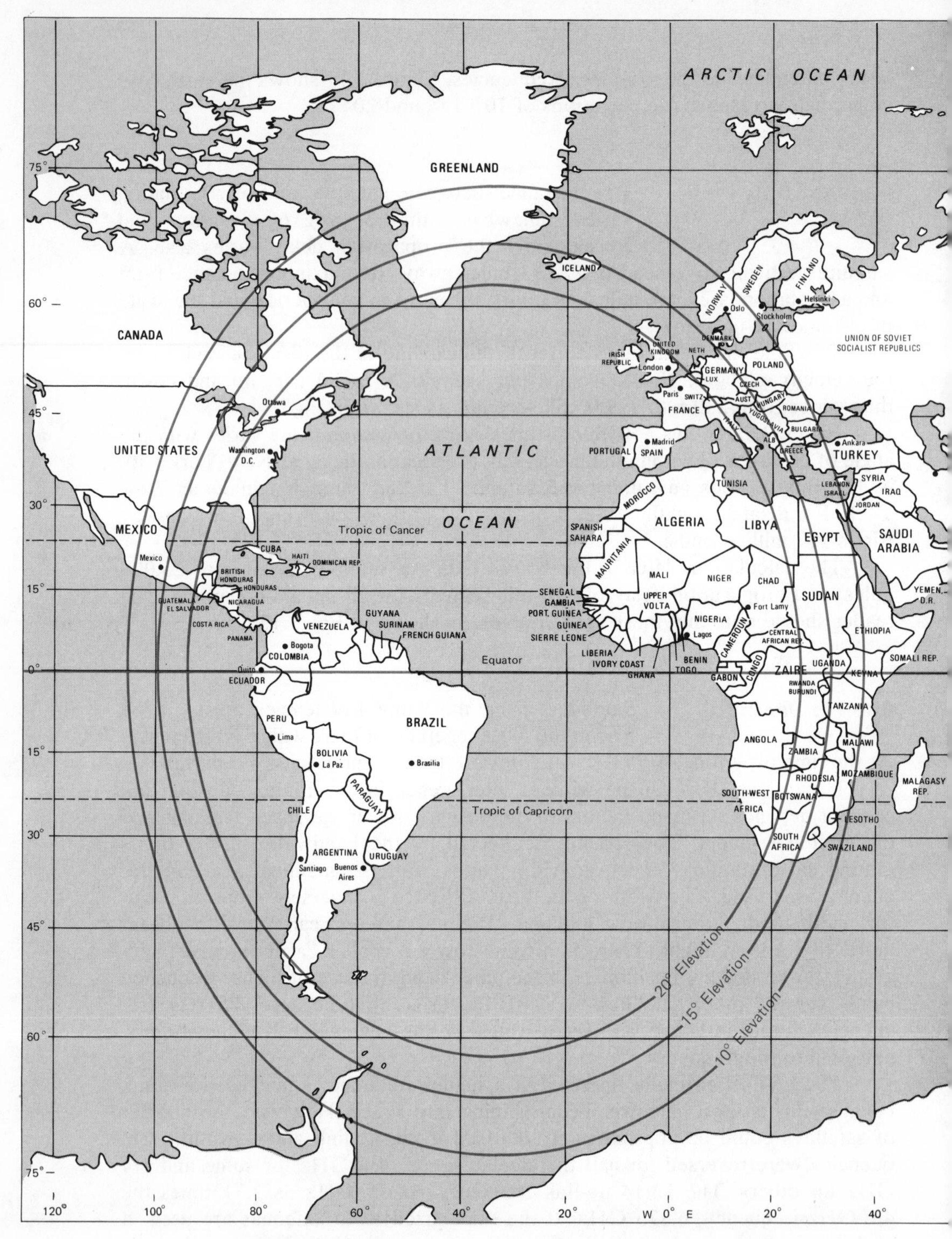

Figure 3.11 Earth coverage contours for a synchronous satellite.

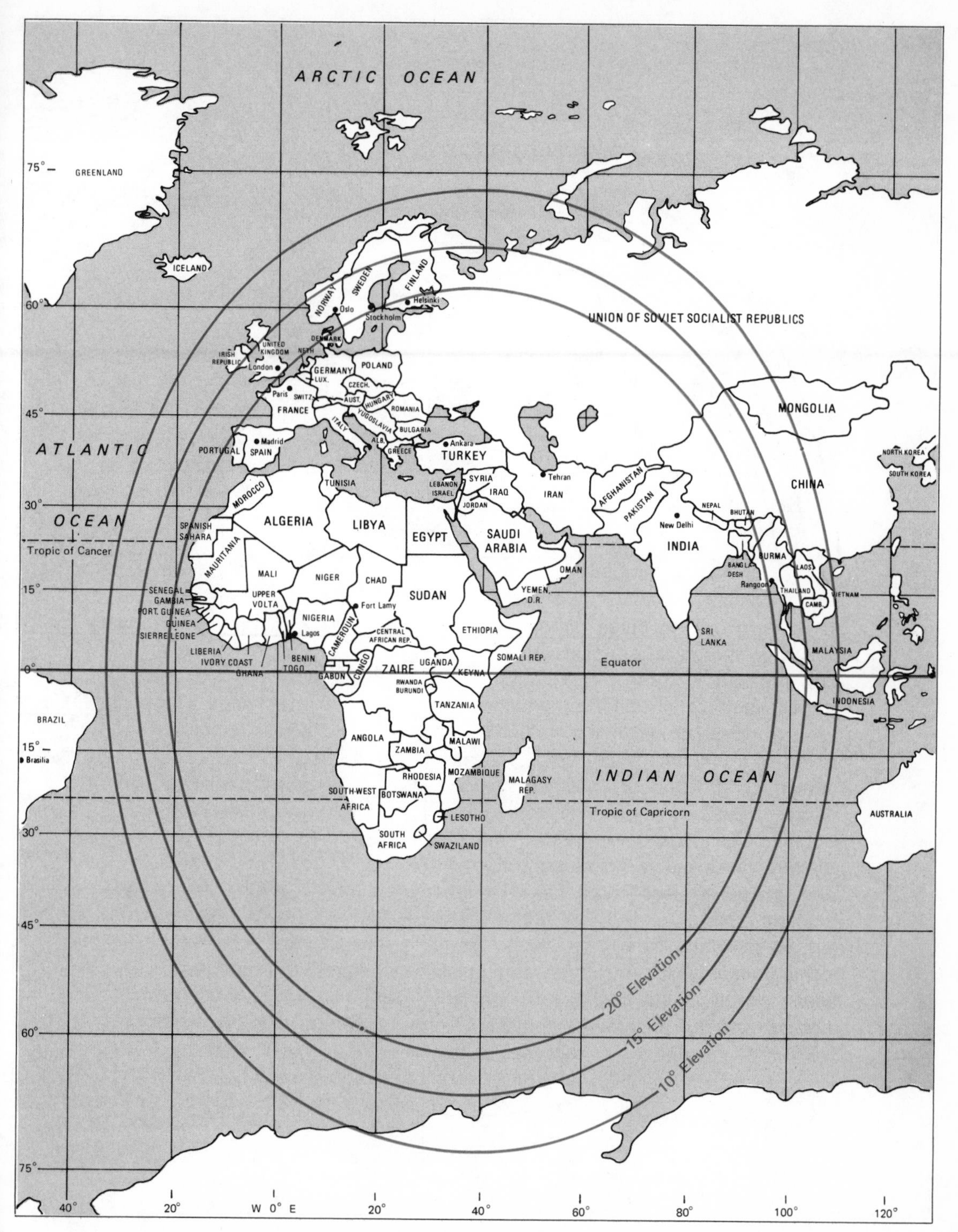

Figure 3.11 (continued).

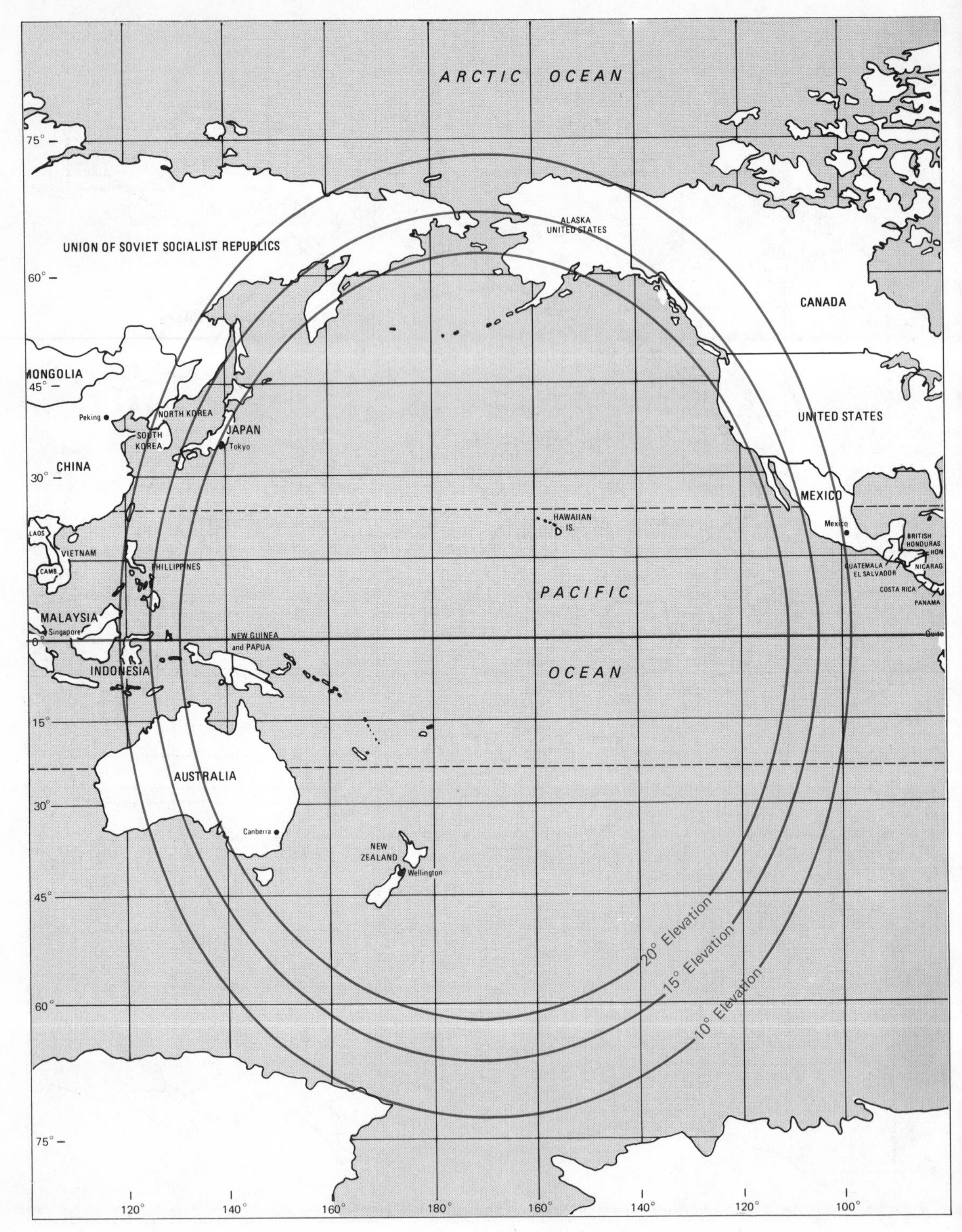

Figure 3.11 (continued).

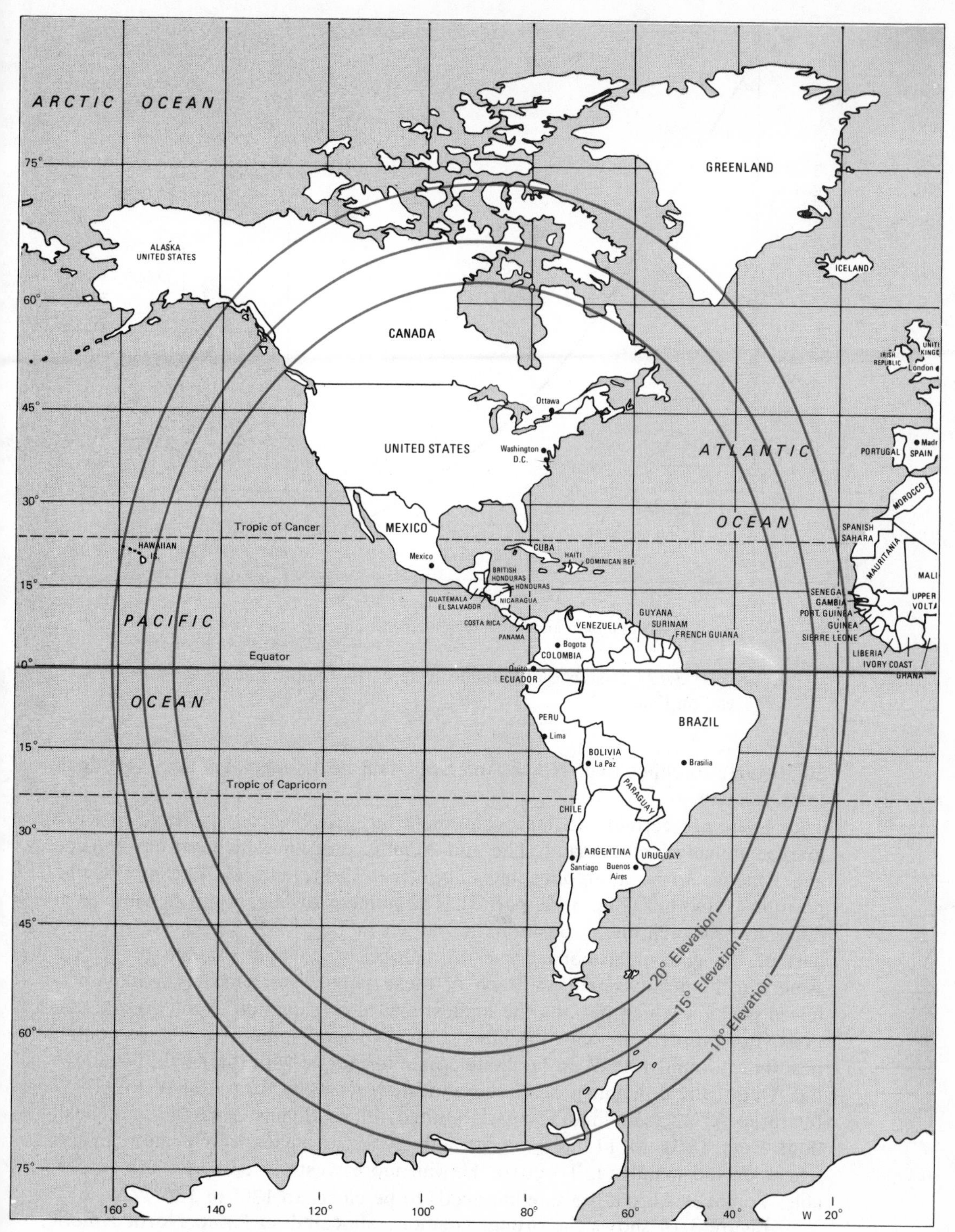

Figure 3.11 (continued).

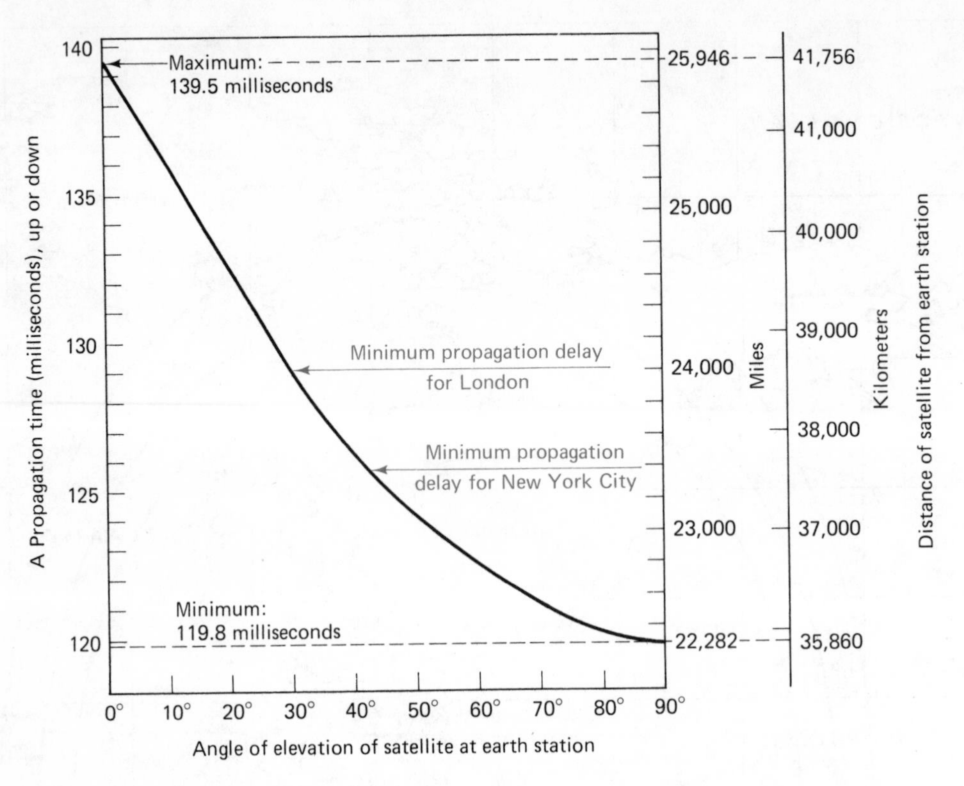

Figure 3.12 Distance of satellite from earth station; and signal propagation time.

20/30-GHz satellites over North America could be 1° apart, giving a very high total capacity.

There are certain satellite positions in the geosynchronous circle that are in exceptionally high demand. The mid-Atlantic position which permits Europe and America to be interconnected is one (Fig. 3.11, part 1). The mid-Pacific position is another (Fig. 3.11, part 3). The number of telephones in view from different geosynchronous positions is plotted in Fig. 3.13. The most crowded part of the geosynchronous circle will probably be that suitable for North American domestic satellites. Each of these popular regions has a most-preferred position which permits the highest angles of elevation at the ground stations. For North American satellites, Canada wants some of the most central positions, around 100°W, to facilitate communications with the north, far above the Arctic Circle. Canada has been allocated positions from 104°W to 114°W for three ANIK satellites 5° apart. United States domestic satellites in positions from 85°W to 115°W give attractively large angles of elevation everywhere on the mainland. To cover Hawaii more westerly positions are preferable. To cover Alaska the satellite needs to be closer to 120° or 130°W.

Figure 3.14 shows the orbital positions allocated so far to North American satellites.

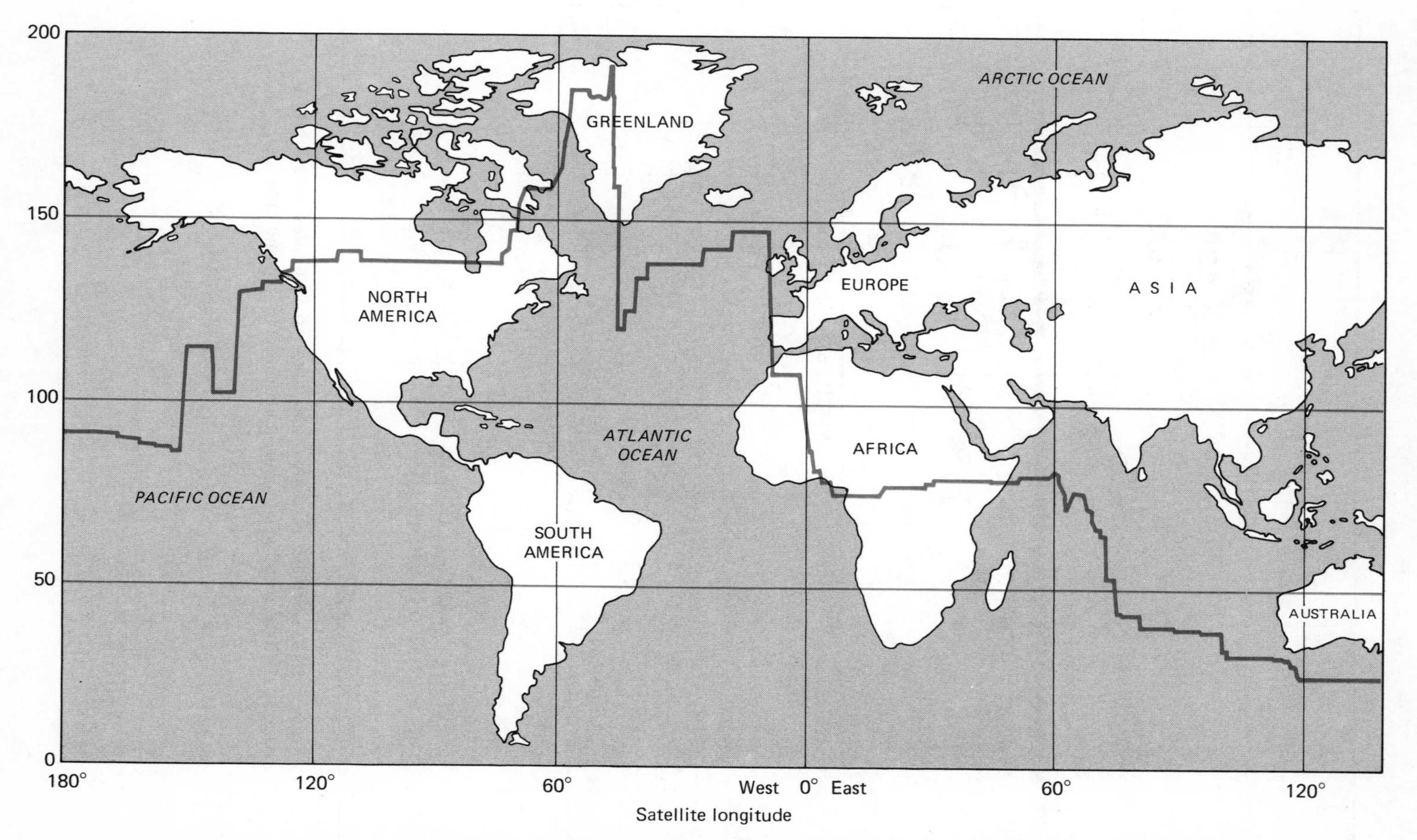

Figure 3.13 Numbers of telephones in line of sight from geosynchronous orbit. *(Data from INTELSAT.)*

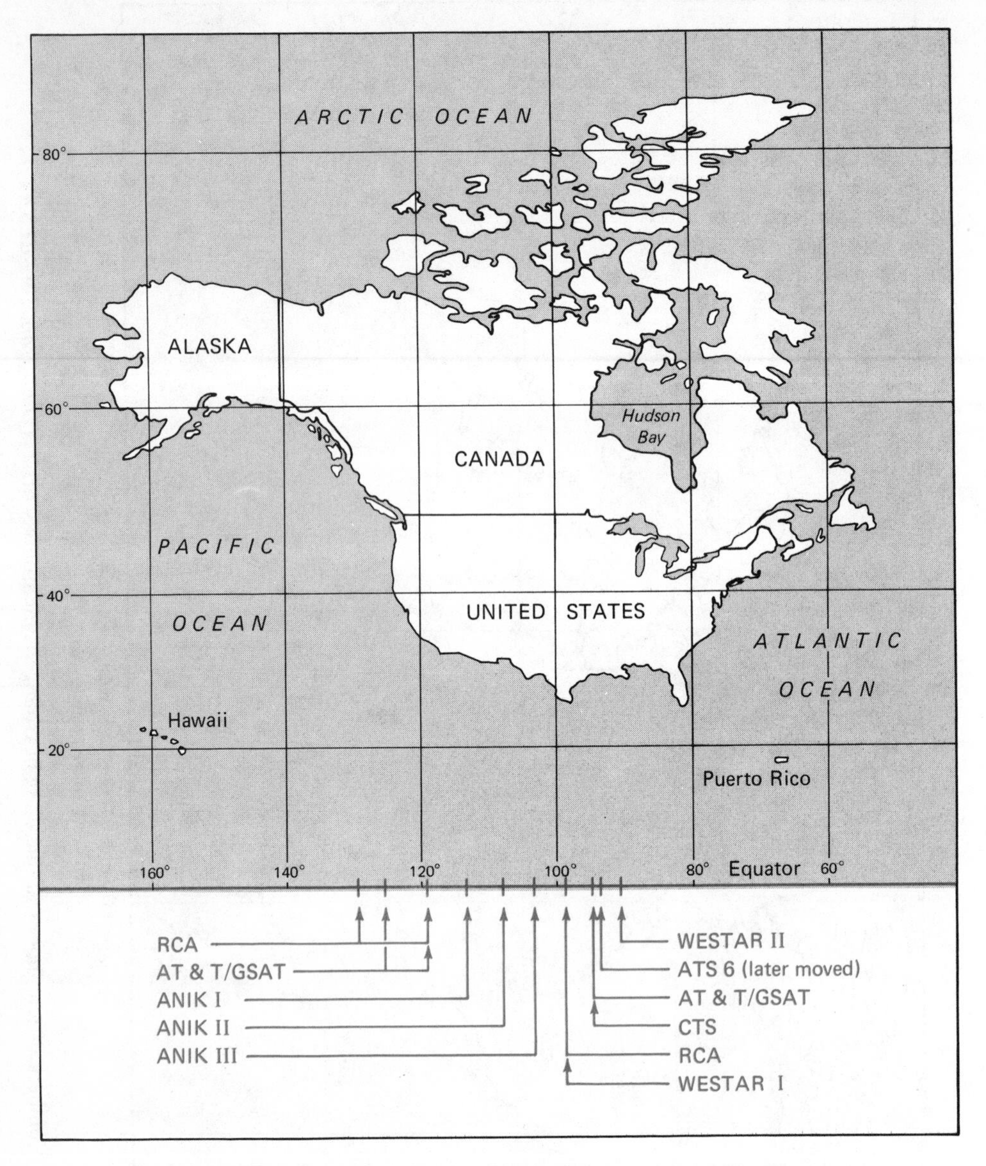

Figure 3.14 Orbital positions allocated to satellites over North America. Parts of the geosynchronous ring will become crowded.

4 SATELLITE CONSTRUCTION

The box in the frontmatter of this book summarizes the first 10 years of geosynchronous satellites and their nonsynchronous forerunners.

SUBSYSTEMS

To do their simple task of relaying signals, geosynchronous satellites need the following subsystems:

1. *An antenna subsystem* for receiving and transmitting the signals.
2. *Transponders* containing the electronics for receiving the signals, amplifying them, changing their frequency, and retransmitting them.
3. *A power generation subsystem* for creating the power to operate the satellite.
4. *A power conditioning subsystem* for converting the power generated into the form required by the electronics.
5. *A command and telemetry subsystem* for transmitting data about the satellite to earth and receiving commands from earth.
6. *A thrust subsystem* for making adjustments to the satellite orbital position and attitude.
7. *A stabilization subsystem* for keeping the satellite antennas pointing in exactly the right direction.

Figure 4.1 illustrates these subsystems. Most of them have varied substantially in design from the satellites launched and proposed to date.

ANTENNA SUBSYSTEMS

To be efficient, a microwave antenna should be designed to focus its signal on the distant location to which it transmits. The more directional the antenna,

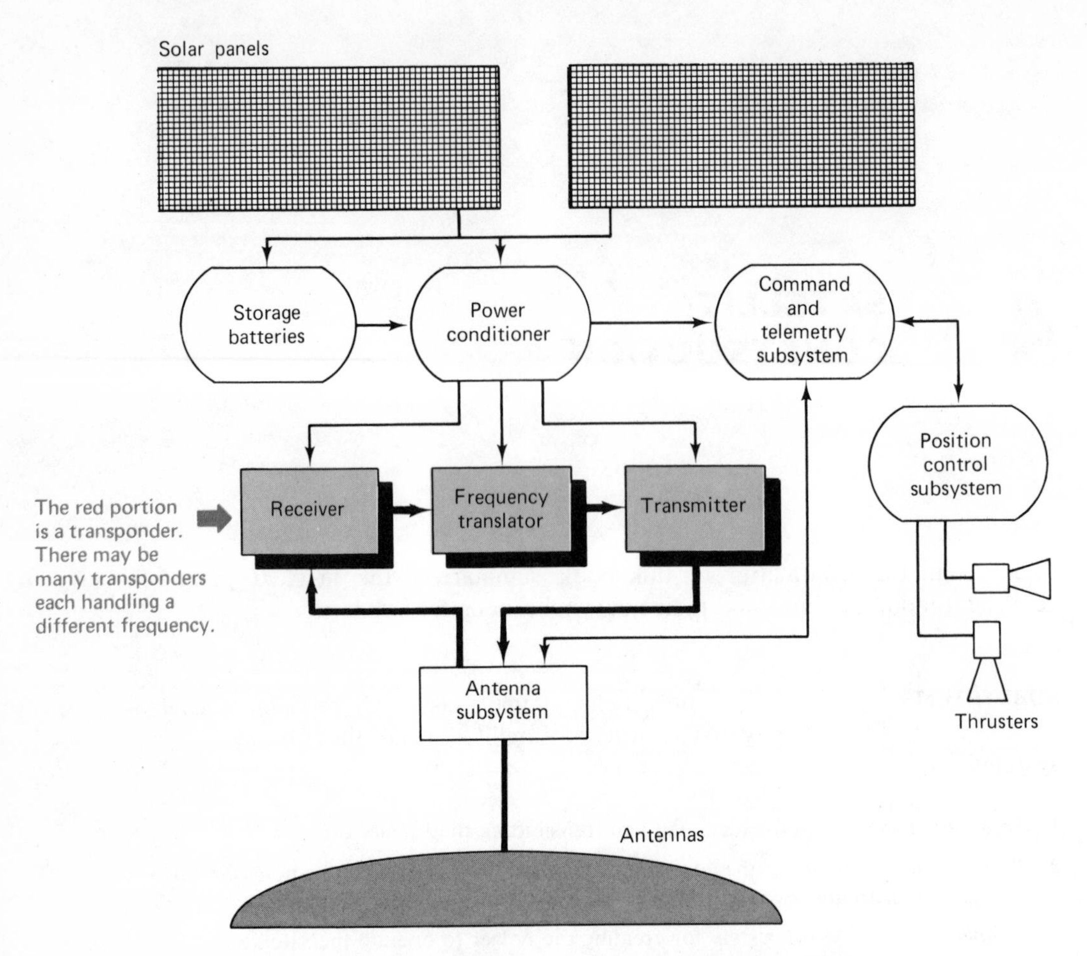

Figure 4.1 The subsystems on board a satellite.

the greater the signal *gain*. A high signal gain in the satellite antennas means either that more information can be transmitted in a given frequency band or that the cost of the earth stations can be lower.

The antennas on the first geosynchronous satellites were not very directional, and most of the signal they transmitted was wasted by being radiated into empty space. INTELSAT III was the first satellite to have a directional earth-coverage horn antenna, aiming the beam down the 17.34° angle subtended by the earth.

As the INTELSAT traffic grew it became clear that most of it was between North America and Europe. Hence INTELSAT IV was designed with spot beams to shine on these two portions of the earth, each spot beam covering an angle of 4.5° (Fig. 2.4). An antenna covering a 4.5° angle has a gain of $(17.34/4.5)^2 = 14.85$ times that of a 17.34° antenna. As the improvement occurs both when receiving and transmitting the signal, the total increase in relayed signal strength is $14.85^2 = 220$.

Figure 4.2 shows the three types of antenna used on INTELSAT IV:

1. Fifty-inch dish antennas, giving 4.5° spot beams, each individually steerable to adjust the area the beam covers on earth.
2. Two receive and two transmit earth-coverage antennas, fixed in position.
3. A nondirectional antenna for receiving commands to the satellite and transmitting telemetry data to earth. The data rate used with this antenna is very low, and so its low gain is acceptable.

The satellite, as with other "spinner" satellites, is stabilized by spinning the main body, at 100 revolutions per minute. To keep the antennas pointing to earth, the antenna subsystem must be *despun,* i.e., the antenna subsystem remains stationary while the body of the satellite spins. To minimize the possible introduction of noise, the transponder electronics are part of the antenna subsystem and are despun with it. Figure 4.13 shows the assembly which is despun.

The ATS-6 antenna reflector shown in Fig. 2.6 is much larger than that on any commercial satellites to date. It opens in space like a huge umbrella. The support structure consists of a 5-foot aluminum hub from which protrude 48 umbrella-like ribs. The ribs are covered with a flexible mesh material made

Figure 4.2 The three types of antennas on INTELSAT IV satellites. The entire antenna subsystem is despun to hold it fixed relative to the earth.

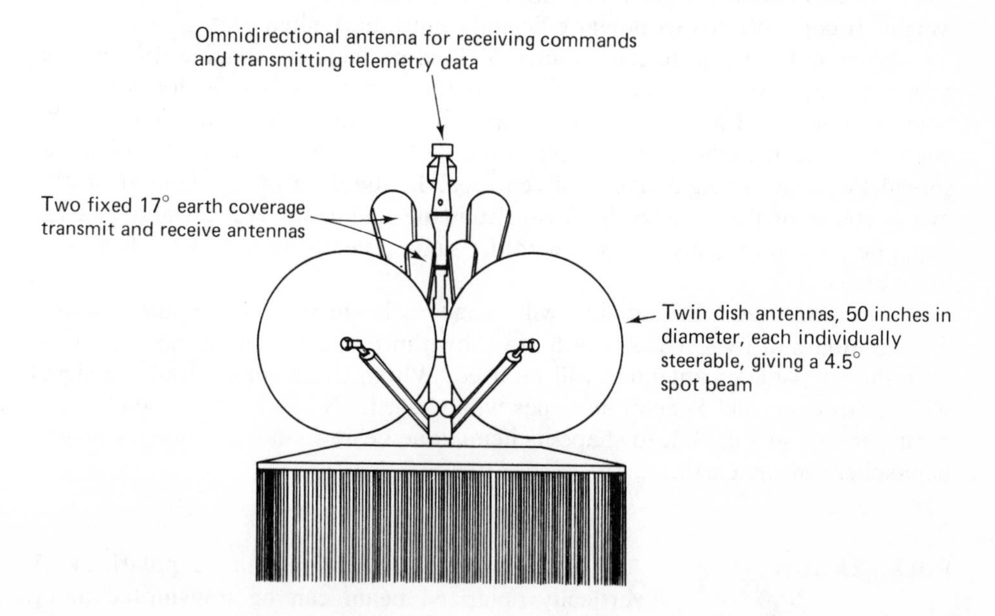

of copper-coated dacron. The radio-reflecting properties of the copper are protected by a coat of silicon to prevent flaking or tarnishing. Before the umbrella opens the ribs are wrapped around the hub and held down by a thin wire. When the wire is severed, in orbit, the coiled energy in the ribs causes the reflector to spring open.

Figure 4.17 shows the ATS-6 satellite deploying its equipment in space, and Fig. 4.18 shows the antenna. The spacecraft carried a small television camera to photograph the reflector in orbit so as to observe whether any flaking or damage was occurring. Figure 4.20 shows a satellite self-portrait taken with this camera.

The high-gain antenna of ATS-6 is not too expensive in terms of weight. The hub, ribs, and mesh weigh 180 pounds.

BEAM SHAPING

A directional microwave antenna uses a reflector to focus the beam, such as the two reflectors on INTELSAT IV. The power from the transmitter is directed toward the reflector by an antenna feed, usually a small horn. The reflector focuses the beam into a narrow cone. When receiving the reflector collects the signal energy which strikes it and directs it toward the antenna feed.

In a domestic satellite system it is desirable that the beam from (and to) the satellite be *shaped* to fit the country it serves. This is especially important if the country has a long thin, or irregular, shape. By concentrating all the available energy onto the country, the satellite can be designed to be smaller and less expensive to launch. The WESTAR satellite, for example, can relay more voice channels than INTELSAT IV although it is less than half the weight. It concentrates its power efficiently onto the United States.

Shaping the beam to the country is done partially by shaping the antenna reflector and, more important, by directing the signal at this reflector with multiple feed horns. Figure 4.3 shows the effect of multiple feeds. The reader might imagine it as being like using three or four bulbs in a car headlight to spread the beam of light over a given area. In the case of the United States, two portions of the country lie away from the mainland, Hawaii and Alaska. Separate feed horns may be used with the same reflector to send spot beams to these areas.

The proposed SBS satellite will shape its beam to fit the United States. Two contours of its signal strength are shown in Fig. 2.6. Within the outer contour shown, 7-meter antennas will be used. Within the inner contour the signal will be stronger, and 5-meter antennas will be used. INTELSAT V used 88 feed horns, shown in Fig. 4.4, to shape its hemi/zone beam to the land masses on one hemisphere of the earth.

POLARIZATION

Any electromagnetic radiation can be polarized. A vertically polarized beam can be transmitted along with a horizontally polarized beam of the same frequency and the two can be

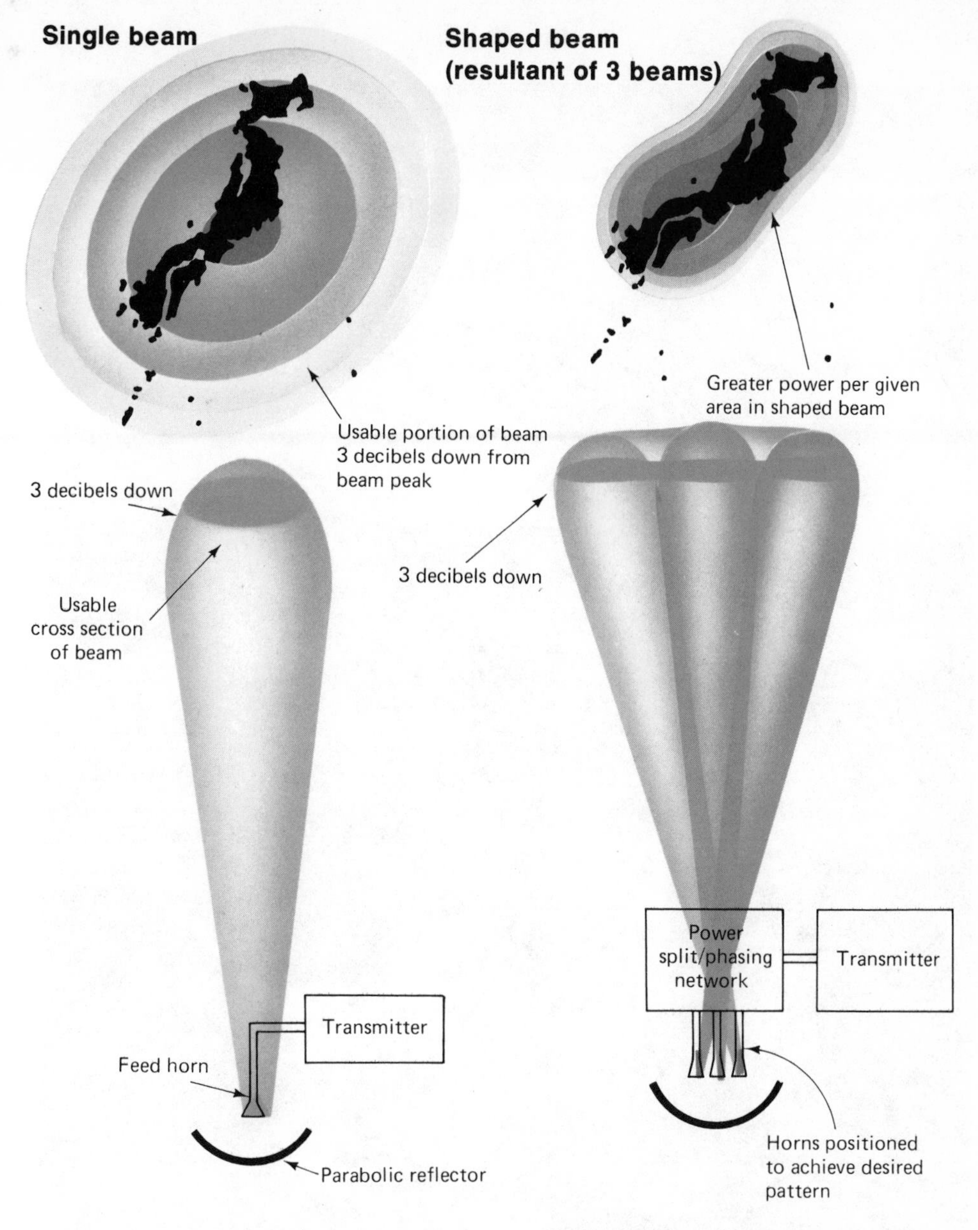

Figure 4.3 With domestic satellites the beam is shaped, as shown, to concentrate as much power as possible on to the country in question. This is done, for example, with the SBS satellite to produce the footprint shown in Fig. 2.6. (*Diagrams courtesy Hughes Aircraft Company.*)

Figure 4.4 The hemi/zone antenna of INTELSAT V has a 2.44 meter reflector and 88 feed horns (bottom). Each feed horn produces a small spot beam which can be set with a feed network to any amplitude and phase. These spot beams in combination form a beam which covers a hemisphere of the earth and is shaped to land areas covered. The entire beam is polarized. (*Photo courtesy* Ford Aerospace.)

detected and received separately. A polarized filter can eliminate one of the beams just as polarized sunglasses can cut out polarized light reflections.

Using two polarized beams in the same frequency range doubles the amount of information that can be sent with that bandwidth. The RCA SATCOM satellite uses twenty-four 36-MHz transponders; the WESTAR and ANIK satellite uses twelve 36-MHz transponders in the same bandwidth. The reason is that the RCA satellite polarizes the beams, and also generates more power with its solar sails.

INTELSAT V reuses the same frequencies in its *hemisphere* coverage beams and in its *zone* coverage beams aimed at smaller areas such as Europe or one side of North America. As shown in Fig. 10.6, INTELSAT V combines beam shaping to fit the continents of the world, with beam polarization to reuse the frequencies twice. It also uses polarized spot beams in a different frequency band, aimed at parts of the earth with a more dense concentration of customers.

The same frequencies can be reused either by polarization or by multiple separate spot beams. As the INTELSAT traffic grows it will be possible to add more spot beams to a subsequent design of satellite. They could serve the same areas as the current spot beams (Fig. 10.6) by using the opposite polarization, or could be pointed at a different part of the earth far enough away to not interfere.

Large satellites of the future will employ multiple spot beams and polarization to reuse the same frequencies many times (see Fig. 14.1).

TRANSPONDER SUBSYSTEM

There are some tight constraints on the design of the transponder electronics. The power amplifier must be highly reliable, light in weight, and efficient because of the smallness of the power supply and must operate over a wide bandwidth (a range of frequencies of 500 MHz). The main contender for such a device is the *traveling wave tube.*

Figure 4.5 shows the essentials of a traveling wave tube. The signal to be amplified travels in a helix down a vacuum envelope. Down the axis of the helix a beam of electrons is fired from the electron gun on the left. The electron beam travels to the collector on the right of the helix and is focused by cylindrical permanent magnets. The spacing of the coils of the helix is such that the radio-frequency (RF) signal to be amplified travels down the tube at the same speed as the electron beam.

The electron beam and the radio-frequency signal traveling down the helix interact with one another. The RF signal creates a sinusoidally varying electric field which travels down the tube at the same speed as the electrons. The sinusoidal variation of the field causes slight corresponding variations in the velocity of the electrons and so tends to bunch the electrons. The bunched electrons traveling down the tube induce a second signal into the helix. This sec-

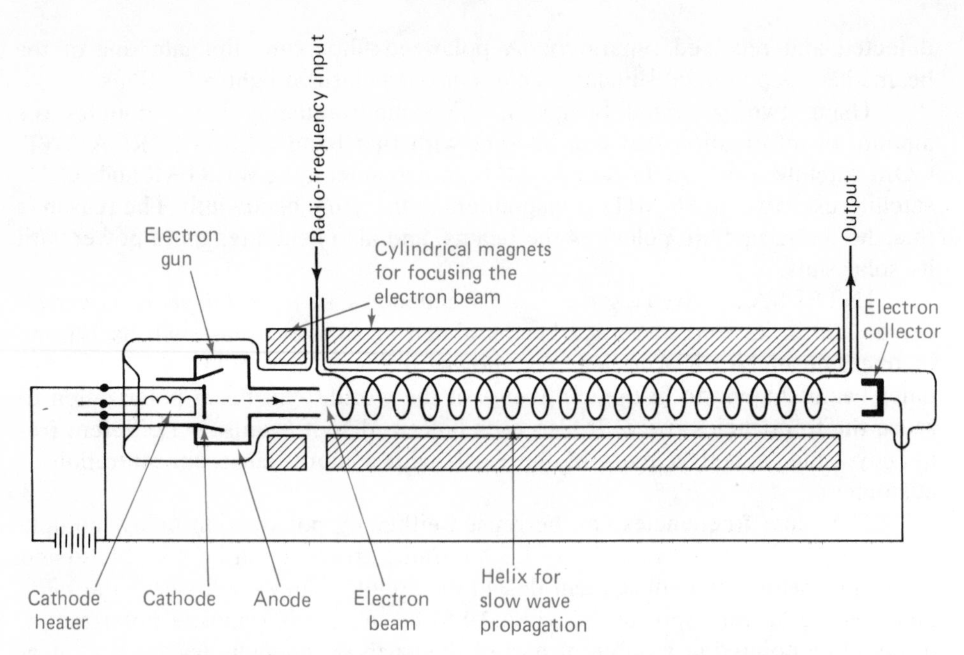

Figure 4.5 Traveling wave tube: the satellite power amplifier.

ond wave, also traveling down the tube, adds to the first and so, in turn, induces greater bunching of the electrons. Greater bunching of the electrons induces a greater signal into the helix, and so on.

The process continues, like compound interest, the further the electrons and RF signal travel down the tube. The longer the tube, the greater the bunching and the greater the sinusoidal variations in the RF signal become, and so a traveling wave tube can produce a large amplification of the oscillations in an RF signal. The amplification grows exponentially with the distance traveled down the tube, just as money grows exponentially with time when it collects compound interest.

The decibel (as explained in Box 6.1) is an exponential unit of measure, and so the gain in decibels is approximately proportional to the length of the tube. The traveling wave tubes in INTELSAT IV are about 2 feet in length and give a power gain of 50 decibels (which means the signal is amplified 100,000 times). They can be seen radiating out from the center in Fig. 4.13.

The traveling wave tubes of future satellites are likely to be substantial improvements on the INTELSAT IV generation. The INTELSAT IV traveling wave tubes have an efficiency of 31% and give a power output of 6 watts. The Canadian Technology Satellite makes experimental use of a traveling wave tube with 50% efficiency and an output of 200 watts. The efficiency of a traveling wave tube is the ratio of RF power output to the dc power input needed to operate the tube. The Canadian Technology Satellite tube requires 400 watts to operate. Future satellites may require tubes producing many kilowatts. Such tubes, however, will need a power supply much greater than that in present satellites.

POWER SUPPLY

The design improvement most needed to make rooftop antennas inexpensive is the provision of more on-board satellite power. With more power, powerful traveling wave tubes could be used, and signals could be beamed to earth strong enough to be picked up with small antennas using relatively inexpensive amplifiers.

The power for today's satellites comes from the sun and is expensive. Approximately 1000 watts of sunshine reaches each meter of satellite surface perpendicular to the sun. One horsepower is equivalent to 745 watts, so an area 20 meters square would intercept enough energy to drive a large American car, if only the energy could be efficiently converted at a reasonable cost.

Today's solar cells, which are used to convert sunshine into electricity, lose more than 85% of the energy, and arrays of them cost $400 per watt. Future technology will make the latter figure drop, perhaps by two orders of magnitude—and then solar energy will be used as an electricity source in houses on earth; today it is used only as a heat source.

Solar arrays on today's satellites are built up by attaching solar cells about 2 centimeters square to a panel. The cells are made of doped silicon crystals. They are connected in parallel and series so as to meet the voltage and current requirements of the satellite. To generate 100 watts, more than 2000 such cells are needed, occupying about a square meter of surface and weighing about 5 pounds.

Most communications satellites have had the cells attached around their cylindrical body as shown in Fig. 4.6. This arrangement is inefficient in that only half the cells are in sunlight and most of those are at an angle to the sun. Only about $1/\pi$ of the sun's radiation can be converted. A more efficient arrangement is to use flat panels of cells which are oriented toward the sun as in Fig. 4.7. Figure 4.7 shows the solar "sails" of the Canadian Technology Satellite. These arrays of solar panels are folded into the satellite until it reaches geosynchronous orbit and are then opened in a concertina-like fashion to their full length of 21.5 feet each (7.1 meters). The satellite needs some selective power prior to this to control its orbital positioning, and so a cylindrical array of cells on the satellite body is also used.

The total area of solar cells on the Canadian satellite's long sails is not much greater than the total area on INTELSAT IV, but it produces more than three times the power. The Canadian satellite has about 16 square meters of cells on its sails and produces 1260 watts. INTELSAT IV has about 15 square meters of cells and produces only 400 watts. INTELSAT V will use solar sails.

The use of flat solar panels, however, introduces two technical complications. First the panels must be moved to follow the sun. In the Canadian satellite a sun sensor and stepping meter are used for this purpose. Because the

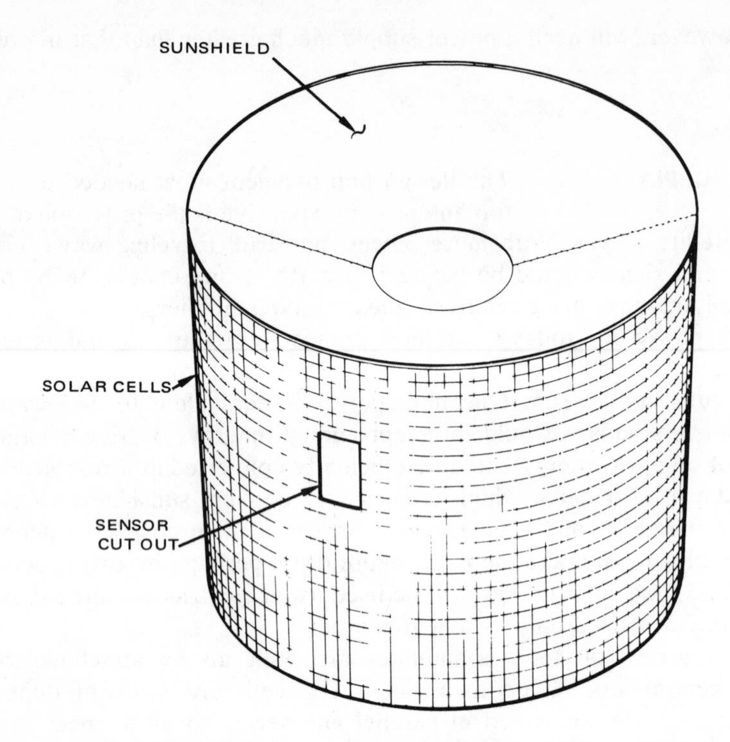

Figure 4.6 Cylindrical solar array (see Fig. 4.9). Only about $1/\pi$ of the sunlight is captured.

antennas must point to the earth and solar panels must point to the sun, three-axis stabilization of the satellite is needed and is achieved using a momentum wheel. With cylindrical solar arrays, stabilization about the axis of the cylinder is sufficient.

Second, flat panels which constantly face the sun have more problems with overheating than cylindrical panels which spin rapidly. It is necessary to maintain suitably low cell temperatures and to limit the mechanical distortions of the sails which can be caused by different thermal expansion.

Silicon solar cells tend to degrade in performance with time. They lose a small percentage of their power output each year. This degradation places a limit on the useful life of the satellite. The main cause of the degradation is bombardment by electrons and to a lesser extent by meteoric dust which travels at high speed (Fig. 5.2). To protect against these factors the silicon cells are covered with very thin glass plates (0.1 to 0.2 millimeters thick).

Solar eclipses occur occasionally, as shown in Fig. 2.14. If the satellite is to continue operating during eclipses, it must carry storage batteries which are charged by the solar panels. WESTAR, for example, has two 8.4-ampere-hour storage batteries with 28 cells per battery.

The satellite power supply is limited to a kilowatt or so with today's techniques. Future satellites will probably have much more power than this partly

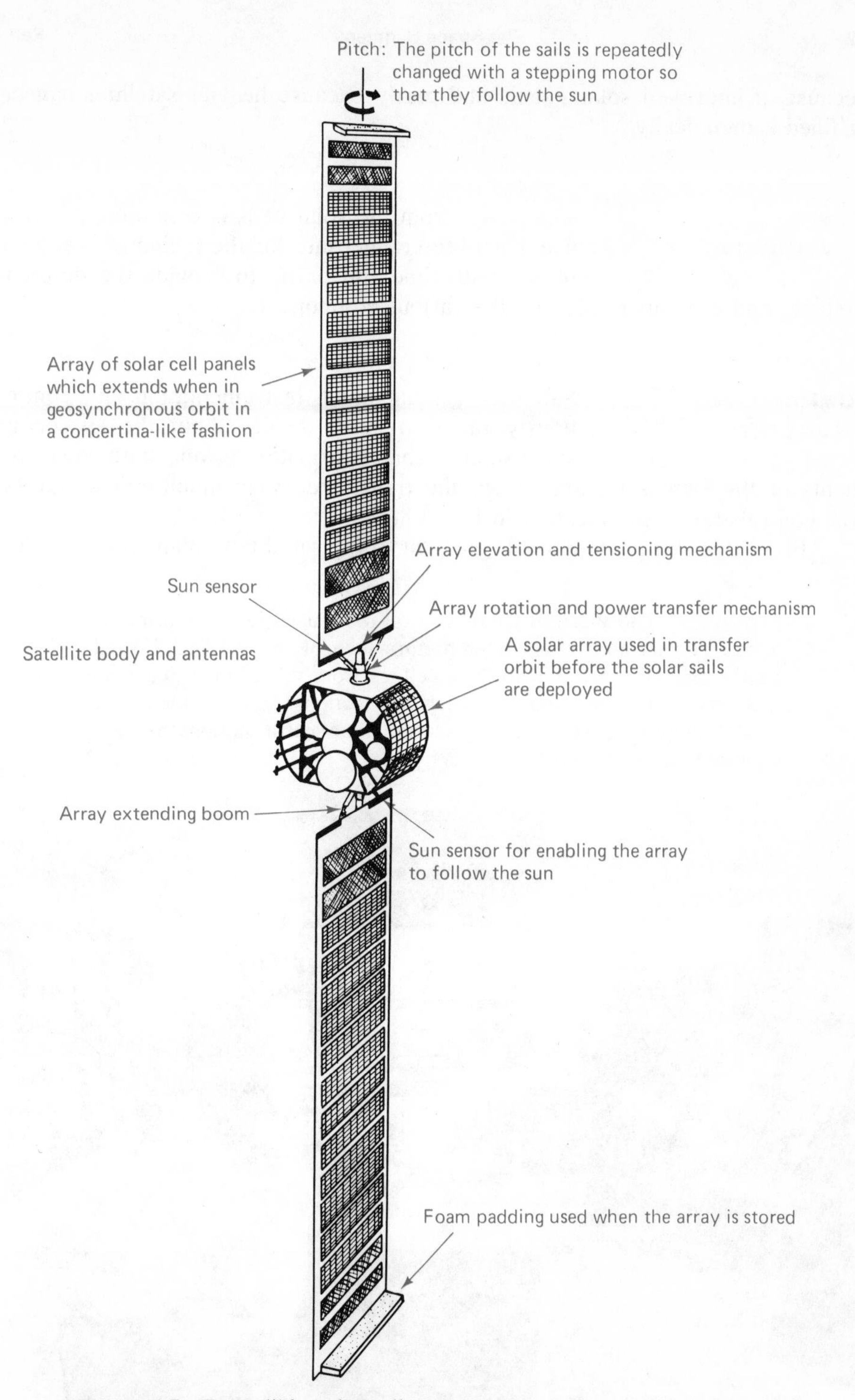

Figure 4.7 Extendible solar sails, each 21.5 ft long, on the Canadian Technology satellite.

because of improved solar panels and partly because heavier satellites can be justified economically.

POWER CONDITIONING

The power from the solar cells is conditioned before use, partly to compensate for the falling of the array output with time and partly to provide the different voltages and currents needed by the various components.

COMMAND AND TELEMETRY

Satellites contain much instrumentation and continuously radio to earth details about the spacecraft subsystems. This information, along with measurements of the signals received from the transponders, is monitored at terrestrial control station such as that in Fig. 4.8.

From this station commands are sent to the satellite to maintain its orbital

Figure 4.8 The Western Union control room used for monitoring the telemetry data and transponder performance of the WESTAR satellites. Commands are sent to the satellites from this room to maintain their orbital position and to switch connections between their electronic equipment. The other Western Union earth stations are also monitored from this room.

position and to keep it functioning correctly. Transponders can be switched in and out of service. Switching between redundant equipment can be performed for reliability purposes. If the satellite has steerable spot beams, they can be adjusted from the control station. The charging and use of storage batteries can be controlled, for example, in preparation for an eclipse. The satellite has several on-board control mechanisms. These are sometimes backed up by terrestrial control mechanisms which can take over if the on-board control fails.

The telementary and command channels carry an information rate much smaller than the main transponders. A separate radio link, with a separate antenna, operating at lower frequency than the main transponders, is often used for telemetry and control. This link can be less efficient than the main radio links because of the low data rate. It is designed to be *highly* reliable and to use as little weight as possible on the satellite. Some of the command functions relate to despinning and pointing the antenna subsystem and must be sent if it is not pointing to earth; hence a separate nondirectional antenna is needed for telemetry and commands.

THRUST SUBSYSTEM Once in position the satellite needs to be given a small push now and then to maintain its orbit. As discussed in Chapter 3, it needs a small north-south push occasionally to compensate for the effects of the sun and moon and a small east-west push to compensate for the elliptical shape of the earth's gravitational field. A spin-stabilized satellite also needs a push to start it spinning.

The small gas jets which give these pushes can be seen projecting through the solar panels in Fig. 4.9. The thurst can be created either from gas under

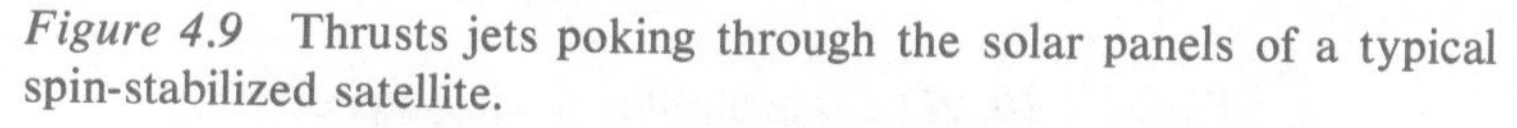

Figure 4.9 Thrusts jets poking through the solar panels of a typical spin-stabilized satellite.

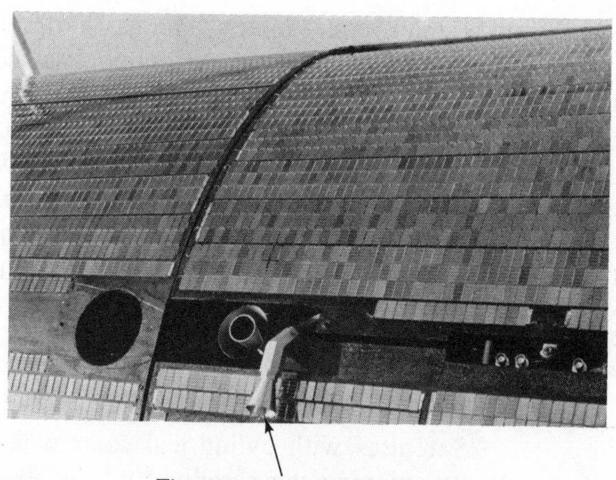

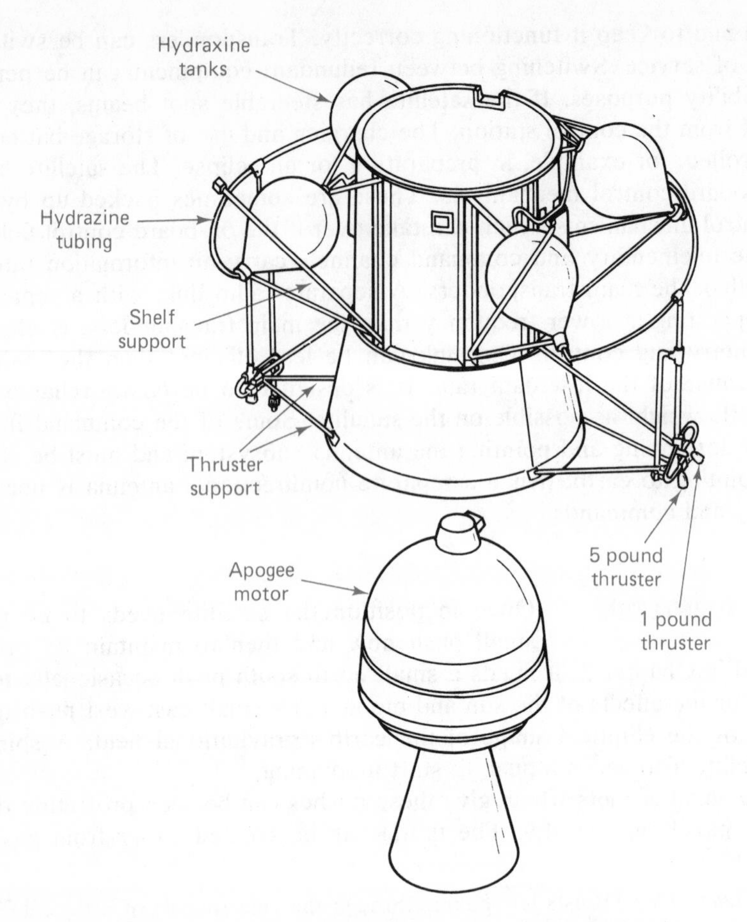

Figure 4.10 The satellite thrust subsystem.

pressure in tanks or from a small rocket motor. The gas hydrazine, which is decomposed with a catalyst in the thrust nozzle, is commonly used.

Figure 4.10 shows the gas tanks, jets, and their supporting structure in a typical satellite. Solenoids operate valves to control the occasional bursts of gas. Figure 4.10 also shows the rocket motor which resides with the satellite and which is used to give the final thrust to the satellite when placing it in orbit. It is fired at the apogee (farthest point) of the elliptical orbit which takes the satellite into space (Fig. 3.5) in order to kick the satellite into its final circular orbit. Prior to firing the apogee motor the gas jets adjust the satellite's orientation.

STABILIZATION SUBSYSTEM

Satellites with cylindrical solar panels are stabilized by spinning the satellite and despinning the antenna subsystem with an electrical motor. The speed at which the antenna subsystem is driven relative to the satellite body is carefully controlled by a servomechanism which keeps the antennas pointing to the earth.

Satellites sometimes tend to wobble, like the nodding of a top. The wobbling is referred to as *nutation* and must be damped out. A nutation damping mechanism is designed to remove wobble energy. INTELSAT IV uses a pendulum-like device, which can be seen near the top of Fig. 4.14. In swinging, the device counteracts the wobble so that INTELSAT IV spins cleanly about its cylinder axis.

Three-dimensional stabilization, which is needed when flat solar panels or large antennas as on ATS-6 are used, is more complex than spin stabilization. It needs a gyroscope-controlled servomechanism. As satellites become large, and need more power, three-dimensional stabilization will become increasingly favored over spin stabilization.

The ATS-6 satellite is designed so that it can be swung in orbit to point at different parts of the earth. It is desirable that these frequent changes in its position should not waste valuable fuel needed for orbital corrections. The positioning is accomplished by momentum wheels in the satellite. Three momentum wheels for the three axes are driven by electric motors. They can be seen in Fig. 4.21. An acceleration or deceleration of one of the wheels causes the satellite to swing. In addition to responding to commands to swing the satellite, the momentum wheels are used for holding the satellite's position steady to $\pm 0.1°$.

Satellite equipment detects the earth and the sun in order to maintain its position in space. Fine positioning on ATS-6 is then done by detecting the star Polaris. Two Polaris sensors are used (37 and 38 in Fig. 4.21). To calculate from this what momentum wheel adjustments are needed, two digital microcomputers are used. There are two of each of these components for reliability.

Figures 4.11 through 4.25 show structural details of WESTAR I, INTELSAT IV and INTELSAT IV A, INTELSAT V, ATS-6, RCA, and Western Union TDRSS satellites.

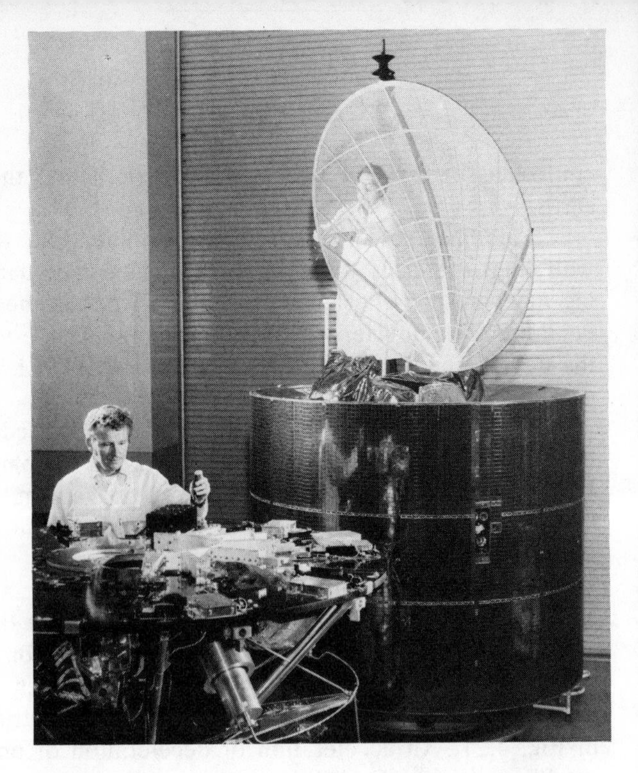

Figure 4.11 Construction of the WESTAR satellite.

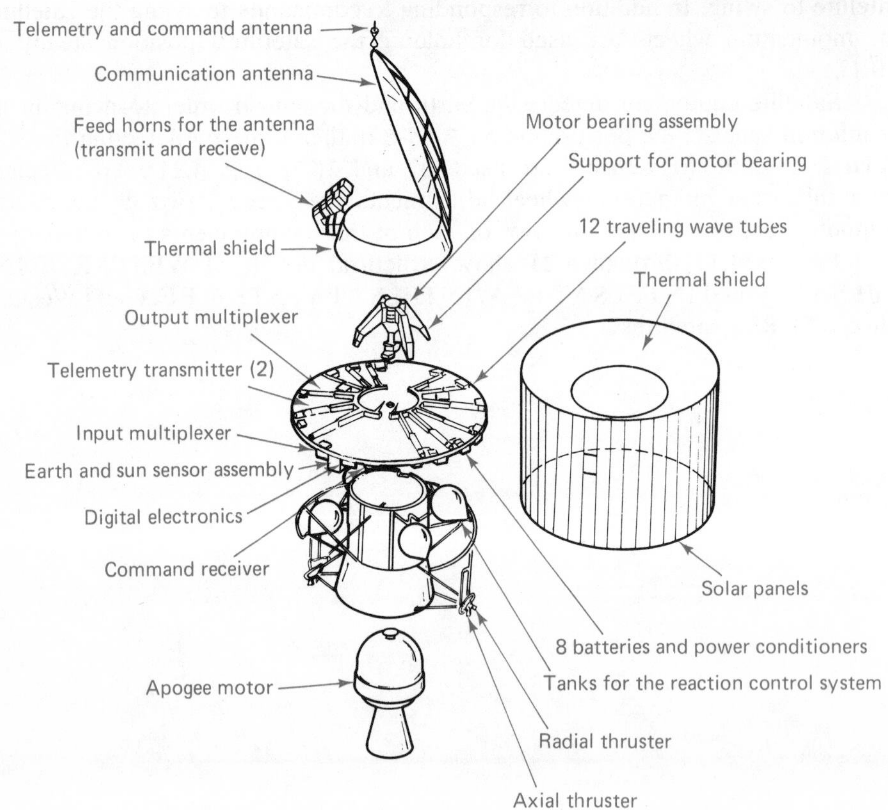

Figure 4.12 Structure of the WESTAR satellite.

Figure 4.13 INTELSAT IV, showing the despun section of the satellite carrying the antennas and transponders.

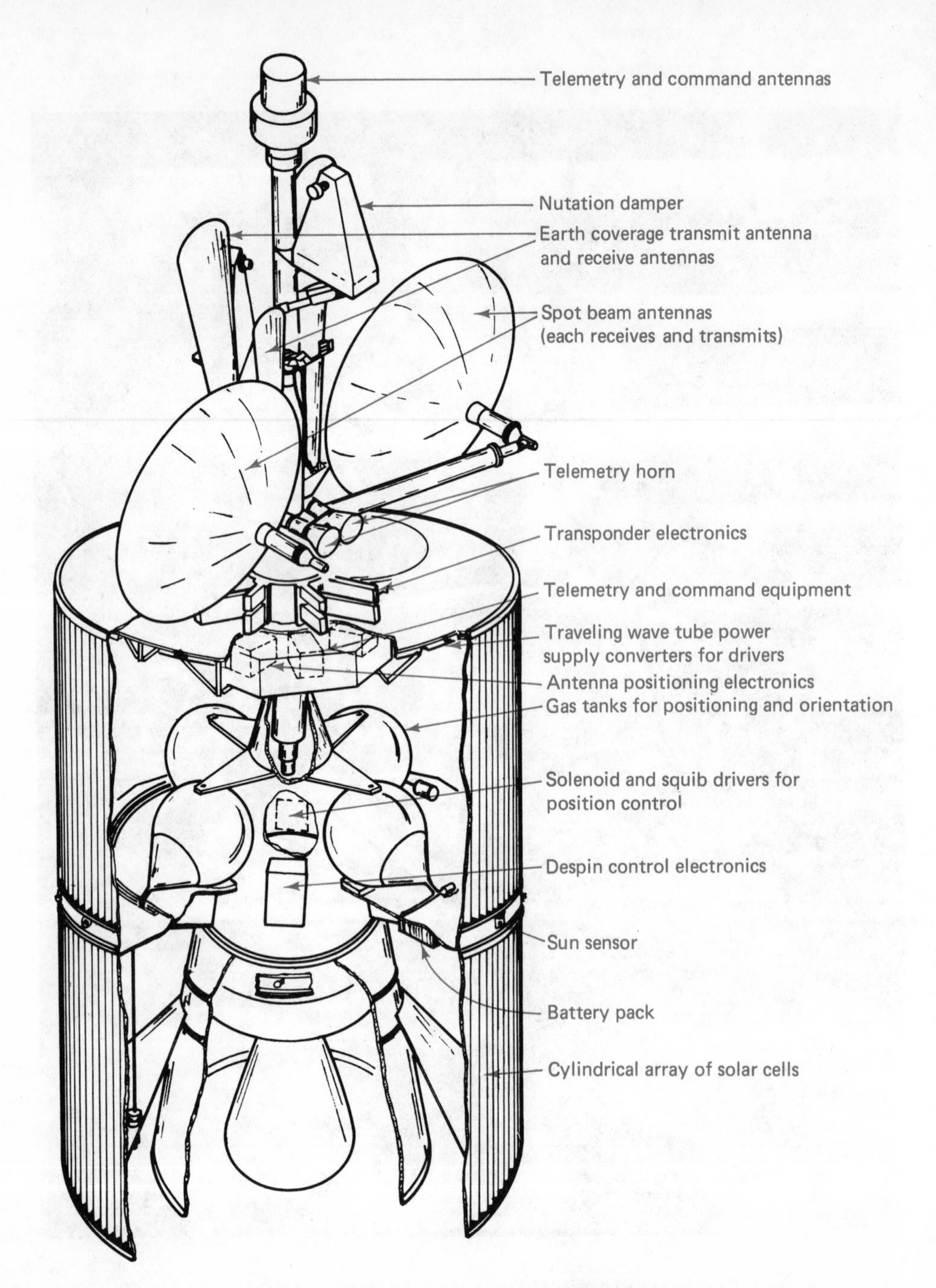
Telemetry and command antennas
Nutation damper
Earth coverage transmit antenna
and receive antennas
Spot beam antennas
(each receives and transmits)
Telemetry horn
Transponder electronics
Telemetry and command equipment
Traveling wave tube power
supply converters for drivers
Antenna positioning electronics
Gas tanks for positioning and orientation
Solenoid and squib drivers for
position control
Despin control electronics
Sun sensor
Battery pack
Cylindrical array of solar cells

Figure 4.14 The structure of INTELSAT IV.

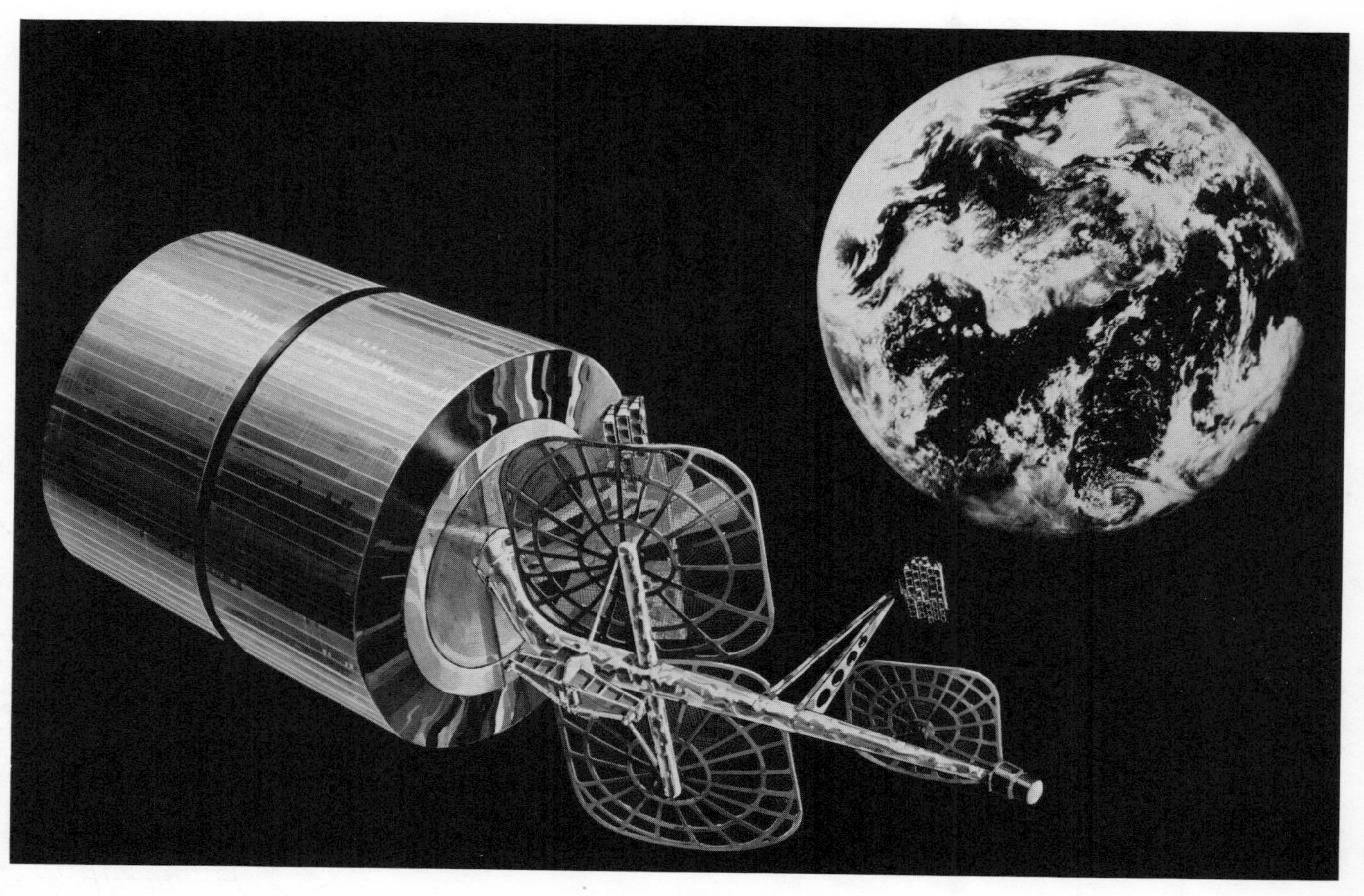

Figure 4.15 INTELSAT IVA.

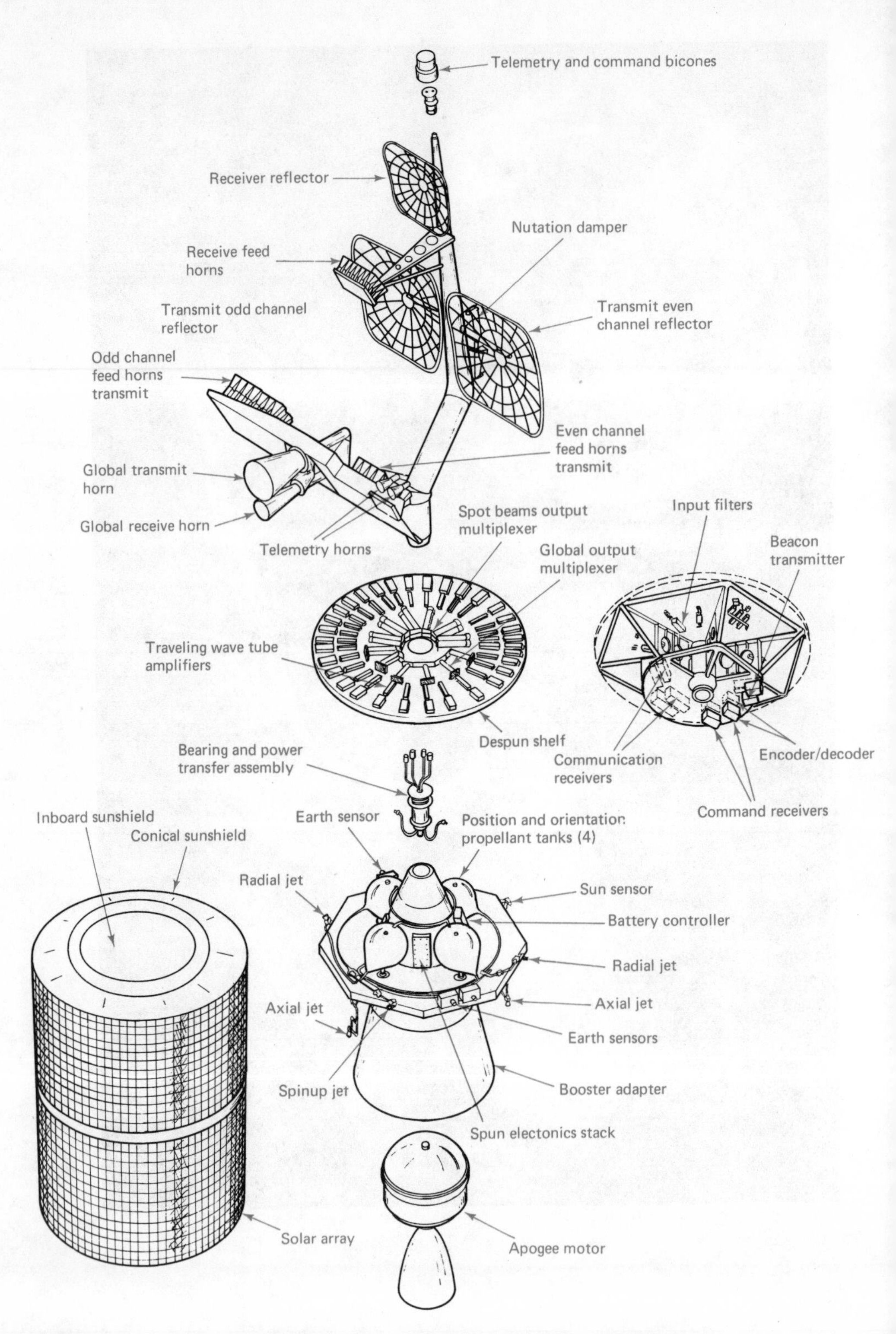

Figure 4.16 The structure of INTELSAT IVA.

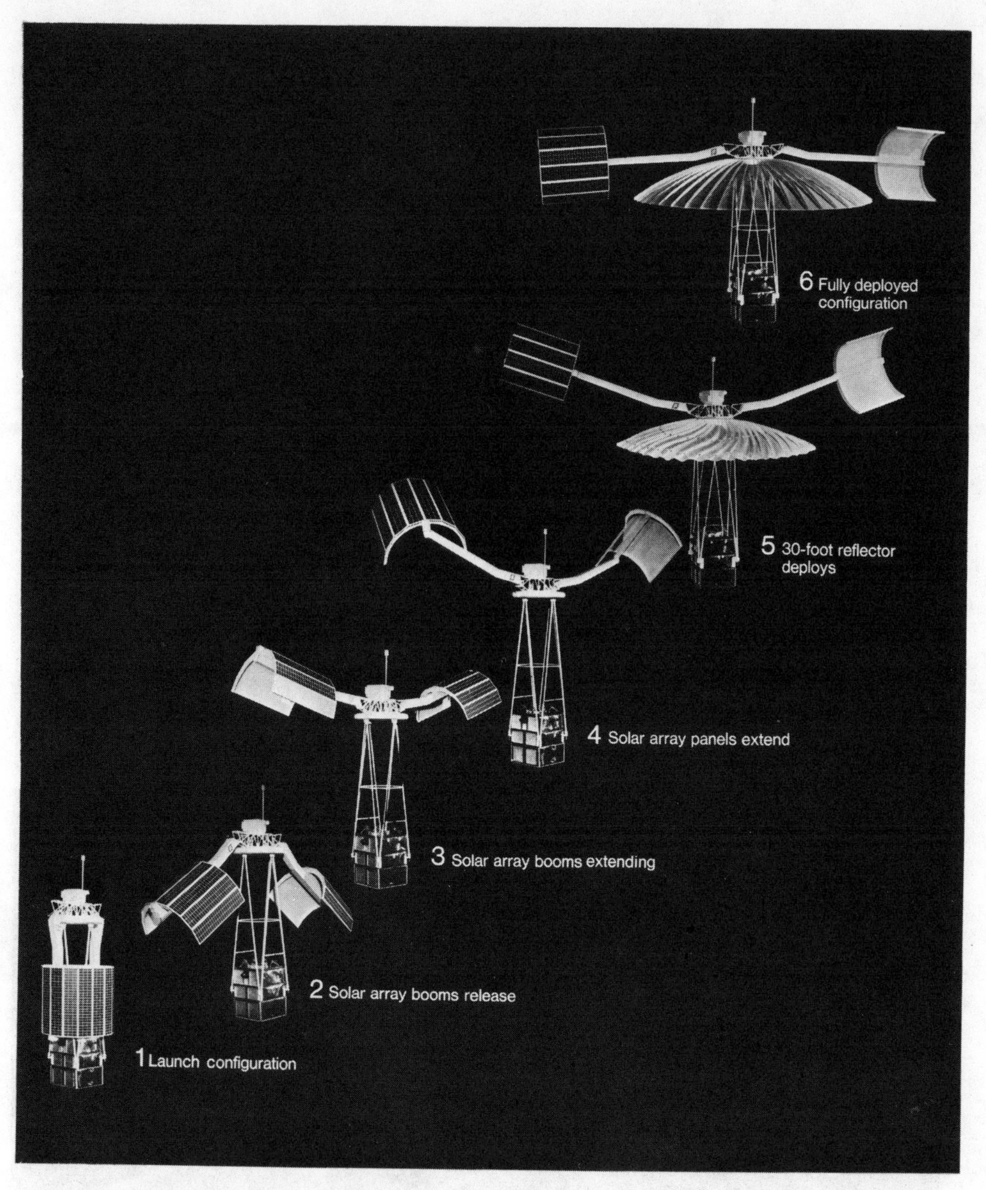

Figure 4.17 The ATS-6 satellite deploying its solar panels and antenna. (*Courtesy Fairchild Industries.*)

Figure 4.18 The NASA ATS-6 satellite, showing its 9-meter antenna. (*Courtesy Fairchild Industries.*)

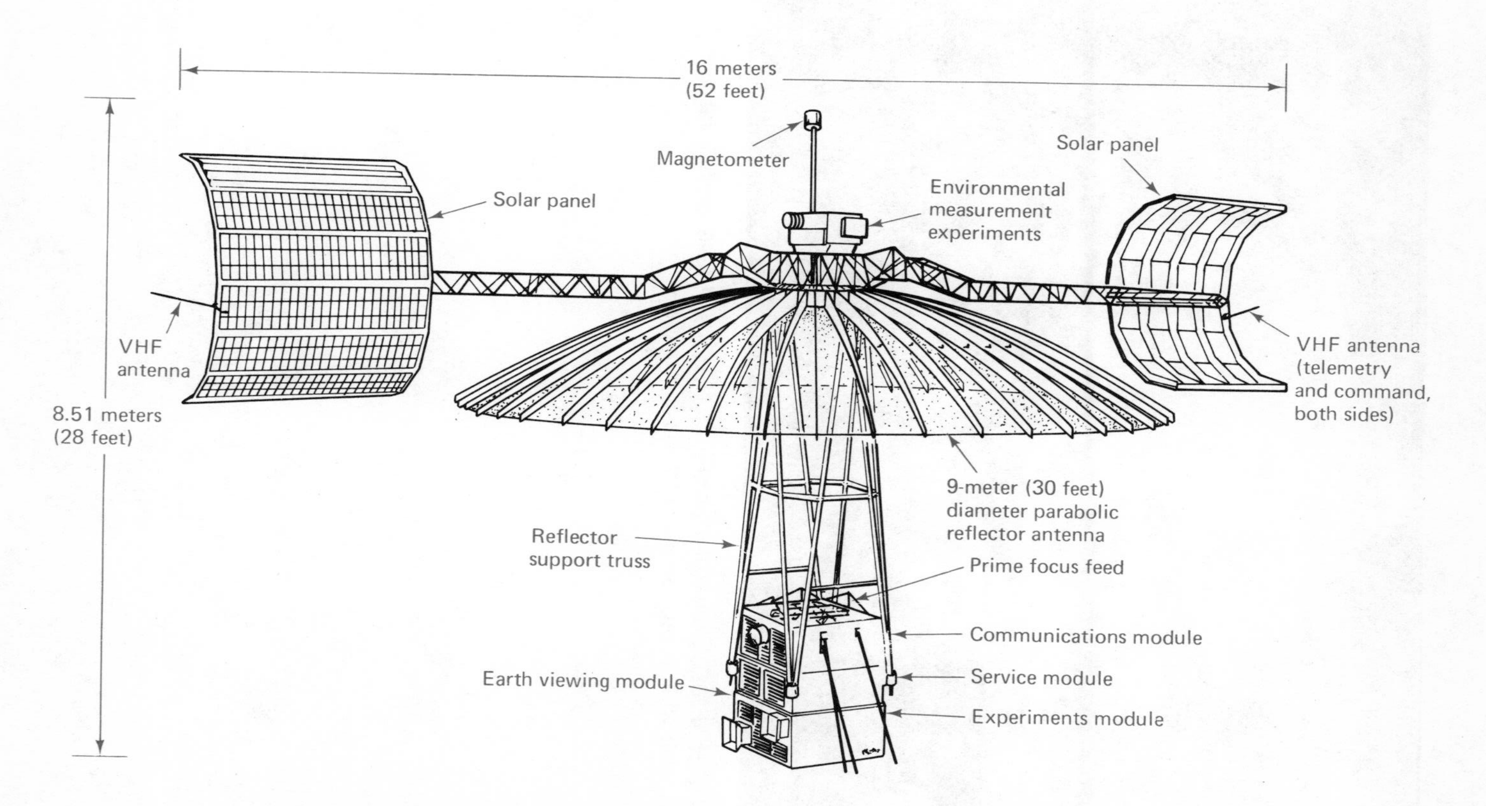

Figure 4.19 The structure of NASA's ATS-6 satellite. Details of the low electronics compartment of the satellite are shown on page 85 (*Courtesy NASA.*)

Figure 4.20 A television picture of NASA's ATS-6 satellite taken in orbit with its own on-board television camera.

The electronics module is fabricated in three separate sections, each of which can be completely tested independently of the others. The upper section is the Communications module and houses the transponders, antenna feeds and associated components. The center section is the Service module. It contains the altitude control, propulsion, telemetry and command, and parts of the power supply subsystems. The bottom section is the Experiment module. It houses a number of antennas and the experiments requiring an earth viewing location.

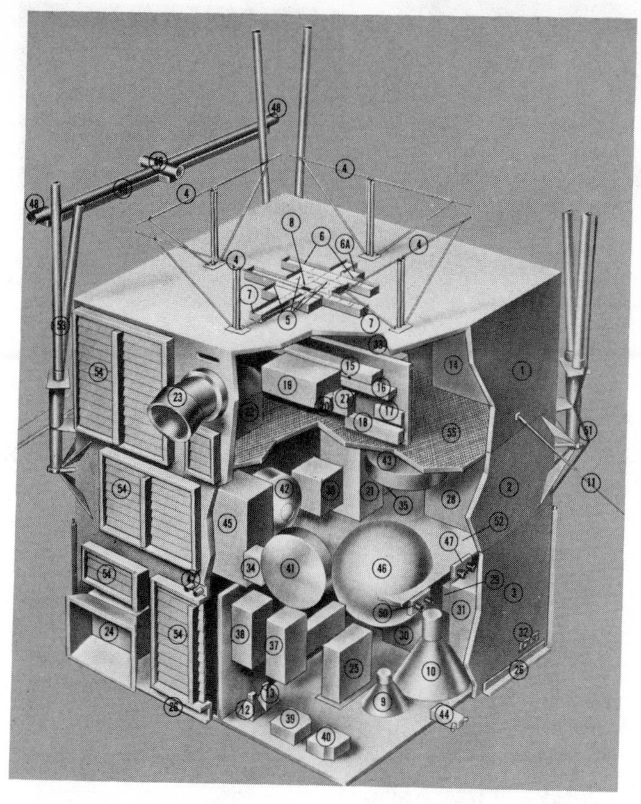

MODULES
1. Communications Module
2. Service Module
3. Experiment Module

30-FOOT PARABOLIC REFLECTOR FEEDS
4. VHF (Monopulse) – Receive & Transmit
5. UHF – Transmit (4 Feeds)
6. L-Band – Receive & Transmit – Fan Beam (7 Feeds)
   6A. L-Band – Receive & Transmit – Pencil Beam (1 Feed)
7. S-Band (Monopulse and Steered Beam) Receive & Transmit (20 Feeds)
8. C-Band (Monopulse) – Receive & Transmit

ANTENNAS
9. Earth Viewing Horn – Receive
10. Earth Viewing Horn – Transmit
11. Radio Beacon Experiment – Transmit
12. Interferometer Reference Horn
13. Interferometer Coarse Horn

COMMUNICATIONS
14. UHF Transmitter
15. S-Band Diplexer
16. S-Band Monopulse Errors Channel Filter
17. S-Band Monopulse Error Channel Preamp.
18. S-Band Preamp. with Redundancy Switches
19. ETV – S-Band Transmitter
20 C-Band Transmitter Output Filter

TELEMETRY
21. Data Acquisition & Control Unit

EXPERIMENTS
22. Ion Engine Electronics
23. Ion Engine Thruster
24. Radiometer Cooler
25. Propagation Experiment Receive Antenna
26. Laser Retroreflector

POWER
27. Shunt Dissipator
28. Power Control Unit
29. Power Regulation Unit
30. Battery No. 1
31. Battery No. 2
32. Battery Test Connectors
33. Load Interface Circuit
34. Shunt Dissipator

COMMAND
35. Command Decoder Distributor

CONTROL
36. Sun Sensor Electronics
37. Polaris Sensor No. 1
38. Polaris Sensor No. 2
39. Earth Sensor Head Ass'y. (Pitch)
40. Earth Sensor Head Ass'y. (Roll)
41. Inertia Wheel (Pitch)
42. Inertia Wheel (Roll)
43. Inertia Wheel (Yaw)
44. Coarse Sun Sensor
45. Actuator Control Electronics
46. Hydrazine Tank
47. Attitude Control Thrusters
48. Orbit Control Thrusters
49. Orbit Control Jets Bar
50. Spacecraft Propulsion System – Fill & Drain Valves
51. Truss Propellant Line Assembly

STRUCTURAL
52. Mounting Plate
53. GFRP Truss

THERMAL
54. Louvers

MISCELLANEOUS
55. RFI Screen

Figure 4.21 Detail of the lower portion of NASA's ATS-6 satellite. *(Courtesy Fairchild Industries.)*

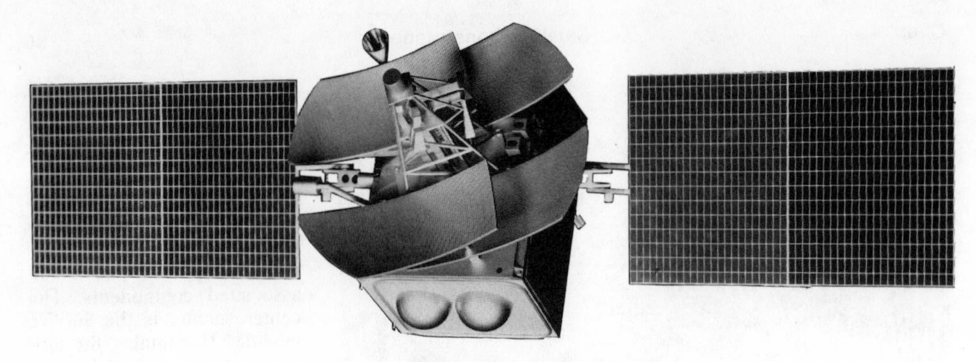

Figure 4.22 The RCA Satcom Satellite.

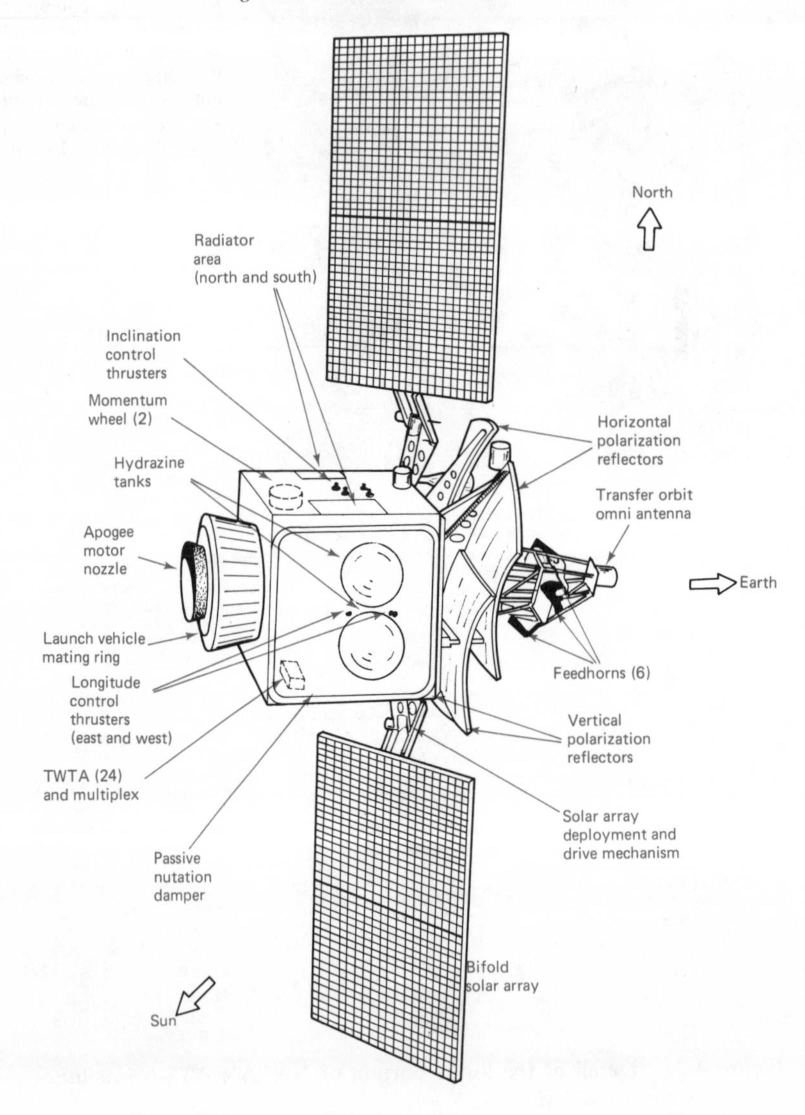

Figure 4.23 Construction of the RCA Satcom satellite.

Figure 4.24 INTELSAT V. 50 feet wing span. (*Photo courtesy* Ford Aerospace.)

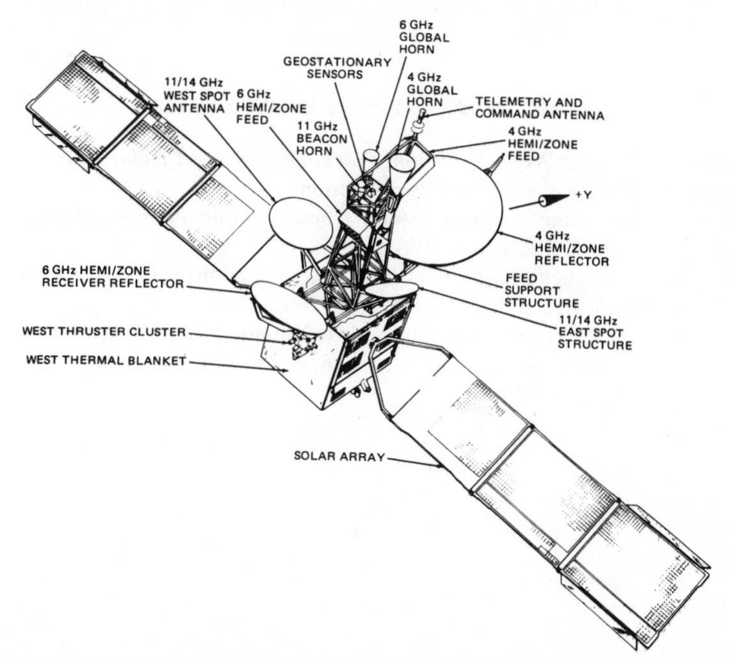

Figure 4.25 INTELSAT V. The 4 GHz hemi/zone feed and reflector are shown in Fig. 4.4. INTELSAT V is made by Ford Aerospace, the prime contractor, and an international team consisting of GEC-Marconi Electronics, Ltd., United Kingdom; Messerschmitt-Boelkow-Blohm, GmbH, Federal Republic of Germany; Mitsubishi Electric Company, Japan; Selenia, Italy; Societe Industrielle Aerospatiale, France.

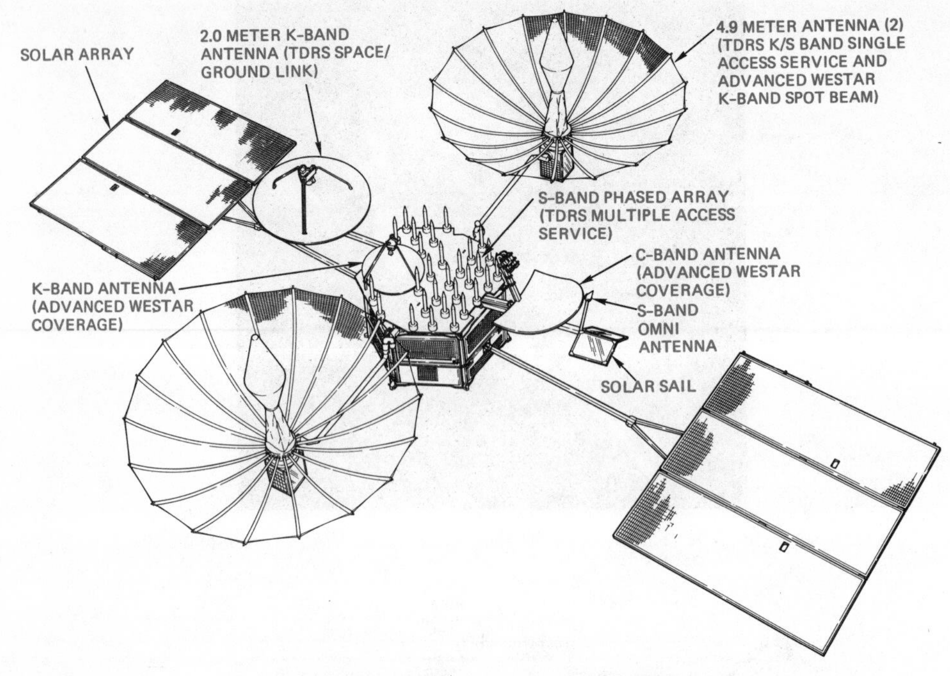

TDRSS/ADVANCED WESTAR SHARED SATELLITE

Figure 4.26 The Western Union *Tracking and Data Relay Satellite System* TDRSS will perform three functions:

i. Data relay services to and from low-orbit NASA satellites including the space shuttle.
ii. Supplementing the 4-6 GHz communications service between large earth stations provided by WESTAR.
iii. 12-14 GHz service between rooftop and parking lot antennas for voice, video and data signals with transmission speeds up to 250 Mbps.

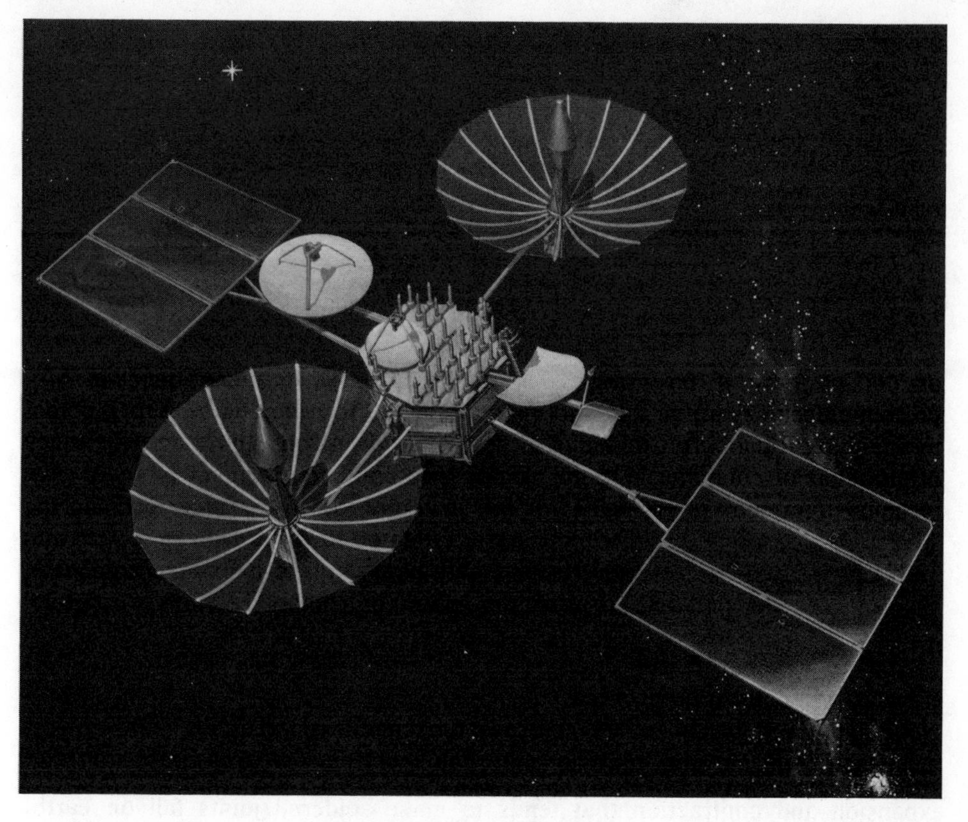

Figure 4.26 (Continued)

The solar arrays provide 1700 watts of power. There are six antennas, three steerable by commands from earth:

- A 30-element S-band phased array for space operations
- Two 16-feet sterrable umbrella-like antennas for S-band space operations and K-band (12-14 GHz) data relay.
- A 6.6 feet steerable antenna for K-band (12-14 GHz) data relay
- A 4.8 feet fixed antenna for terrestrial communications at 4-6 GHz (C-band) like today's WESTAR satellites.
- A 3.7 feet fixed antenna for terrestrial communications with roof-top antennas at 12-14 GHz.

5 THE ENVIRONMENT OF SPACE

Engineering for operation in geosynchronous orbit has some fundamental differences from engineering on earth. One cannot buy components at the local hardware store and expect them to be suitable for space.

At first space seemed a *very* hostile environment, but now it is appreciated that some of its properties can be made use of and that for many techniques it is a better environment than earth. There is no atmospheric corrosion, which causes so much harm on earth. Components exist in the purity of an almost perfect vacuum. There is no continuous temperature variation causing the expansion and contraction that tends to make soldered joints fail on earth, though occasionally the spacecraft plunges into the icy blackness of an eclipse. There is no vibration external to the satellite once the launch is completed. The craft floats in perfect soundless stillness. There are no weather or air-conditioning problems, though the temperature in sunlight becomes high. There are no repairmen, so reliability engineering of the highest order is needed. But there are no human beings to make mistakes, cause damage, drop things, or cause security problems. There is no wind and almost no gravitational force, so large frail structures can be deployed that would be totally impractical on earth.

HIGH VACUUM

Perhaps the most dramatic difference for the engineer is that the equipment operates in high vacuum—a far more complete vacuum than can be produced in the earth's high-vacuum laboratories.

Among the effects of high vacuum are the following:

1. Variations in material strength.
2. Loss of lubrication.
3. Material sublimation.

4. Electrical insulation.
5. Loss of heat convection.
6. Ultra-clean corrosion-force environment.
7. Complete absence of wind stresses.

MATERIAL STRENGTH The mechanical properties of most materials change in the high vacuum of space. Glass is more resistant to fracture. Some metals such as steel and molybdenum have improved creep and fatigue properties because no gases are absorbed by the metal. Weaker metals such as aluminum and magnesium have poorer properties, partly because their surface is not oxidized as it would be on earth, and the oxidization hardens the metal surfaces. Some metals are improved in their properties because surface corrosion does not take place.

LUBRICATION Almost all the lubricants used on earth fail in space. Fluid lubricants are ineffective because the high vacuum causes some of their constituents to vaporize. Powder lubricants such as graphite are ineffective because their moisture content vanishes. Worse, metal bearings operate on earth because of a thin film of air between the surfaces. In space there is no air, and the metal surfaces tend to diffuse into one another in a cold-welding process and bind solidly.

Soft malleable metals or alloys can be used for joints in which there is little movement, like the once-only opening of concertina-like solar sails. The metal surfaces can be soft enough to act as their own lubricant. Such surfaces would not be suitable for continuously moving parts, such as the bearings of the despinning antenna subsystem. For rotating joints, ceramic ball bearings or bearings made from special ceramic-metal compounds such as stellite or nickel-bonded titanium carbide may be used.

MATERIAL SUBLIMATION *Sublimation* refers to a solid substance losing molecules as gas. Some substances, such as sulfur, sublime at normal atmospheric pressure. If sulfur is heated gently, it turns into gas without the intervening stage of becoming liquid; the solid sulfur is redeposited on surrounding surfaces. In the high vacuum of space certain metals sublime very slowly. This can cause severe problems because a thin film of the metal may be deposited on insulating surfaces or bearings. Zinc, cadmium, lead, and magnesium sublime at lower temperatures than many other metals and so are unsuitable for surfaces in spacecraft. The problems of sublimation can be overcome by finding appropriate surface coatings.

Plastics with a higher plasticizer content are avoided in space because some of their components vaporize and leave the plastic in an extremely brittle state. The plastic shrinks and may split. On the other hand, high-polymer resins behave excellently in space. A vast selection of materials suitable for space design can be found, but many are different from materials that would be used on earth.

ELECTRICAL PROPERTIES

The high vacuum of space acts as a good electrical insulator. Electrical surfaces can be closer spaced or carry higher voltages than on earth before arc breakdowns occur. On the other hand, there is a danger of sublimation depositing metallic layers on insulating surfaces.

HEAT TRANSFER

In the vacuum of space there is no convection. Heat transfer occurs by radiation and conduction only. The temperature of a geosynchronous satellite is determined by the balance between the radiation it absorbs from the sun and the heat it reradiates. These factors are determined by its surface shape and texture. A spacecraft can be prevented from becoming too hot by shielding it from sunlight; however, a communications satellite needs to absorb as much sunlight as possible in its solar cells. Long solar sails can become very hot and produce major thermal stresses.

When the satellite passes into the earth's shadow (Fig. 2.13) the energy reaching it becomes very low, and the satellite temperature drops to levels far below temperatures occurring naturally on earth (e.g., −150°C). The equipment, including the storage battery that powers the satellite during eclipses, must be able to withstand these low temperatures as well as the high temperatures of direct sunlight.

A substantial contraction, then expansion, occurs as the spacecraft enters and leaves an eclipse. The materials used and the electrical joints must be able to withstand the mechanical stresses associated with many eclipses.

During normal operation (i.e., not in an eclipse) it is advantageous to regulate the temperature of some components so that they work at their highest efficiency. The ATS-6 satellite (Fig. 4.17) is designed to maintain the components in its main electronics compartment at 20°C ± 15°C. The thermal control is achieved by the use of superinsulation, heat pipes, thermal coatings, and thermal and adjustable louvers. There are some 17 square feet of self-actuating louvers on the north and south faces and about 230 feet of heat pipes bonded into the north and south panels and into the transverse panels between them. The louvers can be seen in Fig. 4.21.

The tubular strusses holding the electronics compartment of the ATS-6 satellite 13 feet below the antenna reflector are fabricated of tough, lightweight, graphite-fiber-reinforced plastic which is impervious to changes in temperature. The aluminum ribs of the umbrella-like reflector have many small holes in them (which can be seen in Fig. 4.18) to help heat balance. Such aspects of the design minimize the twisting or bending of the satellite structure under thermal stresses. This is important because of the stringent pointing requirements of the antenna feeds and reflectors.

SOLAR RADIATION The radiation from the sun enables the satellite to generate its life-giving electricity. However, it also has harmful effects. Almost half of its electromagnetic radiation is in the form of X-rays and γ-rays which penetrate deep in matter, scattering electrons, causing occasional disintegration of nuclei, and causing ionization of matter.

Metals, when bombarded with this high-frequency radiation, have atoms displaced from their crystalline structures. These atomic absentees tend to enhance the strength of metals because the crystals cannot sheer so easily. However, they lower electrical conductivity and so can be detrimental to subsystems which handle the very tiny currents received from earth, for example, the antennas.

Ionization and missing atoms have a more disastrous effect on many organic compounds. Some lose material strength, and some become useless as adhesives, lubricants, or insulators. Also serious is the effect on semiconductors, which can undergo a slow change in their electrical properties.

In addition to electromagnetic radiation, satellites are bombarded by radiation in the form of high-energy particles—alpha particles, protons, and electrons—some from the sun and some from space. The satellite passes through enormous radiation belts consisting of electrons trapped by the earth's magnetic field. The main effect of the flux of electrons is to degrade the performance of the solar cells. The output of solar cells of early satellites launched typically fell by about 6% per year, as shown in Fig. 5.1. Improvements in the design and protective cover of solar cells has now reduced the power degradation.

A satellite is launched with somewhat more power output than it needs, to compensate for the solar cell degradation. At the end of its planned life it should have enough power to keep operating. When its cells are old it may not have enough power to function fully at the same time as charging its storage batteries. The storage batteries may therefore be charged at night or at off-peak periods, with some of the transponders switched off. When the power supply becomes too weak to operate all the equipment the satellite may still be very valuable operating with some of its transponders switched off. The WESTAR satellites, for example, have a hoped-for life of 10 years, but it is thought likely that toward the end of the 10 years only 10 of the 12 transponders will be in use because of power limitations.

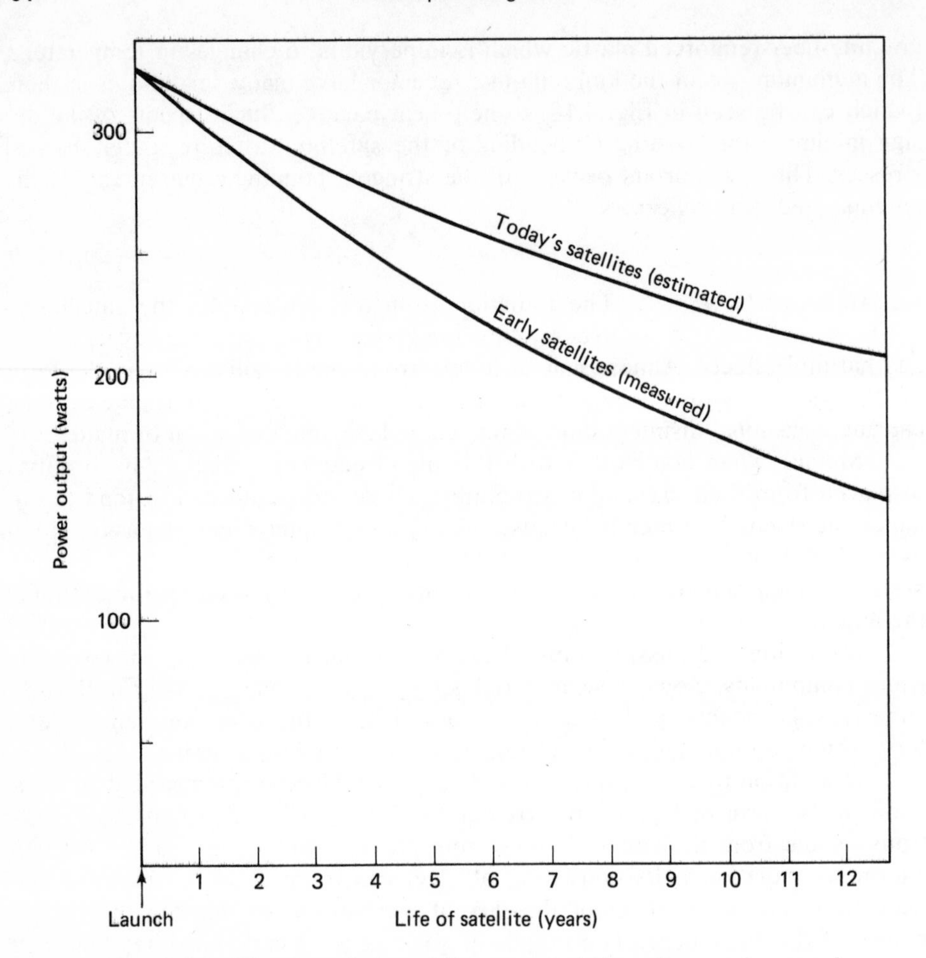

Figure 5.1 The degradation of the output of the solar panels caused by electron bombardment in space.

METEOROIDS AND SPACE DEBRIS

A large quantity of space debris hurtles past the earth at speeds ranging from 7,000 to 45,000 miles per hour. Ten thousand tons of meteoritic material reach the earth daily. Most of the space debris consists of tiny particles which are prevented by the earth's atmosphere from reaching the ground. A satellite has no such protection, and a meteoroid the size of a pinhead can penetrate 2 millimeters of aluminum.

A *meteorite* is a piece of space debris large enough to penetrate the earth's atmosphere and land on the earth. On rare occasions very large meteorites land. The largest one discovered to date weighs more than 50 tons. Very large ones are not found because they explode on impact, causing some of the largest explosions known to man before the atomic bomb. There are craters giving evidence of meteorites weighing more than 200,000 tons.

Meteors are the shooting stars and fireballs seen in the night sky, usually not large enough to reach the earth's surface. Their size ranges from 0.1 millimeter to several meters in diameter. *Meteoroid* refers to all such bodies moving through space, and hence the term includes both meteors and meteorites before they reach the earth's atmosphere. The term *micrometeorite* refers to tiny dust particles below about 0.1 millimeters in diameter.

At one time it was thought that meteoroids would constitute a major hazard for astronauts and space vehicles. Now it is known that large ones are extremely unlikely to hit a relatively small spacecraft, and the spacecraft can survive the impact of small ones. Figure 5.2 shows the mean time between impact of different sizes of meteoroids on a surface 1 square meter in area and shows how deeply they penetrate aluminum. Once per century, on average, a square meter of aluminum will be penetrated to a depth of about 5 millimeters with a hole about 3 millimeters in diameter. This will happen once per year, on average, to a 10-meter square surface of aluminum. The impact rate is random, so there is always the possibility of a larger impact. During the lifetime of a communications satellite it will suffer a little surface erosion and a few pin pricks, but it will be very unlucky if it is hit by a meteoroid the size of a rifle bullet.

RELIABILITY

There are no repairmen in space, and satellite launches cost many millions of dollars, so reliability is of vital importance. The following precautions are taken to increase reliability.

1. Only the most reliable components are used, even though expensive.
2. Physical construction protects the components as much as possible from the vibration and forces of the launch, from the thermal contraction and expansion of solar eclipses, from meteoroids, and from wear in moving parts.
3. The satellite has a prolonged and thorough period of testing before use. Complex equipment usually has a failure rate distribution like that in Fig. 5.3. A satellite is tested until the lower part of the curve is reached.
4. The satellite carries redundant equipment. There are several gas tanks, several storage batteries, multiple solar panels, and so on. If one transponder fails, the others are still useful.
5. The satellite can receive commands from earth instructing it to switch between redundant components, so that failed components can be bypassed. The command subsystem carries a low information rate and is designed to be ultra-reliable.
6. Most systems have more than one satellite in orbit. A second satellite is needed to bridge the periods of outtage when the sun passes behind the satellite and blots out its signal.
7. Most systems have a spare satellite on the ground ready for launching. If a satellite fails in orbit, the earth antennas are pointed to another satellite while a launch of the spare satellite is organized.

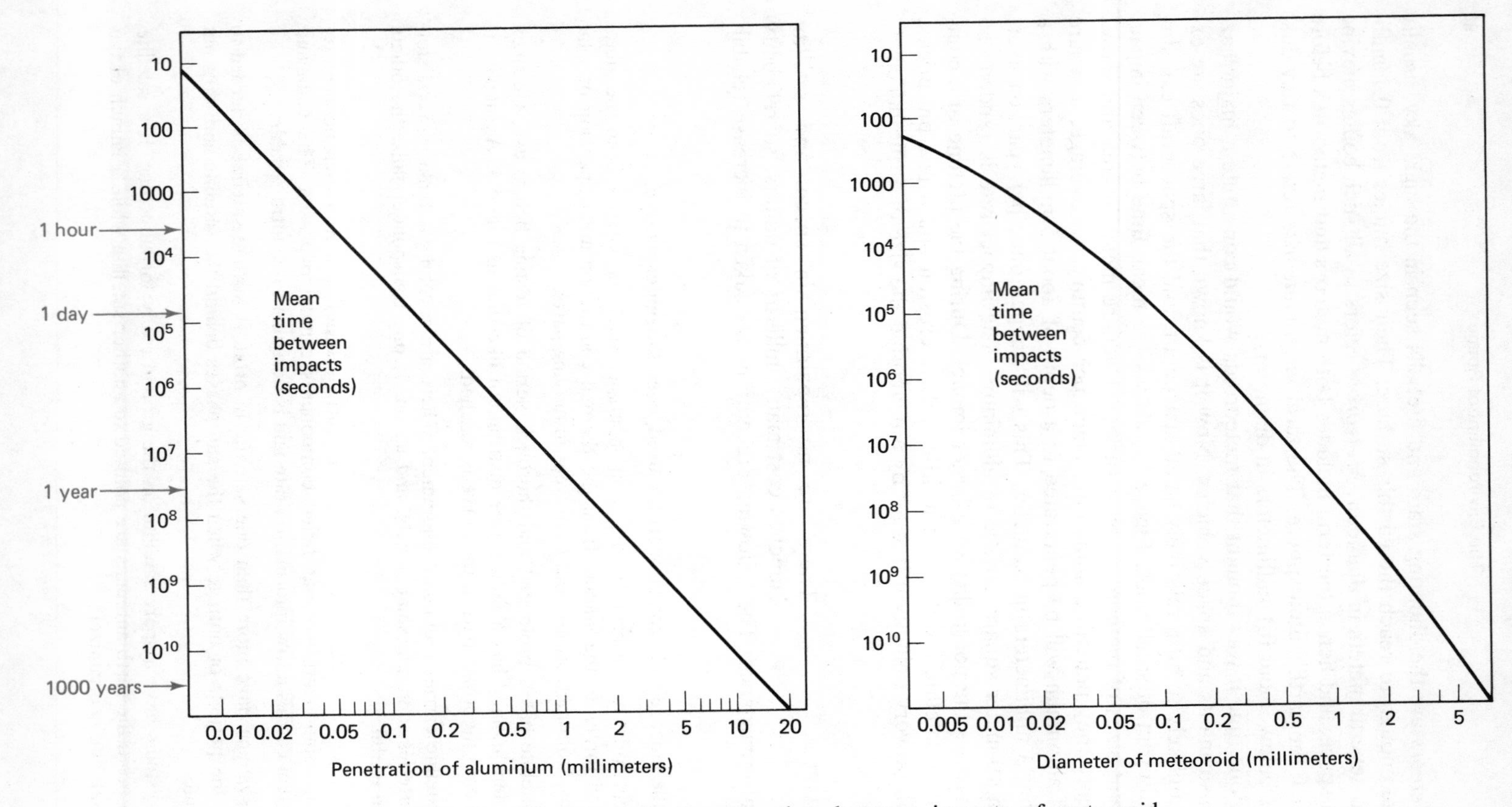

Figure 5.2 Approximate mean time between impacts of meteoroid particles per square meter of surface area of satellite. (*Data from reference* [*1*].)

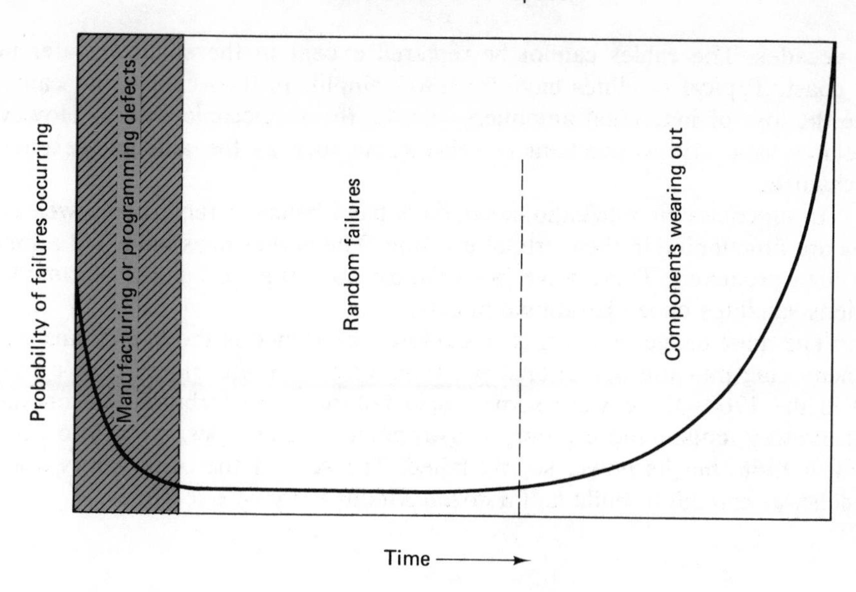

Figure 5.3 The variation in failure rate with time. The shaded portion is avoided by prolonged and thorough testing.

The engineering of spacecraft reliability is well understood. The overall reliability and the trade-offs in the design can be estimated mathematically. The probability of failure during the spacecraft design lifetime (say 10 years) is computed and set at a suitably low figure.

If n components must all be working in order for the satellite to function correctly and the probability of the ith component not failing during the satellite lifetime is τ_i, then the probability of the satellite not failing during its lifetime is the product

$$R = \tau_1 \cdot \tau_2 \cdot \tau_3 \cdots \tau_n$$

If the satellite is to have a failure probability of 1% during its lifetime, then $R = 0.99$. If there are 10,000 components and the values of τ_i are the same, then

$$\tau_i \cdot 10{,}000 = 0.99$$

$$\tau_i = 10^{-6}$$

In other words, each component must have a one in a million chance of failing during the 10-year satellite life. It is clear that very careful selection of components is needed, and subsystem redundancy is an important part of satellite design.

Satellites are not the only telecommunications systems that need extraordinarily high component reliability. The suboceanic cables spanning the Atlantic and Pacific have more than 1000 amplifiers in series and are intended to last for decades. The cables cannot be repaired except in the shallow water near the coast. Typical satellites have far fewer amplifiers than cables and can survive the loss of individual amplifiers—unlike the suboceanic cables. However, satellites have critical mechanical subsystems such as the antenna despinning mechanism.

In practice communications satellites have behaved remarkably well once they are functioning in their orbital position. The earlier ones have had a longer life than predicted. There have been almost no complete failures of communications satellites once operational in orbit.

The most dangerous part of a satellite's existence is the launch and initial maneuvering into the operational position. Launches are more reliable today, but in the 1960s there were some tragic failures. The Orbiting Astronomical Observatory, containing a group of astronomical telescopes, went into perfect orbit in 1966, but its power supply failed. The cost of the observatory and its launch was enough to build half a dozen Mount Palomar telescopes.

6 TRANSMISSION LOSSES

A major difference between satellite links and terrestrial telecommunications is the great distances involved. The strength of a radiated signal diminishes as the square of the distance it travels, so satellite signals become very feeble after their long journey in space. A typical signal strength received from a satellite is only a few picowatts (1 picowatt = 10^{-12} watts).

The transmission path to a satellite is about 800 times as long as the path between two typical terrestrial microwave antennas, so the loss would be $800^2 = 640{,}000$ times greater if antennas of the same size were used. Another such loss occurs on the down-link. As if this is not bad enough, the power available in the satellite for amplifying and retransmitting the signal is limited, and the antennas on the satellite are smaller than many microwave antennas on earth.

First, in this chapter we shall discuss the losses and gains in signal power on a satellite link, and in Chapter 7 the related subject of noise and signal-to-noise levels will be discussed.

DECIBELS

The unit normally used for expressing signal attenuation and gain is the decibel. It measures *differences* in signal strengths, not the absolute strength of a signal, and it is a logarithmic unit, not a linear one. Box 6.1 gives the information the reader should know about decibels.

The decibel was first used as a unit referring to sound. It made sense to refer to sound levels by a logarithmic unit because the response of the human ear is proportional to the logarithm of the sound energy, not to the energy itself. If one noise *sounds* twice as great as another, it is not in fact twice the power; it is approximately 2 decibels greater. The sound energy reaching your ears in the New York subway may be 10,000 times greater than in the room

BOX 6.1 An Explanation of the Decibel Unit

The unit which is normally used for expressing differences in signal strengths in telecommunications is the decibel. The decibel is a unit of power *ratio*. It is not an absolute unit but a unit which is employed to compare the power of two signals. The signal-to-noise ratio is normally quoted, for example, in decibels.

A decibel is equal to 10 times the logarithm (to base 10) of the power ratio:

$$\text{Number of decibels} = 10 \log_{10} \frac{P_1}{P_2}$$

where P_1 is the larger power (normally) and P_2 is the smaller.

The decibel also used to be defined as the unit of attenuation caused by 1 mile of standard No. 19 gage cable at a frequency of 866 cycles, though this definition is now regarded as obsolete. 1 decibel attenuation means that a signal has dropped to 0.794 of its original power. 1 decibel gain means that a signal has increased to 1.259 of its original power.

Voltage and current ratios are also quoted in decibels. Power is proportional to the square of the amplitude of a signal. A power ratio of 100, say, is equivalent to an amplitude ratio of 10. Therefore where the two current levels are a_1 and a_2, or the two voltage levels are v_1 and v_2, we have

$$\text{Number of decibels} = 20 \log_{10} \frac{a_1}{a_2} \quad \text{or} \quad 20 \log_{10} \frac{v_1}{v_2}$$

Decibels are used to express such quantities as gain in amplifiers, noise levels, losses in transmission, and also differences in sound intensity. The decibel is a valuable unit for telecommunications be-

1 – Decibel attenuation means that 0.79 of the input power survives.

3 – Decibel attenuation means that 0.50 of the input power survives.

10 – Decibel attenuation means that 0.1 of the input power survives.

20 – Decibel attenuation means that 0.01 of the input power survives.

30 – Decibel attenuation means that 0.001 of the input power survives.

40 – Decibel attenuation means that 0.0001 of the input power survives.

continued

BOX 6.1 *Cont.*

cause losses or gains in signal strength may be added or subtracted if they are referred to in decibels; thus,

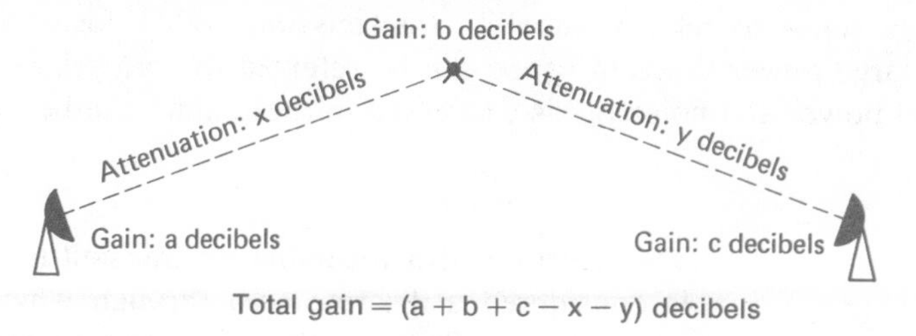

Total gain = (a + b + c − x − y) decibels

Suppose a signal is transmitted over a line which reduces it in power in a ratio 20 to 1. It then passes over another section of line which reduces it in a ratio 7 to 1. The net reduction is in the ratio 140 to 1. Expressing this in decibels, the first reduction is $10 \log_{10} 20 = 13.01$ decibels; the second reduction is $10 \log_{10} 7 = 8.45$ decibels. The net reduction is the sum of these: 21.46 decibels. ($10 \log_{10} 140 = 21.46$ decibels).

Similarly, if we say that line loss is 2 decibels per mile, then the loss at the end of 25 miles of line is 50 decibels. We therefore need an amplifier of gain 50 decibels to produce a signal of the original power.

The chart below will enable the reader to quickly convert power or amplitude ratios to decibels and vice versa.

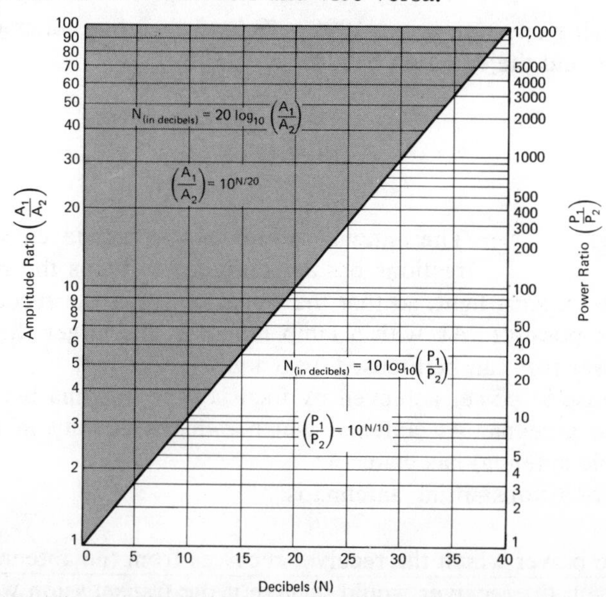

where you are reading this book, but it does not sound 10,000 times greater. It sounds about 40 times greater—you have to shout 40 times harder to make yourself heard to a person the same distance away. Ten thousand times the sound energy is called "40 decibels greater."

It makes sense to refer to satellite transmission with a logarithmic unit because the large power drops in space can be referred to with relatively small numbers, and power and noise ratios expressed in these units can be added and subtracted.

ISOTROPIC RADIATION

The aperture of a receiving or transmitting antenna is considered a discrete area through which all the signal passes. Suppose that the aperture area of the receiving antenna on the satellite is A_r.

If a transmitting antenna radiates equally in all directions, such radiation would be referred to as *isotropic*. Suppost that isotropic radiation is transmitted a distance D and received by an antenna with an aperture of area A_r.

The area of a sphere of radius D is $4 \pi D^2$. The fraction of the isotropic radiation that would be received is therefore $A_r/4\pi D^2$. This fraction is very small if D is 24,000 miles or thereabouts.

$$P_r = \frac{A_r}{4\pi D^2} P_t$$

where P_r = power received
P_t = power transmitted

The receiving aperture is not 100% efficient; therefore an efficiency factor is introduced, z, and the equation becomes

$$P_r = \frac{zA_r}{4\pi D^2} P_t \qquad (6.1)$$

ANTENNA GAIN

The antennas used do not radiate equally in all directions but are designed to focus the radiation, like a car headlight focusing light, so that the signal reaching the receiving antenna is as strong as possible. As with a lamp reflector, the larger the antenna the greater the power that can be focused onto a given receiver.

The increase in power achieved by focusing the antenna is referred to as the *gain* of the antenna. An antenna which radiates equally in all directions (i.e., an isotropic antenna) has gain = 1.

The gain of a transmitting antenna is

$$\frac{\text{The power which the receiver receives from the antenna}}{\text{The power which the receiver would receive if the transmission were isotropic}}$$

Gain is similarly defined by the equation

$$G = \frac{4\pi\phi}{P}$$

where G = gain of the transmitting antenna
ϕ = radiation power per unit solid angle in the required direction
P = total power radiated

It can be shown that the gain of a microwave antenna is

$$G = z\,\frac{4\pi A_t}{\lambda^2}$$

where A_t = aperture area of transmitting antenna
λ = wavelength of transmission
z = the efficiency of the antenna aperture
A typical figure for z for parabolidal antennas is 0.55.

λ, the wavelength of the radiation, is inversely proportional to its frequency, f:

$$\lambda = \frac{C}{f}$$

where C is the velocity of light.

Antenna gain is therefore

$$G = z\,\frac{4\pi f^2 A_t}{C^2} \tag{6.2}$$

$C = 2.99 \times 10^8$ meters per second. A typical figure for z is 0.55. Therefore the antenna gain is

$$G = 60.7 f^2 d^2$$

where f is the frequency in gigahertz and d is the antenna diameter in meters.

For an up-link operating at 6.175 GHz,

$$G = 2315 d^2$$

For a 30-meter antenna the gain is 2.08×10^6 or 63.2 decibels. For a 1-meter antenna the gain is 2315 or 33.6 decibels.

Figure 6.1 plots the gain of an antenna against its diameter for different frequencies.

It will be seen that an advantage of going to higher frequencies is that smaller antennas can give the same gain. If a 20-meter antenna is used at frequencies of 4/6-GHz, a 9-meter antenna would give the same gain in the 12/14-GHz band or a 4-meter antenna in the 20/30 GHz band.

Figure 6.1 Antenna gains at different frequencies. Smaller antennas can be used when the frequency is higher.

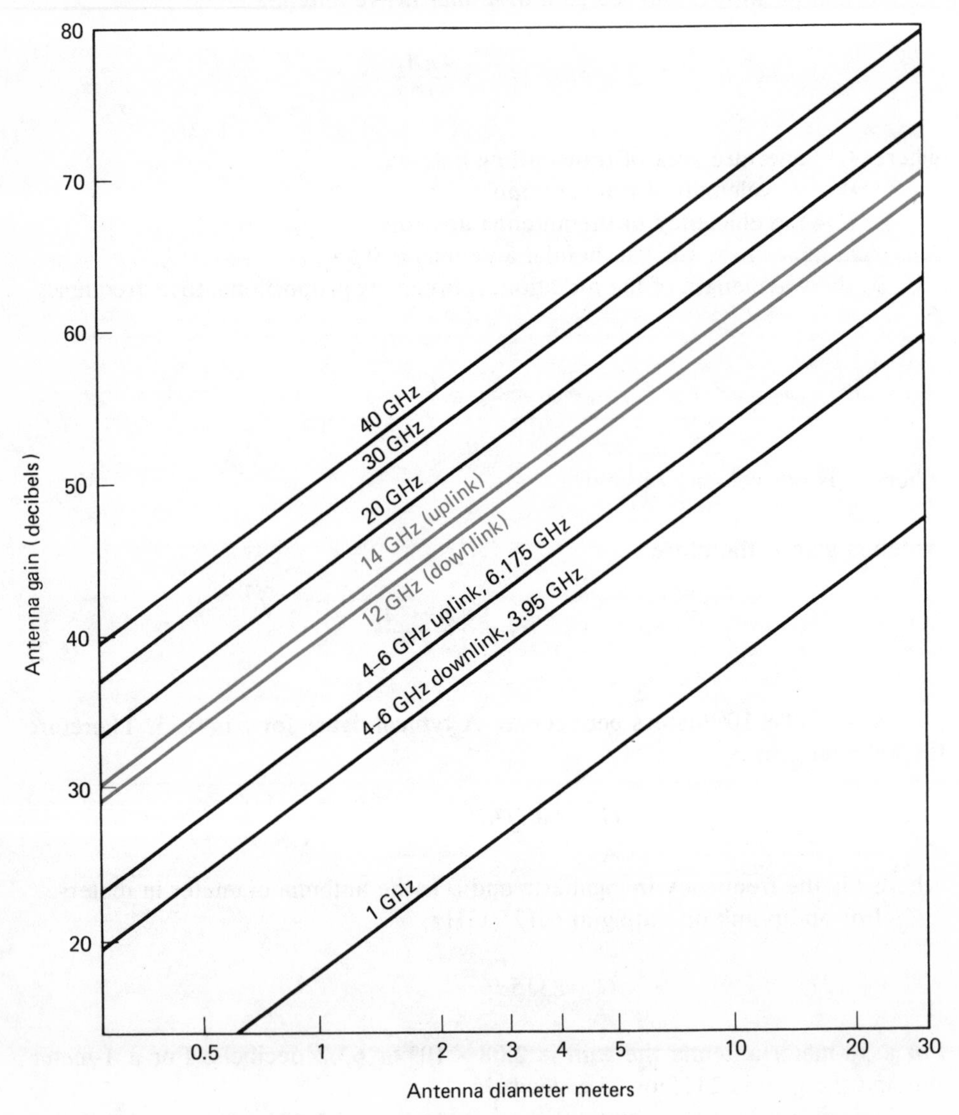

EIRP

The power of the transmitter is often referred to in a manner which takes the gain of the antenna into consideration. It is called *effective isotropic radiated power* (EIRP).

EIRP is the power of a transmitter and isotropic antenna that would achieve the same result as the transmitter and antenna in question.

If a 10,000-watt transmitter has an antenna with a gain of 5000, the resulting EIRP is 50 million watts.

EIRP is used as a measure of the satellite transmission as well as the earth-station transmission. Satellite EIRP has increased with time because both the transmitter power and the antenna gain have increased as satellites could generate more on-board power and deploy bigger antennas. The following table shows the increase:

Year	*Satellite*	EIRP *Watts*
1965	INTELSAT I	14
1967	INTELSAT II	36
1968	INTELSAT III	200
1971	INTELSAT IV	6400
1974	ATS-6	140,000

The big reflector of ATS-6 (Fig. 4.18) adds greatly to the satellite EIRP.

GAIN IN THE RECEIVING ANTENNA

Equation (6.2) is used to describe the gain in the receiving antenna also. An antenna both receives and transmits, with the same gain.

Gain is a measure of the *directionality* of the antenna. Figure 6.2 shows the relative strengths of a signal transmitted from a highly directional antenna. Approximately the same pattern applies to a signal received by the antenna. It is desirable that a receiving antenna should be highly directional so as to exclude as much noise as possible. As we shall discuss in Chapter 7, it is the received signal-to-noise ratio, rather than merely the signal strength, that determines the information-carrying capacity of a link.

Figure 6.3 shows measurements made on earth of the gain from the ATS-6 satellite antenna.

FREE SPACE LOSS

The loss of power of a signal traveling in free space is expressed as

$$L_{fs} = \left(\frac{4\pi D}{\lambda}\right)^2 = \left(\frac{4\pi f D}{C}\right)^2 \tag{6.3}$$

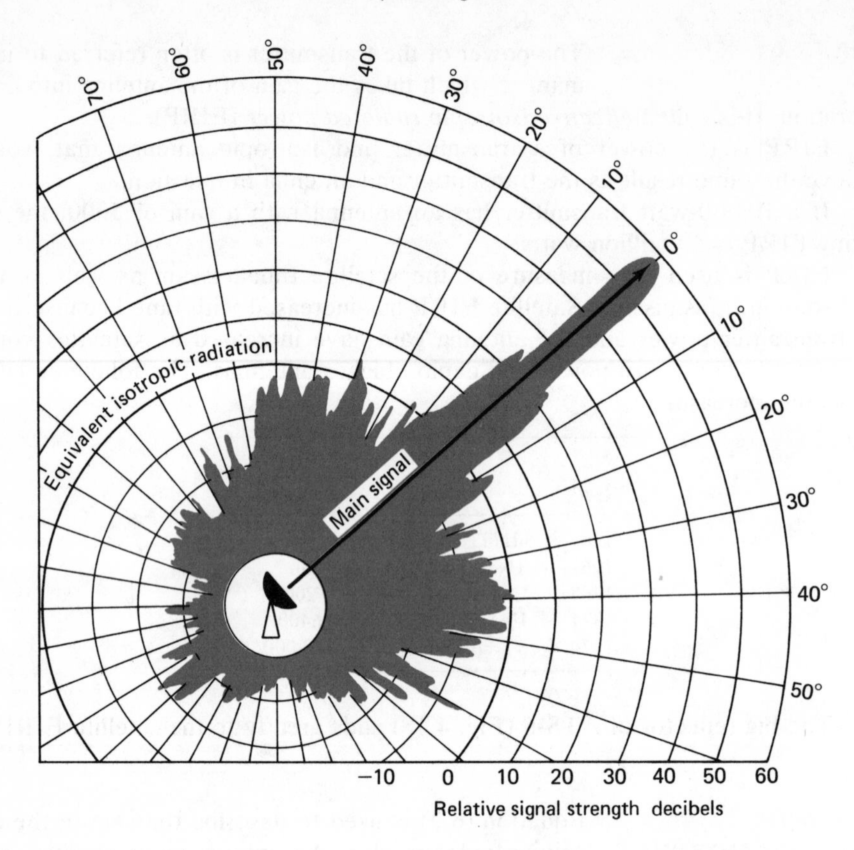

Figure 6.2 Relative strengths of the signal transmitted from an earth antenna with a gain of 57 db. (The signal received is 500,000 times the power of that from an equivalent isotropic antenna.)

where L_{fs} = free space power loss, i.e.,

$$\frac{\text{Power received by an isotropic antenna}}{\text{Power transmitted by an isotropic antenna}}$$

f = frequency
D = distance traveled
C = velocity of light

LINK LOSS

For a receiving antenna of gain G, the received power will be

$$P_r = P_t \frac{G}{L_{fs}}$$

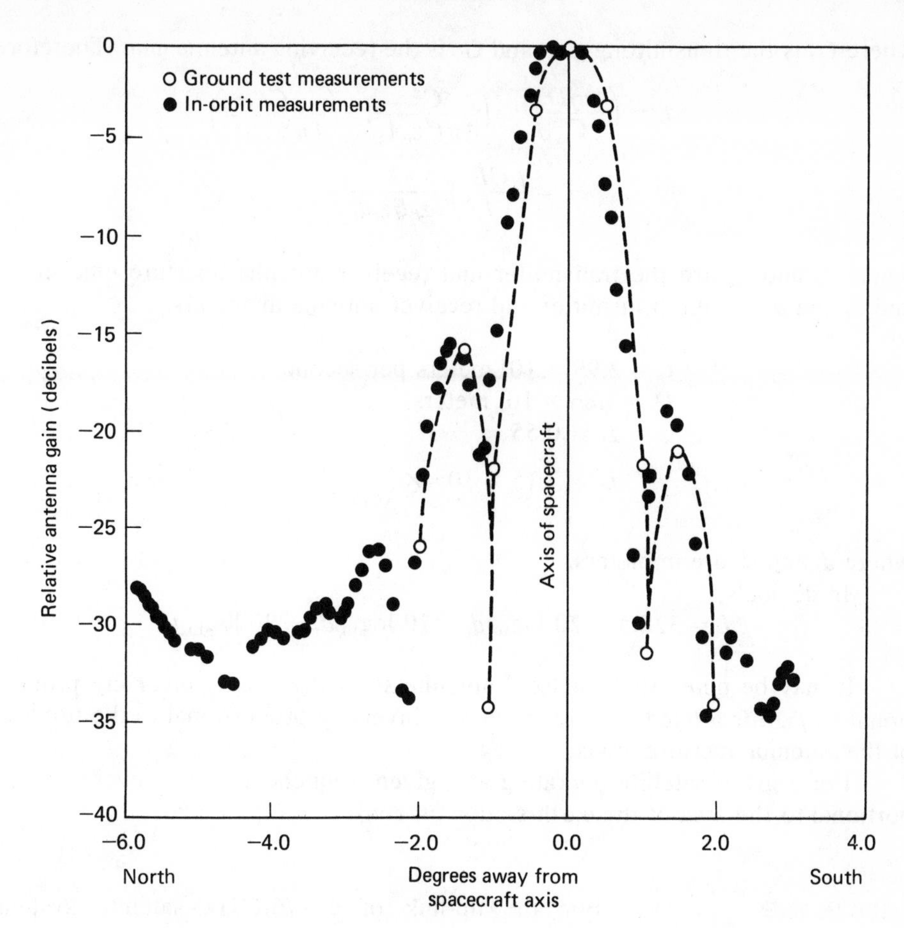

Figure 6.3 Relative antenna gain at points near the axis of the NASA ATS-6 satellite. (*Courtesy NASA.*)

where P_t is the transmitted EIRP.

Substituting from Eq. (6.2) and (6.3),

$$P_r = P_t \cdot z \frac{4\pi f^2 A_r}{C^2} \cdot \left(\frac{C}{4\pi f D}\right)^2$$
$$= \frac{zA_r}{4\pi D^2} P_t$$

where zA_r is the effective area of the receiving antenna. This equation is the same as Eq. (6.1).

If the gains of both the transmitting and receiving antennas are taken into consideration, the total loss between the transmitter and receiver is

$$L = \frac{L_{fs}}{G_t G_r}$$

where G_t is the transmitting gain and G_r is the receiving antenna gain. Therefore

$$L = \left(\frac{4\pi f D}{C}\right)^2 \left(\frac{C^2}{4\pi f^2 z_t A_t}\right) \left(\frac{C^2}{4\pi f^2 z_r A_r}\right)$$

$$= \left(\frac{CD}{f}\right)^2 \frac{1}{z_t d_t^2 z_r d_r^2}$$

where z_t and z_r are the transmitter and receiver antenna aperture efficiencies and d_t and d_r are the transmitter and receiver antenna diameters.

$$C = 2.99 \times 10^8 \text{ meters per second}$$
$$D \simeq 3.86 \times 10^7 \text{ meters}$$
$$z_t = z_r \simeq 0.55$$
$$\therefore \quad L \simeq 7.15 \times 10^{32} \times \left(\frac{1}{d_t d_r f}\right)^2$$

where d_t and d_r are in meters.

In decibels,

$$L = 328.5 - 20 \log_{10} d_t - 20 \log_{10} d_r - 20 \log_{10} f \tag{6.4}$$

It may be noted that for fixed antenna sizes the loss is inversely proportional to f^2. For a fixed frequency, loss is inversely proportional to the product of the antenna aperture areas.

For a given satellite operating at a given frequency, loss is inversely proportional to the area of the earth-station antenna.

LOSS IN THE UP-LINK

For the up-link of a 4/6-GHz satellite system, $f = 6.175 \times 10^9$. Hence

$$L = 132.7 - 20 \log_{10} d_t - 20 \log_{10} d_t \tag{6.5}$$

A large earth antenna is 30 meters in diameter. A typical satellite antenna is 1.5 meters in diameter. Using these figures,

$$L = 132.7 - 29.5 - 3.5 = 94.7 \text{ decibels}$$

In other words, the power transmitted is 2.95×10^9 times that received at the satellite—a spectacular loss.

LOSS IN THE DOWN-LINK

On the up-link the gain of the transmitting antenna can be large. A 30-meter earth antenna has a gain of a million or so. On the down-link the transmitting antenna is smaller and so its gain cannot be so great, but the receiving antenna is the large earth-station dish and so the total loss is comparable to that of the up-link.

The satellite antenna beams the signal so that it covers a certain portion of the earth. The more directional the antenna, i.e., the smaller the area of its earth coverage, the greater will be its gain.

The antenna may be designed to cover the entire earth, the projected area of which is $\pi \times 3962^2$ square miles. If an earth-coverage antenna could be perfect so that no signal spilt over beyond the earth, the antenna gain would be

$$\frac{4\pi \times 24{,}000^2}{\pi \times 3962^2} = 147$$

$$= 22 \text{ decibels}$$

In reality an antenna cannot be this perfect, and the gain of a good earth-coverage antenna is about 17 decibels.

If the antenna is designed to cover a smaller portion of the earth, for example, an area 3000 miles across to give coverage of the continental United States, the gain would be

$$\frac{4\pi \times 24{,}000^2}{\pi \times \left(\frac{3000}{2}\right)^2} = 1024 = 30 \text{ decibels}$$

The gain of the signal received at different parts of the earth varies, and so a contour map can be drawn of the satellite antenna gain. Figure 6.4 is such a map for the WESTAR I satellite. The contours of high antenna gain are referred to as a *footprint* of the satellite. As indicated in Fig. 4.3, the contours can be shaped to fit a given area by appropriate antenna design.

To make an antenna more directional, either its size must be increased or the frequency of its transmission must be increased. The WESTAR I antenna giving the footprint of Fig. 6.4 is 1.5 meters in diameter. A satellite antenna 4 times the width could give an equally sharply defined footprint one-fourth of the size of that in Fig. 6.4 covering, say, Japan or New Zealand. Its gain would be 16 times as great (12 decibels larger). The weight penalty on the satellite would be higher.

For a 4/6-GHz satellite (such as WESTAR) the down-link frequency is 3.95 GHz. Equation (6.4) then becomes

$$L = 136.6 - 20 \log_{10} d_t - 20 \log_{10} d_r \tag{6.6}$$

For an earth-station diameter of 30 meters and a satellite antenna diameter of 1.5 meters,

$$L = 136.6 - 3.5 - 29.5 = 103.6 \text{ decibels}$$

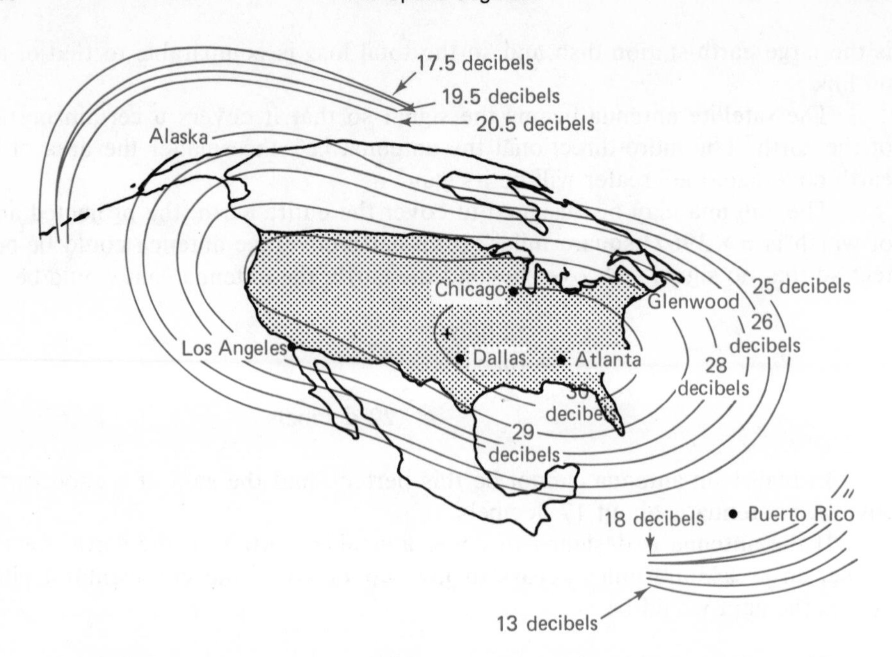

Figure 6.4 A "footprint" of WESTAR I over the continental United States. The contours show the gain of the transmitting antenna on the satellite. (*Reproduced with permission from reference* [*1*].)

In other words, the power received is about a ten-billionth of that transmitted.

ATMOSPHERIC ABSORPTION

In addition to the losses inherent in the large distances of space, the earth's atmosphere also causes propagation losses.

Atmospheric absorption is caused by six main factors:

1. Molecular oxygen.
2. Uncondensed water vapor.
3. Rain.
4. Fog and clouds.
5. Snow and hail.
6. Free electrons in the atmosphere.

The first two of these are relatively constant; the latter four vary greatly with weather and atmospheric conditions. Figures 6.5 and 6.6 show the absorption due to oxygen and water vapor. The absorption is greater for larger angles of elevation because the radio beam has a longer path through the atmosphere.

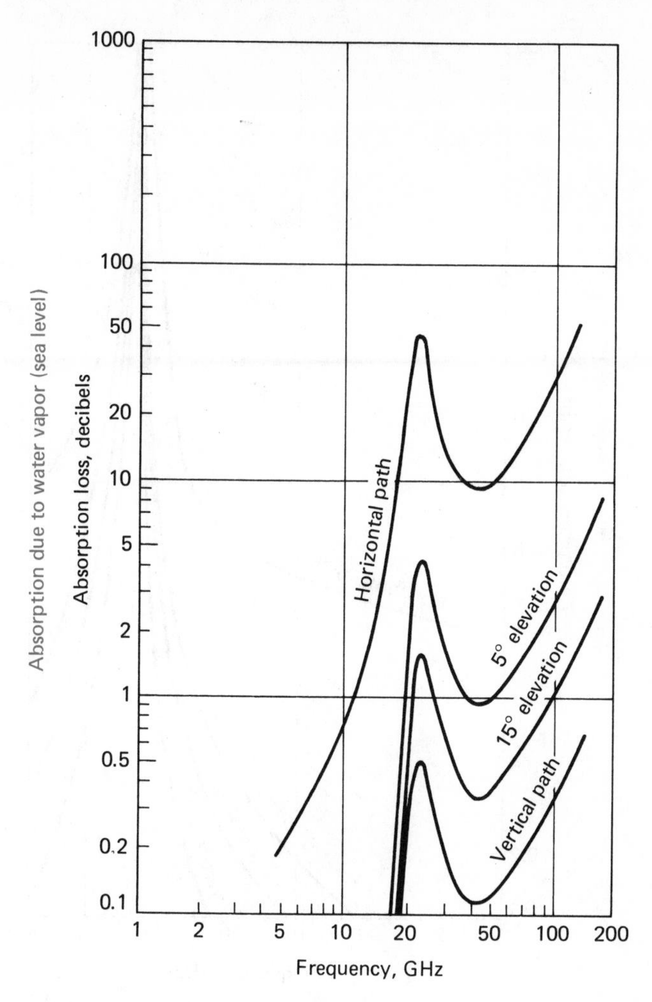

Figure 6.5 Absorption in the atmosphere caused by uncondensed water vapor.

The absorption by molecular oxygen has a sharp peak at 60 GHz, and the absorption by water molecules has a peak at 21 GHz. The absorption is caused by the radio wave changing the rotational energy levels of the molecules, and resonance effects occur at the above frequencies. Atmospheric nitrogen does not have a resonance peak of this type; carbon dioxide has one above 300 GHz.

When there are free electrons in the earth's atmosphere the radio waves collide with them. This causes absorption because radio energy is transferred to the electrons. The electron density of the ionosphere is greatly reduced during the dark hours, so the absorption drops considerably at night. Electron ab-

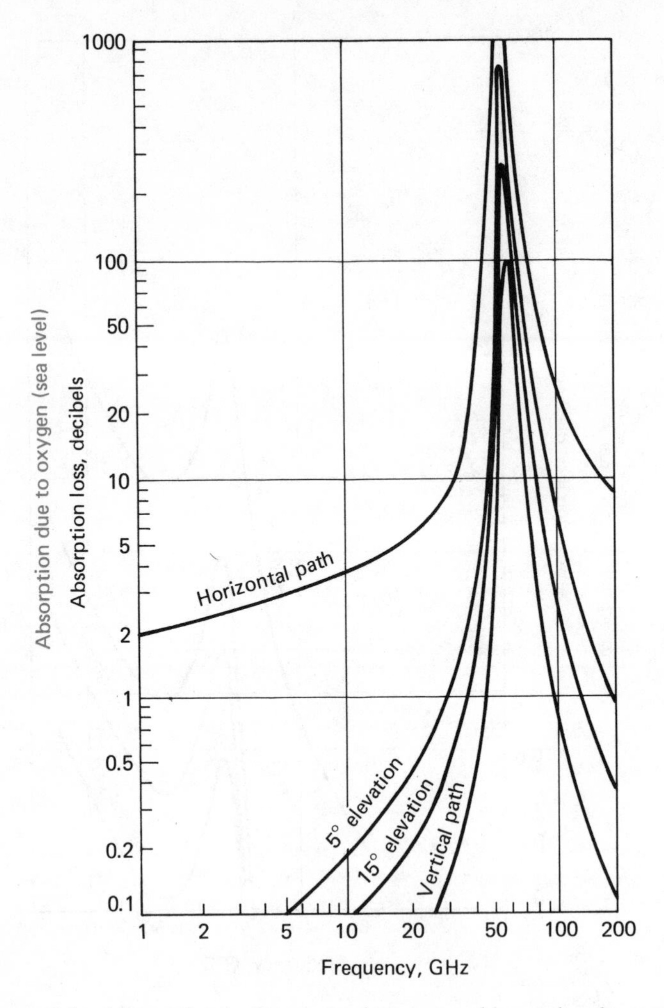

Figure 6.6 Absorption in the atmosphere caused by molecular system.

sorption mainly affects radio at frequencies below 100 MHz. It has negligible effect on the UHF and SHF bands.†

Figure 6.7 is a composite diagram showing the absorption caused by electrons, oxygen, and water vapor. It will be seen that there is a window in the UHF and SHF bands between 300 MHz and 10 GHz. About 40 GHz, atmospheric absorption is high. There is a rather opaque window around 30 GHz between the water vapor and oxygen absorption peaks. Satellite experiments have been proposed for investigating the transmission properties in this window. The WARC 20/30-GHz frequency allocation (see Fig. 8.3) would enable a satellite to straddle the water vapor absorption peak, with an up-link at 27.5

†The designation is explained in Box 2.1.

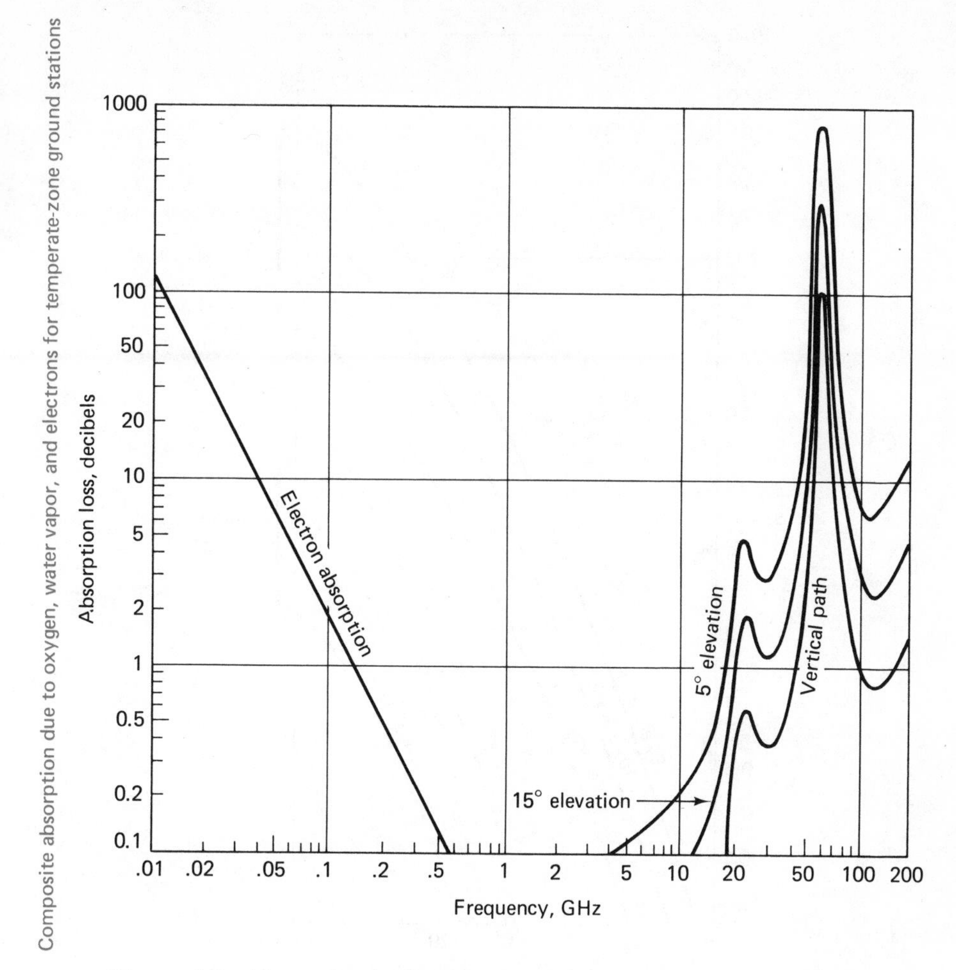

Figure 6.7 Absorption in the atmosphere caused by electrons, molecular oxygen, and uncondensed water vapor.

to 29.5 GHz and a down-link at 17.7 to 19.7. Its alternative down-link allocation, 19.7 to 21.2, lands right on the water vapor peak.

The curves shown are for attenuation at sea level. If the earth station is on a mountain, the losses can be reduced by about a half.

BAD WEATHER The losses discussed so far in this chapter have been relatively constant and predictable. Losses caused by rain and fog vary greatly. Snow and hail cause much less attenuation than rain and fog.

Figures 6.8 and 6.9 show curves of the attenuation typical from heavy rain or fog and clouds. Only rarely does the attenuation become worse than the

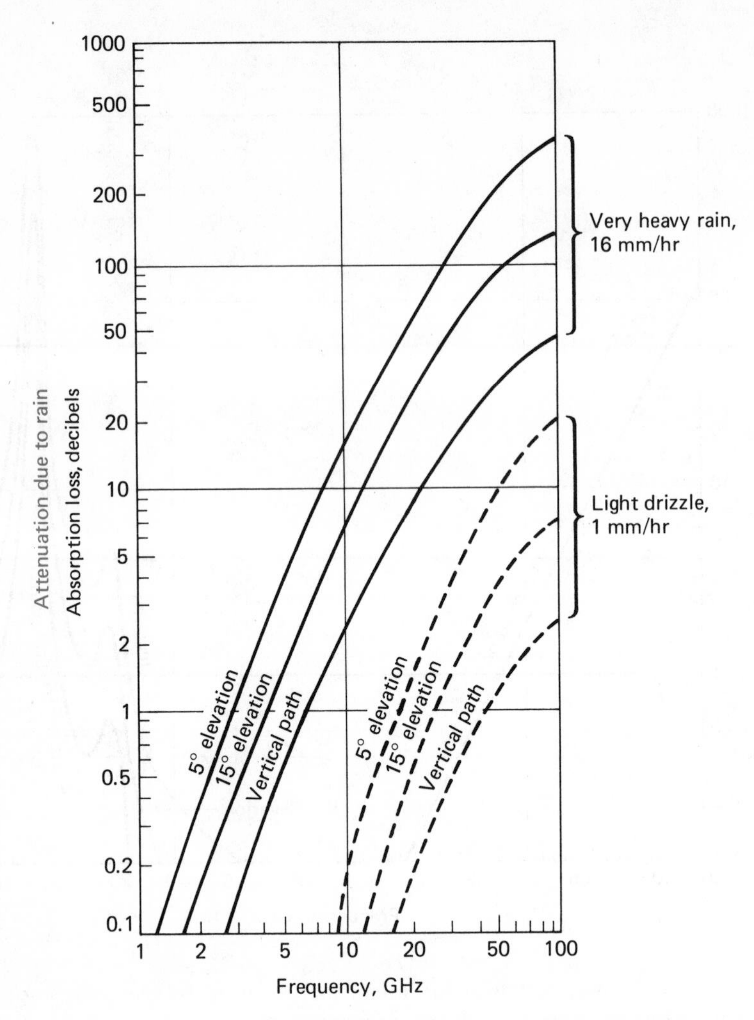

Figure 6.8 Typical absorption due to rain in the atmosphere.

solid lines in these figures. The reader may compare the figures with Fig. 6.7. Most rain occurs below an altitude of 2 kilometers, so again there will be an advantage in having an antenna on a mountaintop.

Gentle rain may spread over a wide area, but severe storms, which cause the chief problem, are almost always small in extent. There would be considerable benefit, therefore, in being able to switch the transmission path between earth stations some miles apart. Stations using two antennas that are several miles apart, linked to the same control center, are called *diversity* earth stations. They may come into use by common carriers using frequencies above 10 GHz. In future networks, some transmission will not be "real-time," as are telephone calls, but will include the transmission and *storage* of data, movies,

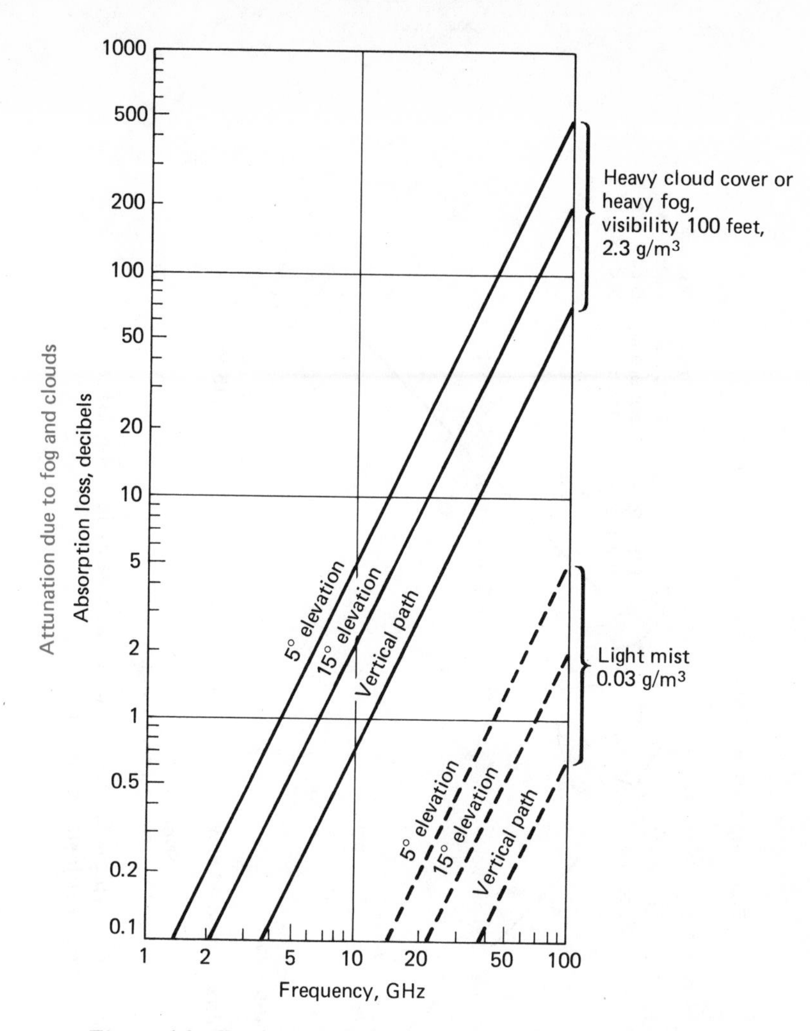

Figure 6.9 Typical absorption due to fog, mist, and clouds.

facsimile mail, and so on. The drop in capacity caused by a storm will merely delay the non-real-time transmission.

EQUATION FOR RECEIVED POWER

An equation can be written which evaluates the power received from a satellite link. It adds up the various sources of signal gain and loss. All of these are predictable except the losses caused by rain, clouds, or fog, which are highly variable and can cause severe degradation of the link, especially at frequencies of 12/14 GHz and higher. Satellite engineers come to work like sailors with one eye on the weather.

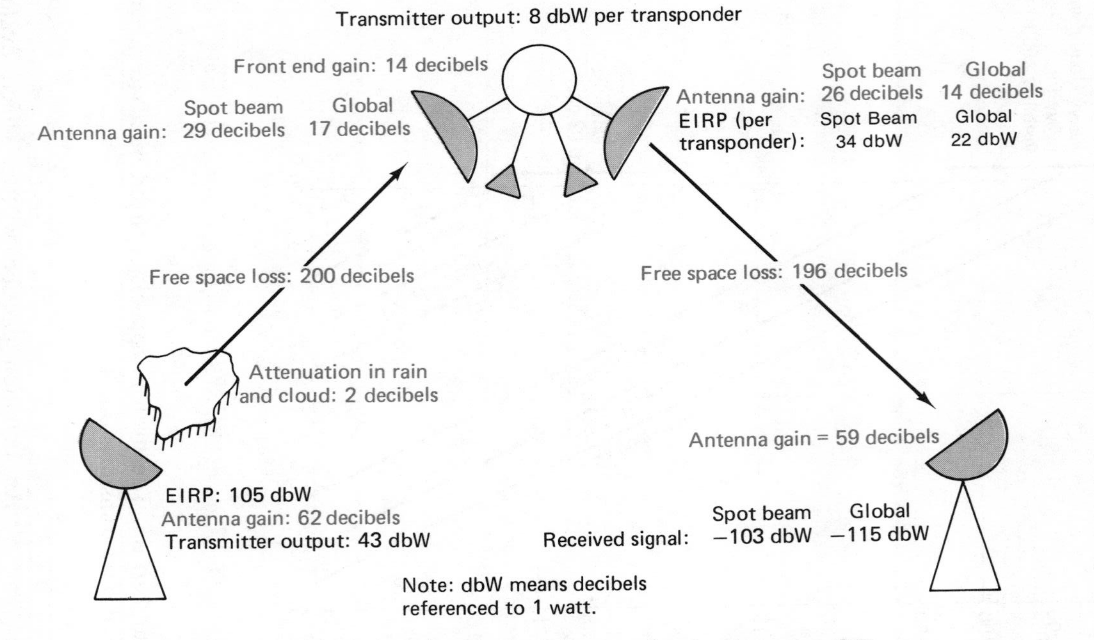

Figure 6.10 Typical figures for losses and gains in a 4-6 GHz system with large earth station antennas.

TYPICAL FIGURES Figure 6.10 shows some typical figures for the losses and gains in a satellite system. Figure 6.11 also illustrates losses and gains showing that the relative importance of loss caused by very bad weather is more severe at the higher frequencies.

REFERENCES

1. *The Western Union Communicator,* Summer 1973. The whole issue is about WESTAR and is available from Western Union Telegraph Company, N.J.
2. "Future Communications Systems via Satellites Utilizing Low-Cost Earth Stations," prepared by the Electronics Industries Association for the President's Task Force of Communications Policy, Washington, D.C., 1969.

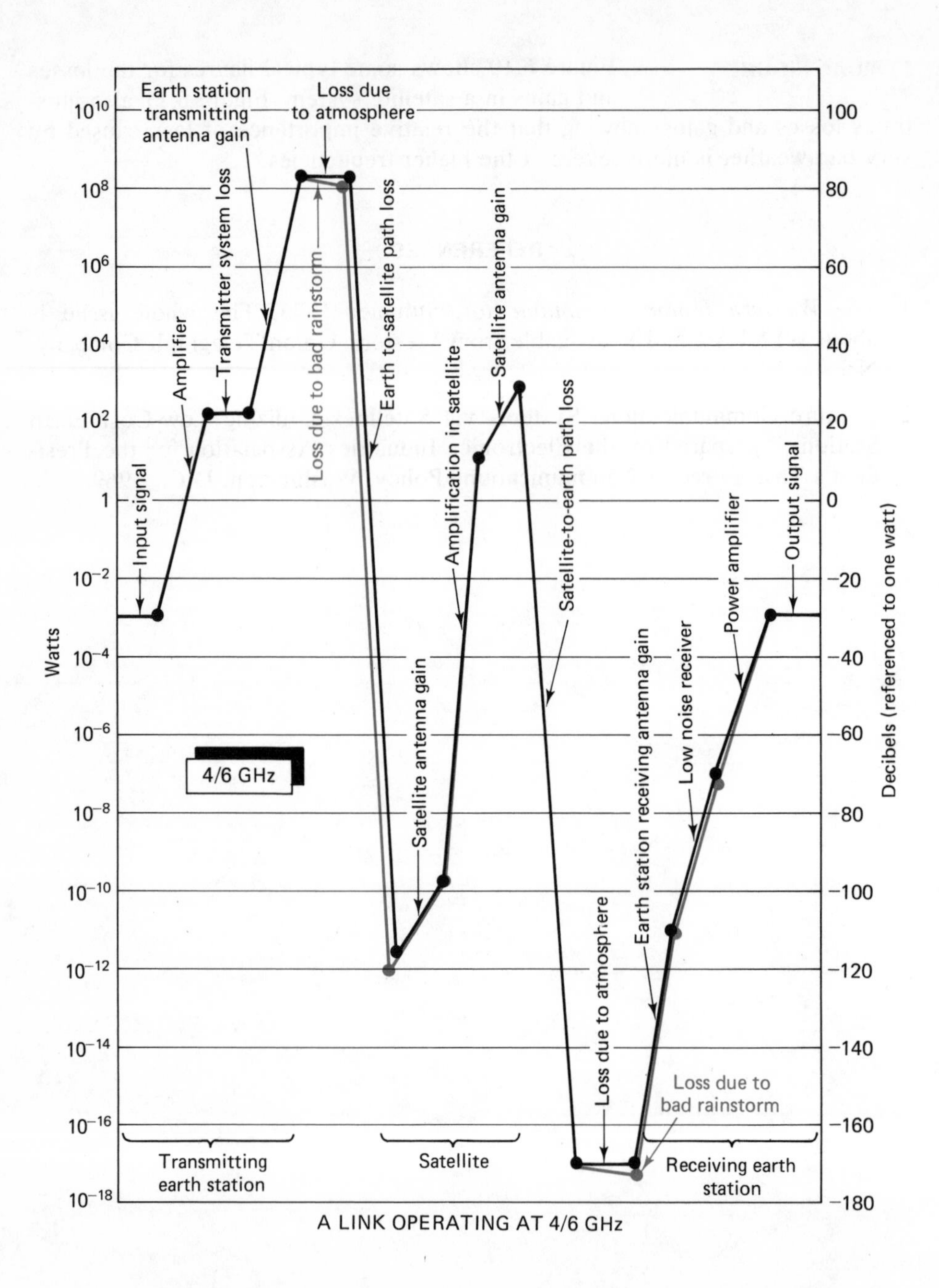

Figure 6.11 Gains and losses in power of signals being relayed by satellite. The signal falls to about one hundred billion billionth of its strength (10^{-20}) on each of its 25,000-mile journeys through space. This great loss is balanced by the gains of the antennas and amplifiers.

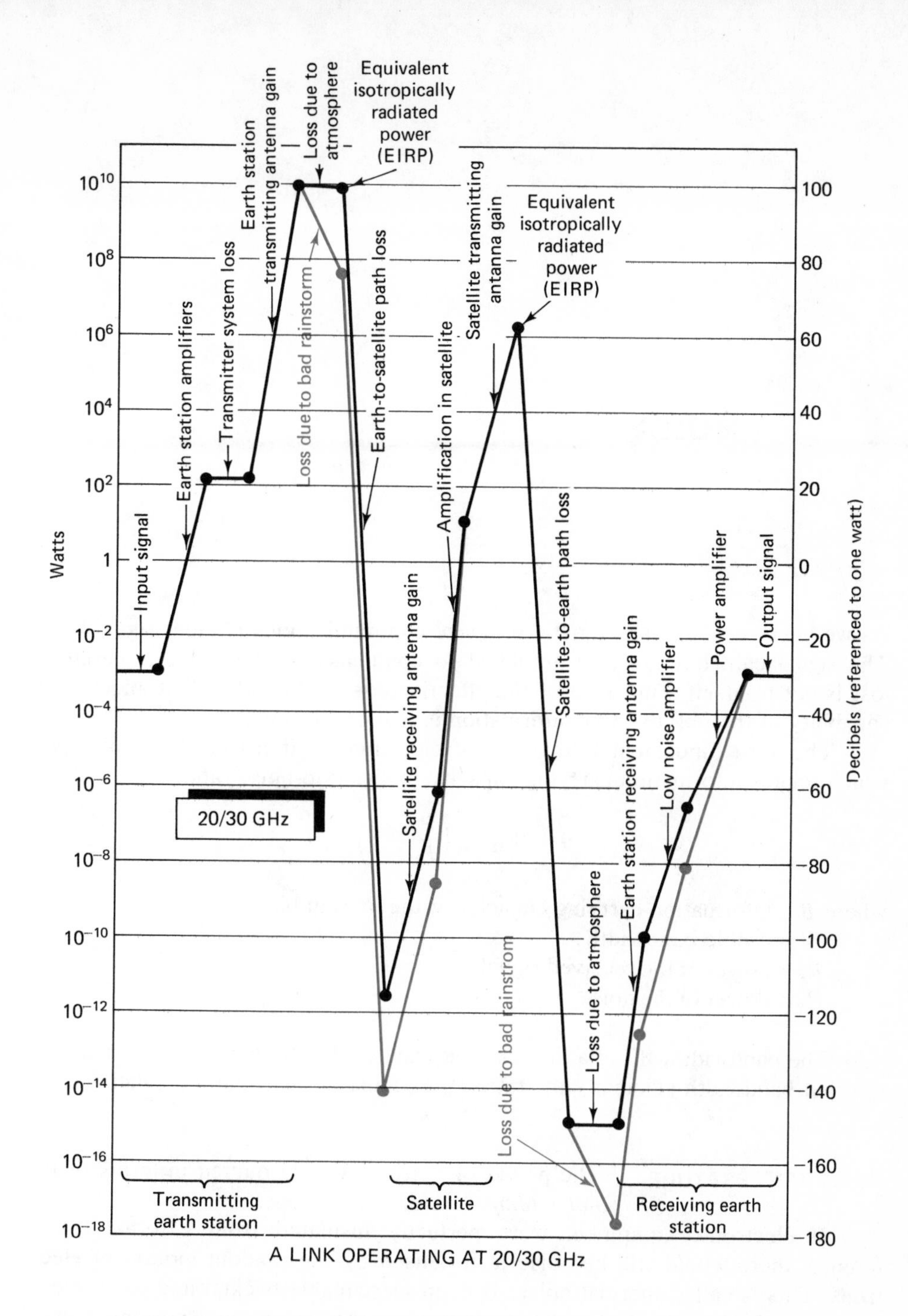

At the higher frequencies of the above illustration, antennas provide more gain in signal strength. However, the loss caused by bad rain storms is worse. As discussed in the following chapter, bad weather causes not only loss but also noise which degrades the signal.

7 NOISE

As we have seen, a characteristic of space communications is great signal loss. The signal can, however, be amplified to compensate for the loss; therefore loss is not by itself a measure of the effectiveness of the link. What affects the capability of the link to carry information is the signal-to-noise ratio.

The most fundamental equation in telecommunications is Shannon's law, which relates information-carrying capacity to signal-to-noise ratio:

$$B = W \log_2 \left(\frac{P_R}{P_N} + 1\right) \tag{7.1}$$

where B = information-carrying capacity of the link, in bits
W = usable bandwidth
P_R = power of the received signal
P_N = power of the noise

The bandwidth, W, of a link may be fixed by the frequency allocation. To use that bandwidth efficiently the ratio P_R/P_N must not become too small.

NOISE TEMPERATURE

The power of noise is usually quoted in terms of its *noise temperature.*

If electronic equipment were perfectly insulated from external interference, there would still be noise in it caused by the random motion of electrons. This is called thermal noise. It is an inescapable background to all electronic processes. The higher the temperature, the faster the electrons move, and the higher the power of the thermal noise.

The power of the thermal noise affecting a given range of frequencies is proportional to the absolute temperature and to the bandwidth of frequencies in question:

$$P_N = kTW \tag{7.2}$$

where P_N = noise power, in watts
k = Botzmann's constant, 1.380×10^{-23} watt-seconds/°K
T = temperature, in °K
W = bandwidth, in hertz

If thermal noise were the only form of noise affecting the signal, the maximum information-carrying capacity of the signal would be [substituting into Eq. (7.1)]

$$B = W \log_2 \left(\frac{P_R}{kTW} + 1 \right) \tag{7.3}$$

In fact there are many other forms of noise, but it is convenient to refer to them with the same equation, so an imaginary temperature is used called the noise temperature.

The noise temperature of a noise source is that temperature which produces the same noise power over the same frequency range.

Thus if a noise source creates noise of power P_N, its noise temperature, sometimes called *equivalent noise temperature,* ENT, is

$$T = \frac{P_N}{kW} \tag{7.4}$$

NOISE DENSITY

The term *noise density* refers to the noise per hertz of bandwidth:

$$\text{Noise density} = \frac{P_N}{W} = kT \tag{7.5}$$

C/kT CARRIER-TO-NOISE RATIO

A ratio often used to state the quality of a satellite channel is

$$\frac{\text{Received carrier power}}{\text{Noise density}} = \frac{P_R}{\left(\frac{P_N}{W}\right)} = \frac{P_R}{kT}$$

Carrier power is often referred to with the symbol C. The above ratio is referred to as C/kT and is called *carrier-to-noise ratio.* The carrier-to-noise ratio is related to the quantity of information that can be transmitted in unit bandwidth, as shown by Shannon's law:

$$B = W \log_2 \left(\frac{C}{kT} + 1 \right)$$

C/kT is plotted against earth-station EIRP in Fig. 7.1 for the up-link of a typical American domestic satellite. It will be seen that the ratio cannot be improved above a certain level because channel saturation is reached.

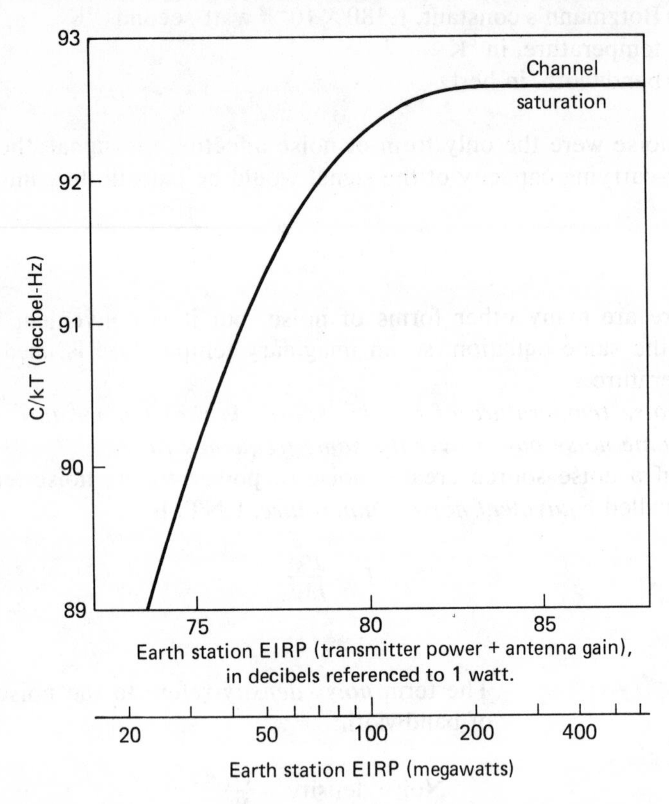

Figure 7.1 C/kT for a typical American domestic satellite.

EXTERNAL SOURCES OF NOISE

The following are external sources of noise in a radio link:

- The sun
- The moon
- The earth
- Galactic noise
- Cosmic noise
- Sky noise
- Atmospheric noise
- Man-made noise

These sources differ in their intensity, frequencies, and location in space.

If a satellite antenna points toward the *sun,* the signal will be effectively blotted out because the noise temperature of the sun is 100,000°K or more. The noise from the sun varies with solar activity, as shown in Fig. 7.2. The noise temperature of the *sky* is about 30°K. The directivity of the antenna thus not only focuses the beam being transmitted but also protects the signal being received from noise sources.

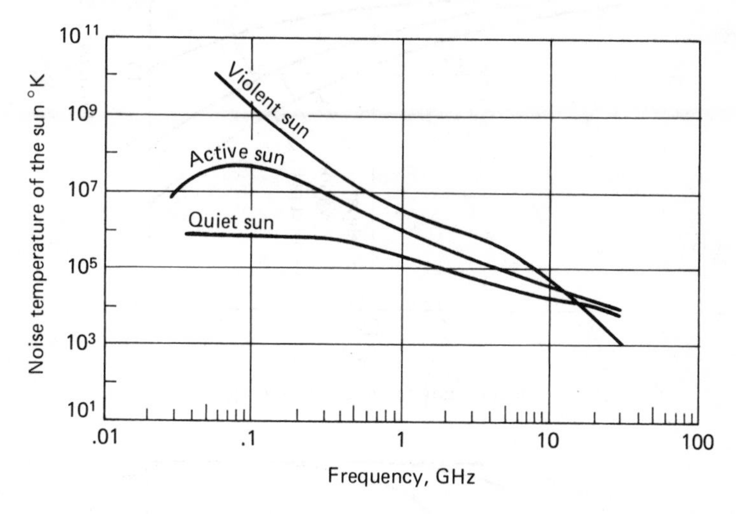

Figure 7.2 Solar noise.

The noise temperature of *earth,* viewed from space, averages 254°K. A satellite antenna with a beam width equal to the projected width of the earth would receive this quantity of noise as a background to the signals from earth stations. Because of local terrain variations, spot beams directed at some portions of the earth receive a noise temperature slightly greater than 254°K.

Galactic noise refers to noise from radio stars in the galaxy. This noise falls off rapidly at higher frequencies and has negligible effect on transmission over 1 GHz. *Cosmic noise* refers to other radio noise from outer space, and this also becomes low above 1 GHz.

Lightening flashes and electrostatic discharges in the atmosphere are a major source of radio noise below 30 MHz. Fortunately they radiate negligibly at the frequencies used by satellites. *Atmospheric noise* originates mainly from the oxygen and water vapor molecules which absorb radiation, as discussed in Chapter 6, and then reemit it. Consequently the frequencies at which atmospheric absorption is high (Figs. 6.3 and 6.4) are the same as those at which atmospheric noise is high, and the noise, like absorption, is greater for lower angles of elevation. Figure 7.3 shows the noise temperatures of atmospheric oxygen and water vapor.

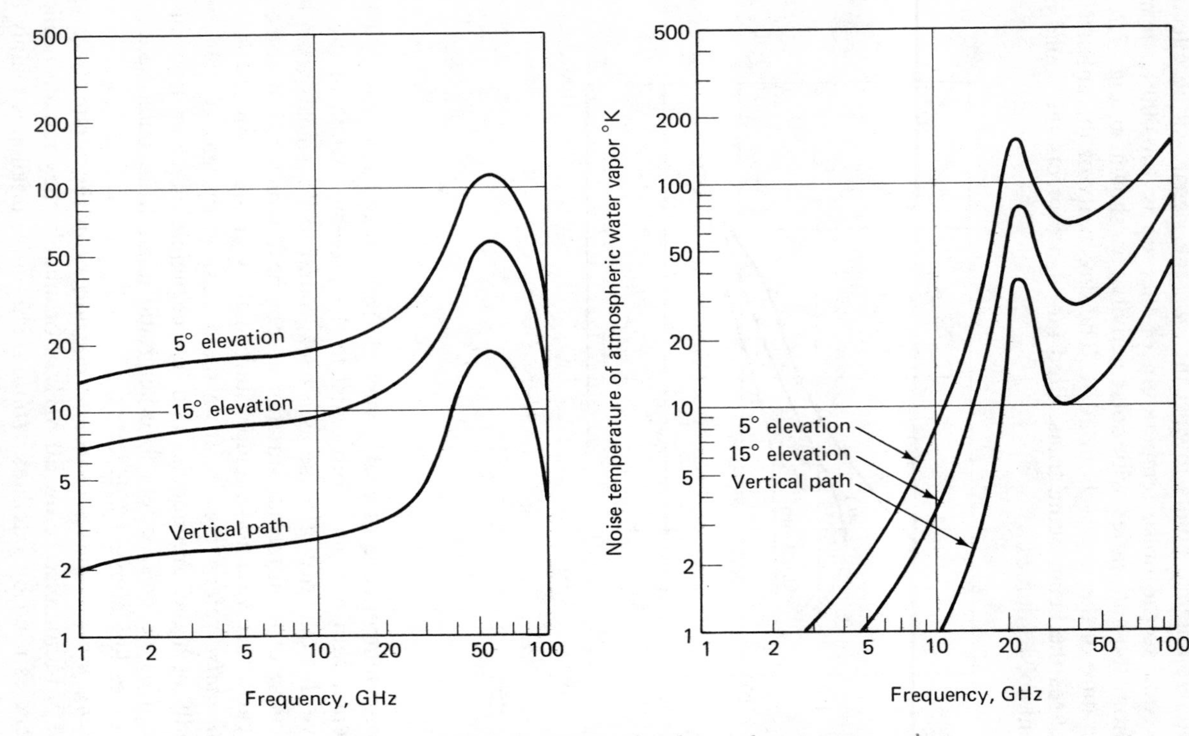

Figure 7.3 Noise caused by oxygen molecules and water vapor in the atmosphere (at sea level in temperate zones).

Man-made noise, which plagues the lower radio frequency on earth, has little effect above 1 GHz. It arises mainly from electrical machinery and is much greater in industrial areas. If present, it can be reduced by shielding the antenna. It is virtually absent in space. Interference from terrestrial radio links is a different, and serious, problem, as discussed in Chapter 8.

Figure 7.4 shows how the combination of these different types of noise affect the received signal. Noise received at the satellite is dominated by the noise temperature of earth (the dotted line). At the earth station there is a noise window between the cosmic noise effect and the effect of water vapor.

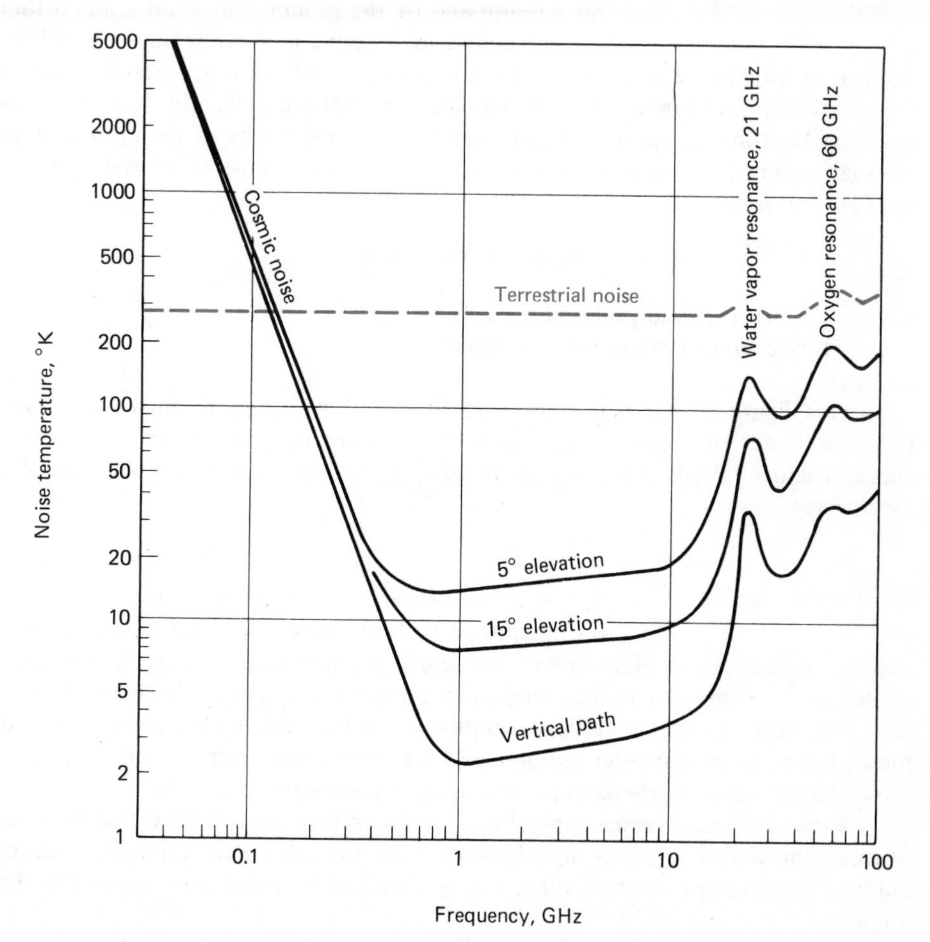

Figure 7.4 A composite noise diagram. The solid lines show typical noise received by an earth antenna, excluding the effects of rain, fog, and cloud. The dotted line shows the noise received at the satellite, caused mainly by the earth.

BAD WEATHER Very heavy rain causes more noise at the earth station than all the other noise sources combined. As with absorption, its effect is worse at the higher frequencies. Figure 7.5 shows the effect of heavy rain and bad fog or heavy cloud cover.

At frequencies above 10 GHz the noise received can be highly variable because of weather conditions. It is desirable to avoid, if possible, a low angle of elevation at these frequencies.

FIGURE OF MERIT, G/T Because of the very feeble signal received, both by the satellite and by the ground station, it is important that the receiving antenna and electronics should introduce as little noise as possible. To avoid losses and noise in lines connecting the receiving antenna to the electronics, the antenna usually has the preamplifier built into it, as shown in Figure 7.6. The efficiency of the combination is usually quoted as the ratio of the gain to the noise temperature and is called the *figure of merit*

$$\text{Figure of merit} = \frac{G}{T}$$

where G = antenna and preamplifier gain
T = receiver system noise temperature

This figure of merit is related to the resulting signal-to-noise ratio, and hence indicates the relative capability of the receiving subsystem to receive a signal. Figure 7.7 plots some typical G/T values for receivers with uncooled electronics.

EQUIPMENT NOISE T, the noise temperature of the receiving equipment, is contributed to both by the antenna structure itself and by its associated electronics. The early earth stations used cryogenically cooled preamplifiers to reduce the noise temperature. Today, with more powerful satellites, cheaper receiving equipment can be used, with a lower figure of merit, having both a smaller antenna and a higher noise temperature. The more powerful the satellite, the cheaper the receiving equipment can be.

Table 7.1 shows some typical noise temperature figures for different types of electronics. The resulting signal-to-noise ratio is calculated assuming that the antenna noise temperature is 60°K and a signal of 10 picowatts enters the preamplifier.

A signal-to-noise ratio of 10 decibels is typical on satellite links, whereas a signal-to-noise of 30 decibels is more likely in terrestrial telecommunications. We can feed these two figures into Shannon's equation and compare the theoretical throughputs of a typical satellite and terrestrial link of the same bandwidth.

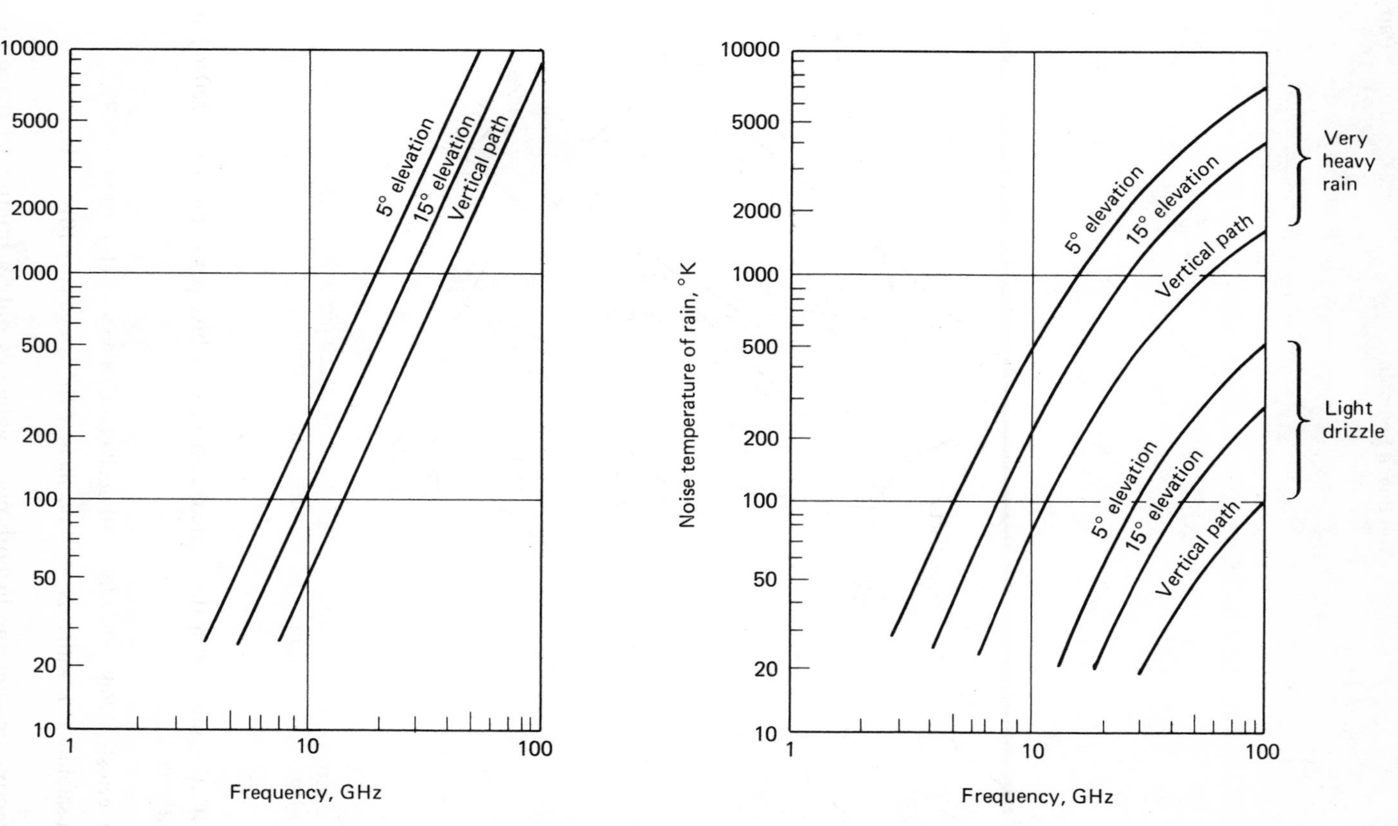

Figure 7.5 Noise caused by heavy cloud, fog, and rain.

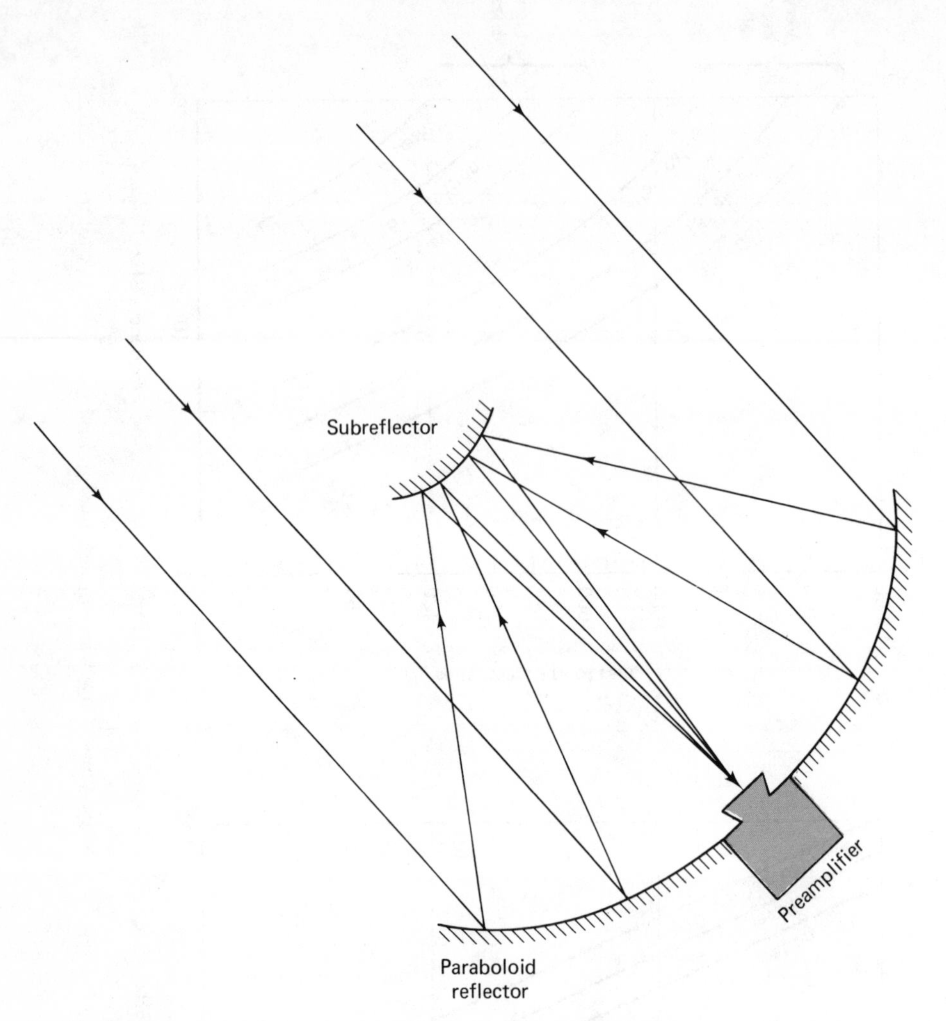

Figure 7.6 To minimize noise in the receiving subsystem, the preamplifier is built into the antenna dish construction.

C/W, the maximum theoretical capacity in bits per second divided by the bandwidth, is

1. For a terrestrial link with $S/N = 30$ decibels: $C/W = \log_2 (1 + 1000) = 9.97$.
2. For a satellite link with $S/N = 10$ decibels: $C/W = \log_2 (1 + 10) = 3.46$.

Cost-effective engineering in both cases achieves only a fraction of these theoretical figures. The 36-MHz transponder on a satellite carries, typically, 50 million bits per second ($C/W = 1.39$). A 3-kHz voice channel on earth carries

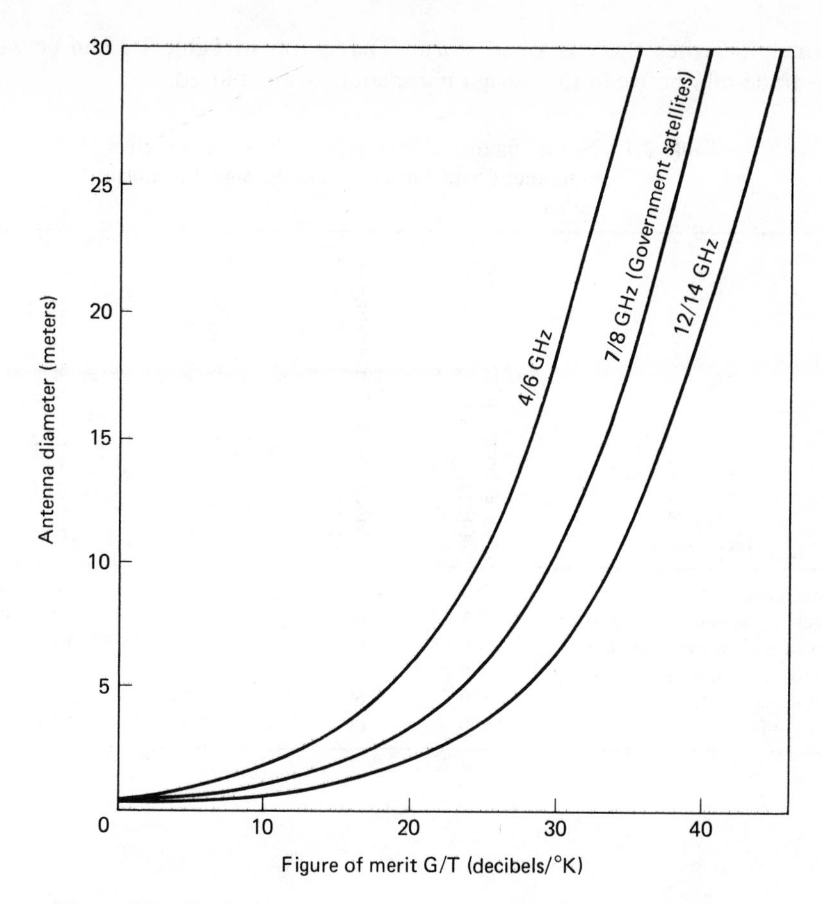

Figure 7.7 Typical earth station receiver figures of merit, G/T, calculated assuming an uncooled receiver with a noise temperature of 316°K.

9600 bits per second ($C/W = 3.2$). A 60-MHz waveguide channel on earth carries 274 million bits per second ($C/W = 4.6$).

One color television channel occupies 6 MHz of channel bandwidth on a terrestrial link and a 36-MHz transponder on a satellite. Nine hundred voice channels occupy 4 MHz of bandwidth on earth (on facilities which transmit a CCITT mastergroup) but occupy a whole 36-MHz transponder in space.

LINK DESIGN SUMMARY

A satellite link designer will produce a balance sheet showing the losses, gains, and noise in the up-link and the down-link. Table 7.2 shows such a summary. It evaluates the received signal-to-noise ratios, SNR, when the weather is normal and when there is a bad storm. The figures in Table 7.2 can be varied as the trade-offs between the design parameters are explored.

Table 7.1 Noise figures for typical types of receiver equipment and typical resulting signal-to-noise ratios.

Type of Electronics	Noise Temperature of Electronics	Typical Noise Temperature of Antenna	Combined Noise Temperature of Antenna & Electronics	*S/N* for a Received Signal of 10 picowatts (10^{-11} watts)
Maser (cooled to 4.2°K)	10	60	70	14.6 decibels
Parametric amplifier (cooled to 25°K)	35	60	95	13.3 decibels
Uncooled parametric amplifier	120	60	180	10.5 decibels
Inexpensive parametric amplifier	300	60	360	7.5 decibels
Tunnel diode amplifier	530	60	590	5.3 decibels
Schottky mixer	1000	60	1060	2.7 decibels

Table 7.2 A summary of the gains, losses and noise in the satellite link. The figures can be varied as trade-offs between the design parameters are explored.

	A 4/6-GHz Link. Satellite Antenna: Earth Coverage. Earth Antenna: 12 meters. Moderately Low-Cost Electronics in Earth Station.	A 12/14-GHz Link. Satellite Antenna: 1.8 Meters. Earth Antenna: 1.8 Meters. Low-Cost Earth-Station Receiver.	A 20/30-GHz Link for Common Carrier Use. Satellite Antenna: 2 Meters. Earth Antenna: 27.5 Meters. Cryogenically Cooled Receiver.	A 12/14-GHz Link on a Broadcast Satellite. Satellite Antenna: 9 Meters. Earth Antenna: 1.8 Meter Receiver-only; 12-Meter Transmit. Low-Cost Earth Station.
Up-Link				
Transmitter power, dBw*	35	25	20	20
Transmitter system loss, decibels	−1	−1	−1	−1
Transmitting antenna gain, decibels	55	46	76	62
Atmospheric loss, decibels	0	−0.5	−2	−0.5
Free space loss, decibels	−200	−208	−214	−208
Receiving antenna gain, decibels	20	46	53	60
Receiver system loss, decibels	−1	−1	−1	−1
Received power, dBw*	−92	−93.5	−69	−68.5
Noise temperature, °K	1000	1000	1000	1000
Received bandwidth, MHz	36	36	350	36
Noise, dBw*	−128	−128	−118	−128
Received SNR, decibels	36	34.5	49	59.5
Loss in bad storm, decibels	2	10	25	10
Received SNR in bad storm, decibels	34	24.5	24	49.5
Down-link				
Transmitter power, dBw*	18	20	8	10
Transmitter system loss, decibels	−1	−1	−1	−1
Transmitting antenna gain, decibels	16	44	49	58
Free space loss, decibels	−197	−206	−210	−206
Atmospheric loss, decibels	0	−0.6	−2	−0.6
Receiver antenna gain, decibels	51	44	72	44
Receiver system loss, decibels	−1	−1	−1	−1
Received power, dBw*	−114	−100.6	−85	−96.6
Noise temperature, °K	250	1000	250	1000
Received bandwidth, MHz	36	36	350	36
Noise, dBw*	−131	−128	−121	−128
Received SNR, decibels	17	27.4	36	31.4
Loss in bad storm, decibels	2	10	25	10
Received SNR in bad storm, decibels	15	17.4	11	21.4

*dBw means decibels referenced to one watt. I. E. 1 watt = 0 dBw; 100 watts = 2 dBw, etc.

8 FREQUENCIES

The choice of frequency at which a satellite operates is determined by two main factors. First there are many contenders for the useful parts of the radio spectrum, and they would interfere with one another if not carefully controlled. The frequency must be chosen to avoid harmful interference. Second, as we have seen, transmissions at some frequencies are more susceptible to loss, absorption, and noise. The frequency should be chosen to minimize the cost of the transmission or maximize its information-carrying rate.

FREQUENCY ALLOCATIONS

Because space transmissions transcend national borders, it is very important that there should be international agreement on the use of space frequencies. Such agreements originate at Administrative Radio Conferences held under the auspices of the International Telecommunications Union (ITU) and comprise the *Radio Regulations*. These regulations have the force of a treaty to which each signatory is bound under international law. Two conferences have allocated frequencies for space usage: the Extraordinary Administrative Radio Conference of 1963 (EARC) and the World Administrative Radio Conference for Space Telecommunications of 1971 (WARC).

The frequencies of interest for satellites are the higher UHF frequencies, the SHF band, and the lower EHF frequencies (see Box 2.1). These are the frequencies of the radio "window" illustrated in Figs. 6.6 and 7.3.

The lowest frequency allocated for geosynchronous communications satellites is 2.5 GHz. Figure 8.1 shows the allocation of radio frequencies of 2.5 GHz and above made at the EARC and WARC conferences.

The radio regulations divide the world into three regions, as shown in the map at the top of Fig. 7.1. Region 1 consists of Europe, Asia Minor, Africa, all U.S.S.R. territory outside Europe, and the Mongolian Peoples Republic. Region 2 consists of the Western Hemisphere: the Americas, including Hawaii and Greenland. Region 3 consists of Australia, New Zealand, Oceania, and the parts of Asia not in Region 1. There are some differences among the allocations for the three regions.

There are many frequency allocations shown in Fig. 8.1 for geosynchronous (fixed) communications satellites and a few for broadcasting satellites. Various other uses of space are allocated.

Various allocations for radio in space are made for frequencies below 2.5 GHz, and these are listed in Fig. 8.2. Most of them are for mobile satellites or satellites having specific functions. Most are for bandwidths which are low compared with the bandwidths of today's main communications satellites. Some communications satellites low in the UHF band have been used, with low bandwidths, and can be operated with relatively inexpensive earth stations. Other UHF frequencies are used for sending commands to geosynchronous satellites. Some military satellites use UHF frequencies not allocated internationally for space purposes.

Frequencies in the lower part of the UHF band have the advantage that the electronic equipment is relatively efficient and inexpensive. The military use UHF satellites primarily for tactical communications from equipment on vehicles, ships, and planes. Some small, highly mobile, antennas on vehicles are used. Although very limited in their bandwidth compared with SHF satellites, they are nevertheless of great value in tactical military communications.

The frequency bands which are allocated internationally are in turn reallocated for national use by national administrations. Not all the international allocations for a given service are portioned out nationally; some may be kept in reserve. Some of the national frequency allocations are for government or military usage, some for nongovernment usage, and some for both. In the United States, nongovernment usage is regulated by the Federal Communications Commission (FCC) Rules and Regulations, and government usage is regulated by the Interdepartmental Radio Advisory Committee (IRAC).

COMMUNICATIONS SATELLITE BANDS

Most of today's communications satellites carry enough transponders to utilize 500 MHz of bandwidth. Commercial satellites use the *4/6-GHz band,* with an up-link of 5.925 to 6.425 GHz and a down-link of 3.7 to 4.2 GHz. Government and military satellites in many countries use the *7/8-GHz band,* with 7.9 to 8.4 GHz up and 7.25 to 7.75 GHz down. Satellites are now being designed for the *12/14-GHz band,* using 14 to 14.5 GHz up and either 11.7 to 12.2 GHz down or 10.95 to 11.2 and 11.45 to 11.7 down.

Higher in the spectrum there are satellite allocations of greater bandwidth. Experiments are being conducted at these higher frequencies, but more development work is needed before they will be commercially usable. 20/30 GHz will probably be of great commercial value one day because of the 3.5-GHz bandwidth allocated for these frequencies.

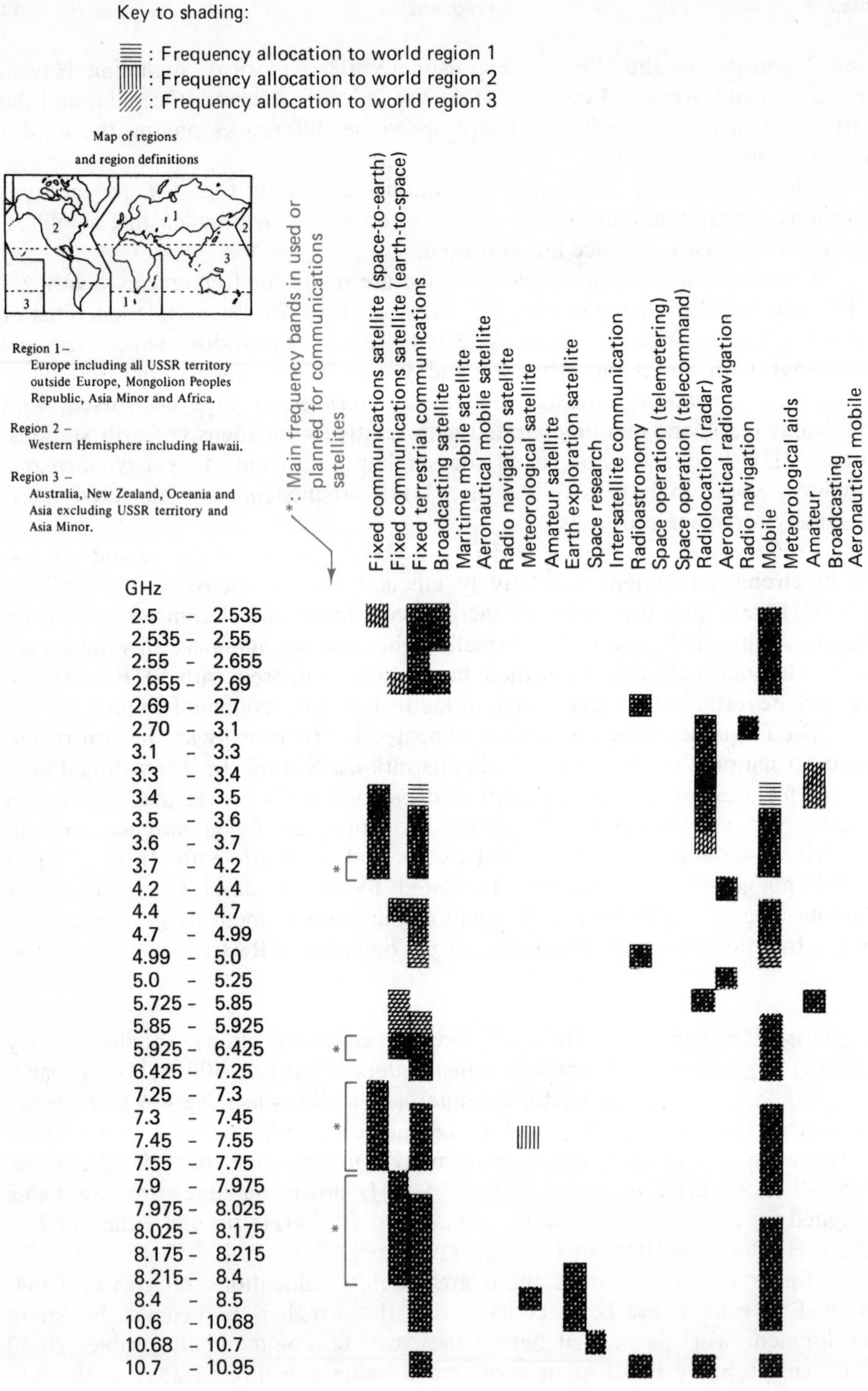

Key to shading:

- : Frequency allocation to world region 1
- : Frequency allocation to world region 2
- : Frequency allocation to world region 3

Map of regions
and region definitions

Region 1 –
Europe including all USSR territory outside Europe, Mongolion Peoples Republic, Asia Minor and Africa.

Region 2 –
Western Hemisphere including Hawaii.

Region 3 –
Australia, New Zealand, Oceania and Asia excluding USSR territory and Asia Minor.

*: Main frequency bands in used or planned for communications satellites

Fixed communications satellite (space-to-earth)
Fixed communications satellite (earth-to-space)
Fixed terrestrial communications
Broadcasting satellite
Maritime mobile satellite
Aeronautical mobile satellite
Radio navigation satellite
Meteorological satellite
Amateur satellite
Earth exploration satellite
Space research
Intersatellite communication
Radioastronomy
Space operation (telemetering)
Space operation (telecommand)
Radiolocation (radar)
Aeronautical radionavigation
Radio navigation
Mobile
Meteorological aids
Amateur
Broadcasting
Aeronautical mobile

GHz
2.5 – 2.535
2.535 – 2.55
2.55 – 2.655
2.655 – 2.69
2.69 – 2.7
2.70 – 3.1
3.1 – 3.3
3.3 – 3.4
3.4 – 3.5
3.5 – 3.6
3.6 – 3.7
3.7 – 4.2
4.2 – 4.4
4.4 – 4.7
4.7 – 4.99
4.99 – 5.0
5.0 – 5.25
5.725 – 5.85
5.85 – 5.925
5.925 – 6.425
6.425 – 7.25
7.25 – 7.3
7.3 – 7.45
7.45 – 7.55
7.55 – 7.75
7.9 – 7.975
7.975 – 8.025
8.025 – 8.175
8.175 – 8.215
8.215 – 8.4
8.4 – 8.5
10.6 – 10.68
10.68 – 10.7
10.7 – 10.95

Figure 8.1 International radio frequency allocations above 2.5 GHz.

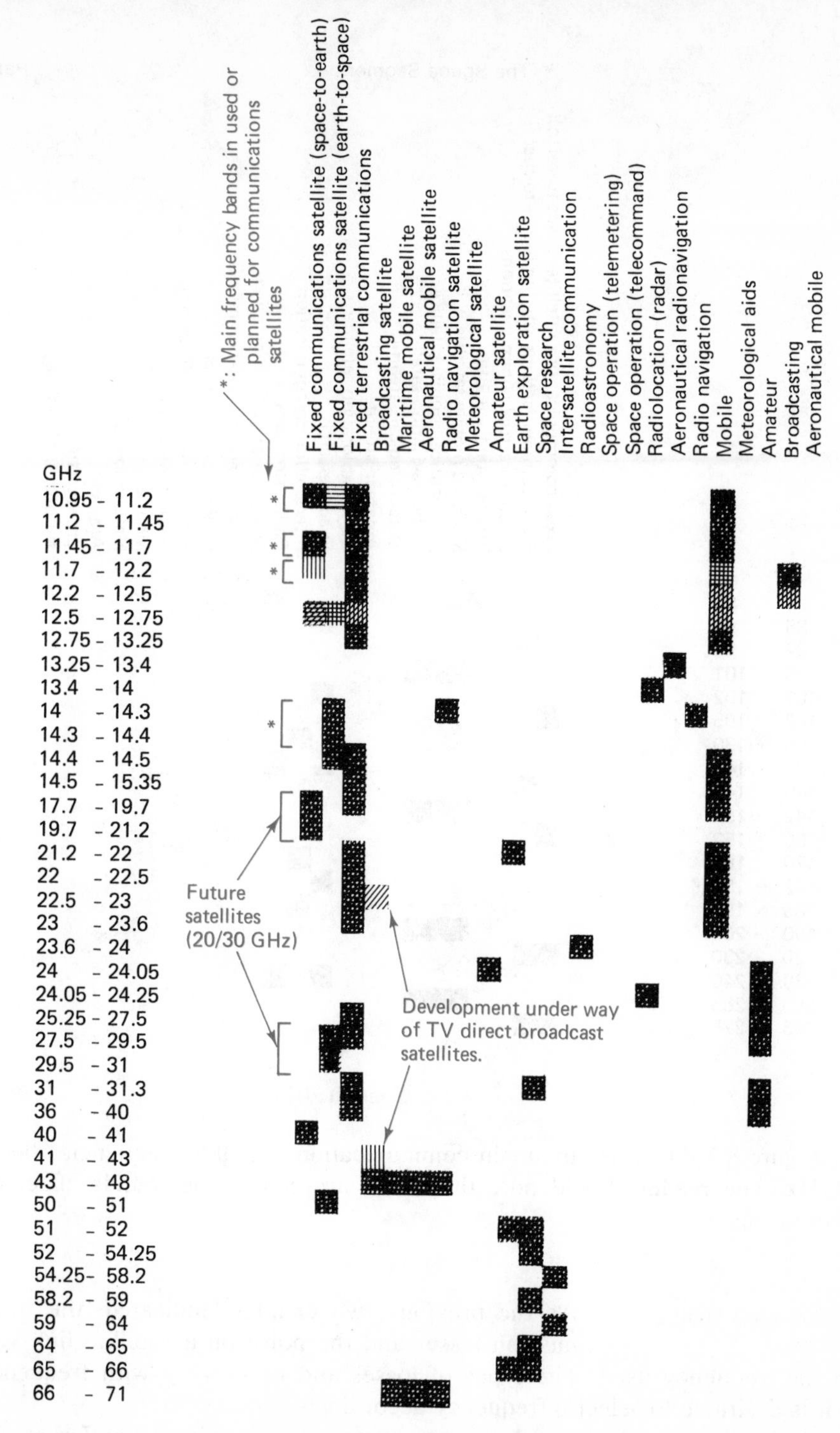

Figure 8.1 (Continued).

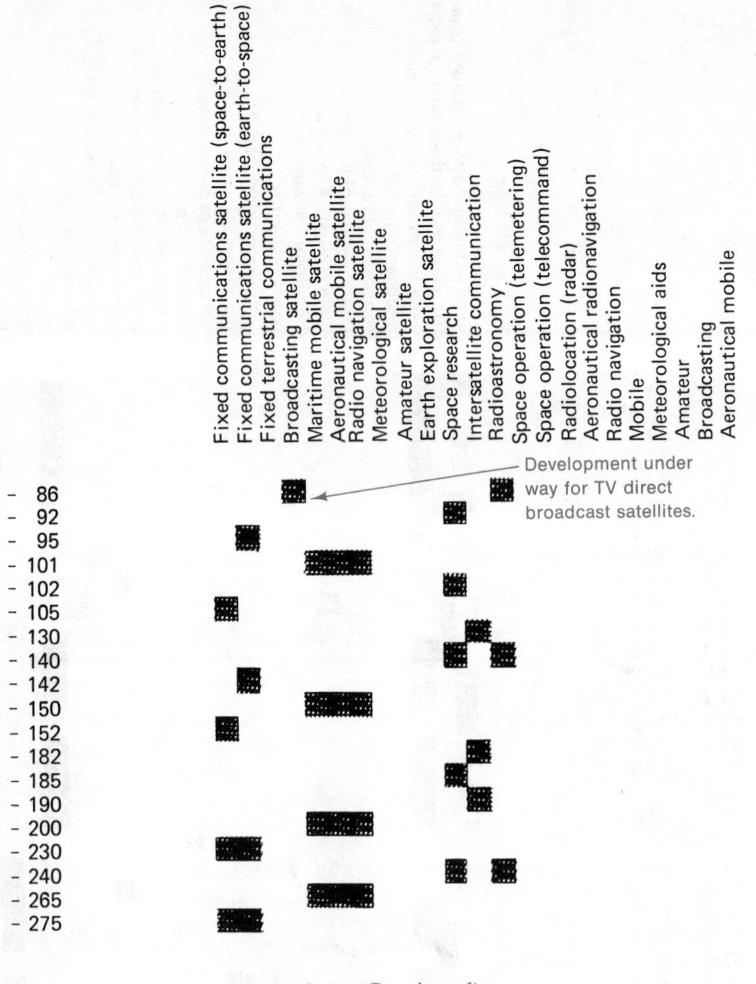

Figure 8.1 (Continued)

Figure 8.3 illustrates the main communications satellite frequencies below 40 GHz. The reader should note that there are some other bands allocated, shown in Fig. 8.1.

LOSSES AND NOISE

As the previous two chapters indicated, the transmission losses and the noise on a satellite link vary with the frequency used. The types of losses and noise vary with frequency, and it is desirable to select a frequency accordingly.

Figure 8.4 summarizes the relative losses and noise at different frequencies. The gain of the antennas is proportional to the square of the frequency. If there were no absorption, the overall loss in the space link would be

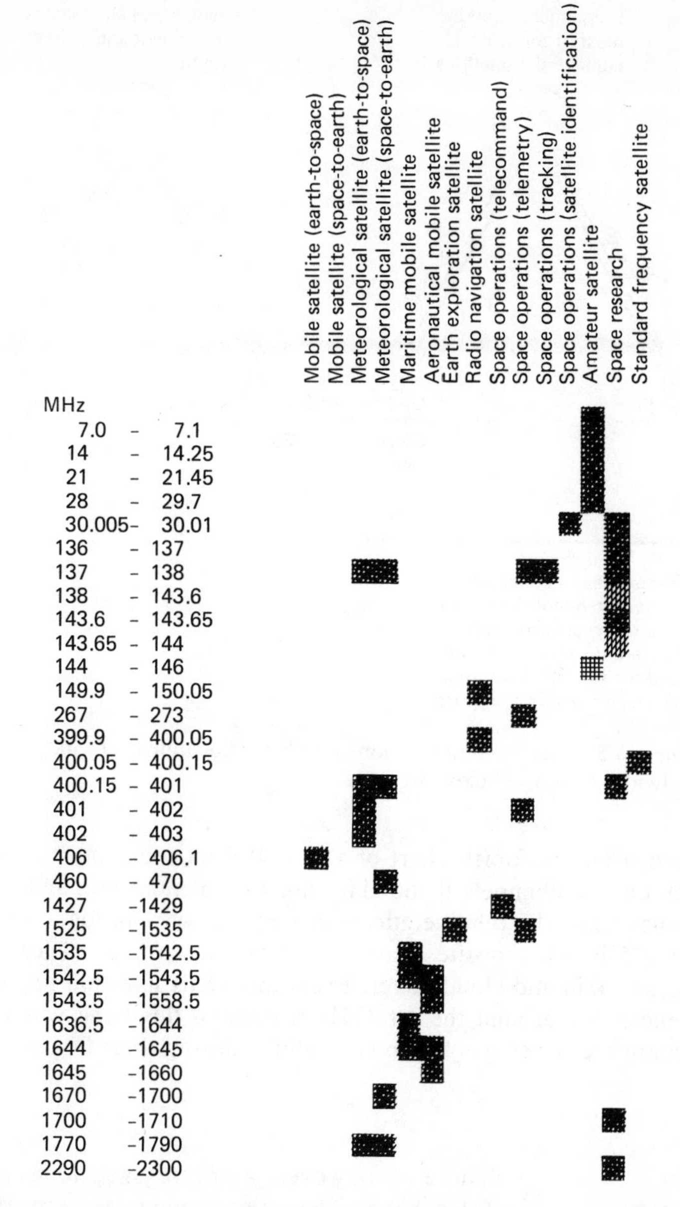

Figure 8.2 Radio frequencies below 2.5 GHz which are allocated to space uses.

inversely proportional to the square of the frequency. Most forms of radio interference are less at higher frequencies. However, these factors are counterbalanced by noise and absorption, which seriously affect frequencies about 10 GHz.

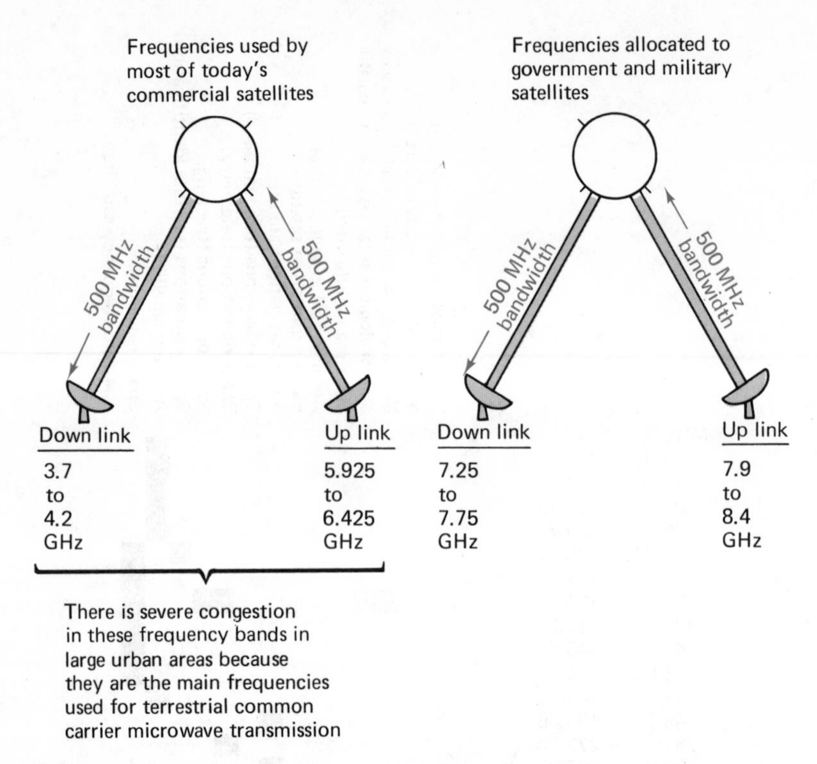

Figure 8.3 The communication satellite allocations of 500 MHz bandwidth or more, below 40 GHz.

As seen from the fourth chart of Fig. 8.4, frequencies up to 40 GHz can provide high-quality channels if the skies are free of rain, fog, and clouds. The last four charts show the relative effects of very heavy rain, fog, and clouds.

Figure 8.5 is a composite chart adding all the sources of loss and noise with very heavy rain and cloud cover. From this chart it would appear that the ideal frequencies are around the 4/6-GHz band, and this band was used exclusively by commercial geosynchronous satellites during their first decade of operation.

MICROWAVE INTERFERENCE

There is, however, a severe snag to using the 4/6-GHz band. The same frequencies are allocated to terrestrial microwave links and there is a serious problem of radio interference. The major cities of the world are becoming highly congested with traffic at this frequency. Figure 8.6 shows microwave traffic in the New York City area.

Four types of interference are theoretically possible:

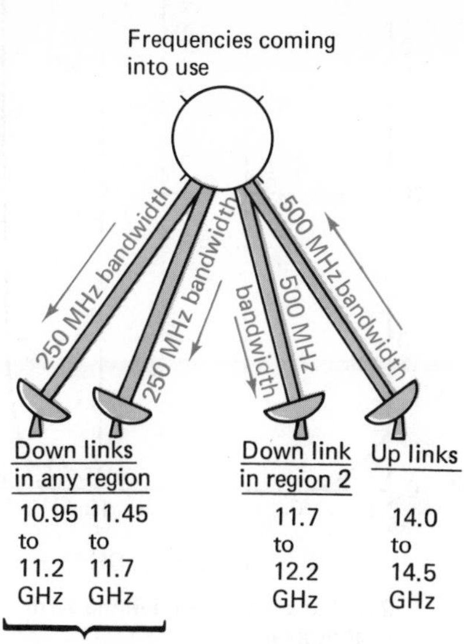

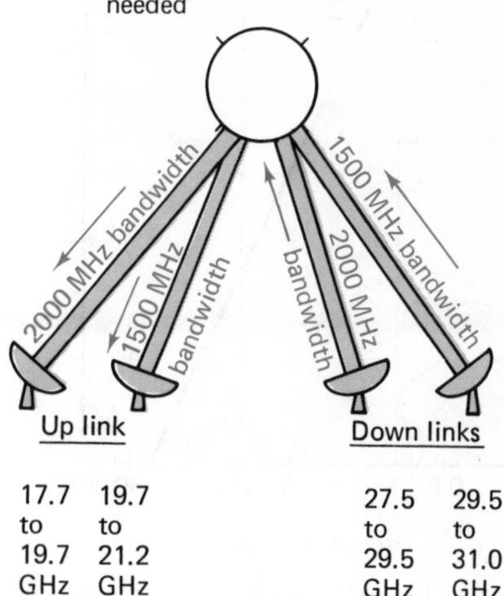

These frequencies are also allocated to mobile radio and terrestrial common carrier microwave, but are not congested

(These frequencies are allocated to the western hemisphere, North and South America, only)

Frequencies allocated for communication satellites at the 1971 WARC (World Administrative Radio Conference)

1. Transmission from the earth station interferes with the terrestrial link receiver.
2. Terrestrial link transmission interferes with reception from the satellite.
3. Transmission from the satellite interferes with the terrestrial link receiver.
4. Terrestrial link transmission is received by the satellite.

The first is by far the most serious. An earth station must transmit a powerful signal to compensate for the vast distance of the satellite. The antenna transmits a highly directional beam toward the satellite, but nevertheless some of the signal spills in other directions (as shown in Fig. 6.2) and may interfere with a microwave receiver. The earth-station transmitter must therefore not be too close to a microwave antenna.

The second of the above types of interference is the next most serious. To avoid it, an earth station should not be located close to a terrestrial microwave path so that part of the terrestrial beam shines into the receiving antenna.

Because of microwave congestion, earth stations using the 4/6-GHz band

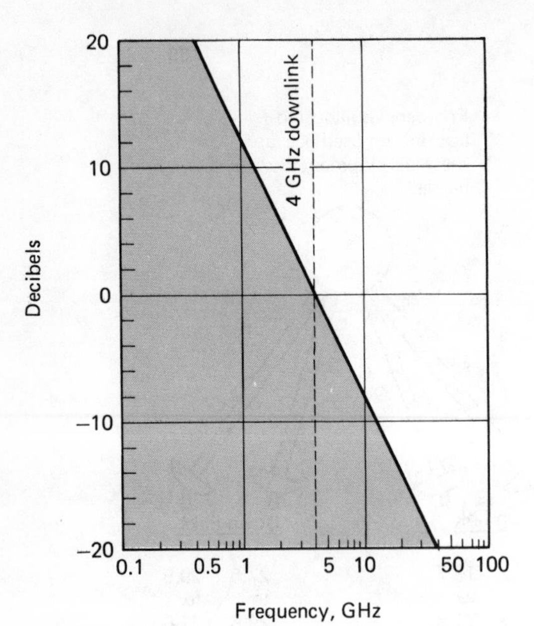

1. Relative loss in the space link (taking into consideration the gain of the transmitting and receiving antennas).

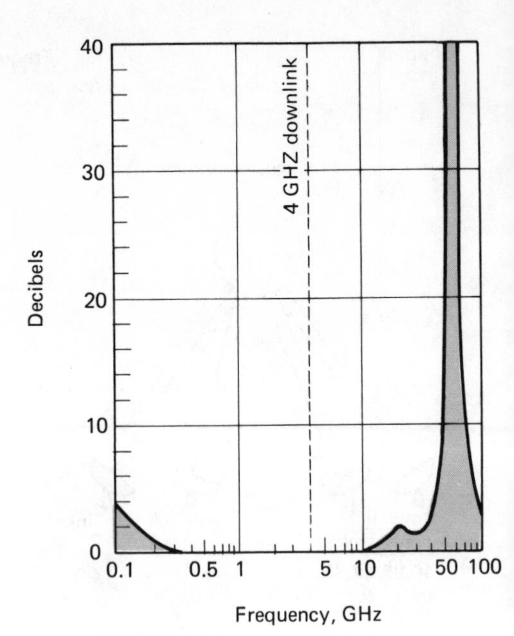

2. Relative attenuation in the earth's atmosphere.

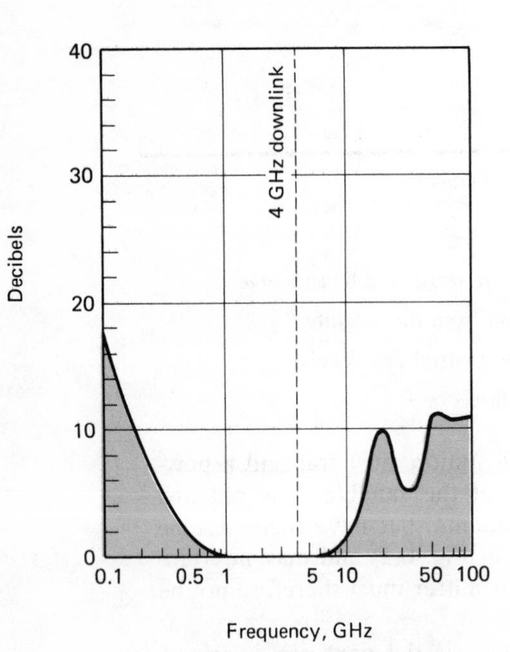

3. Relative power of atmospheric noise

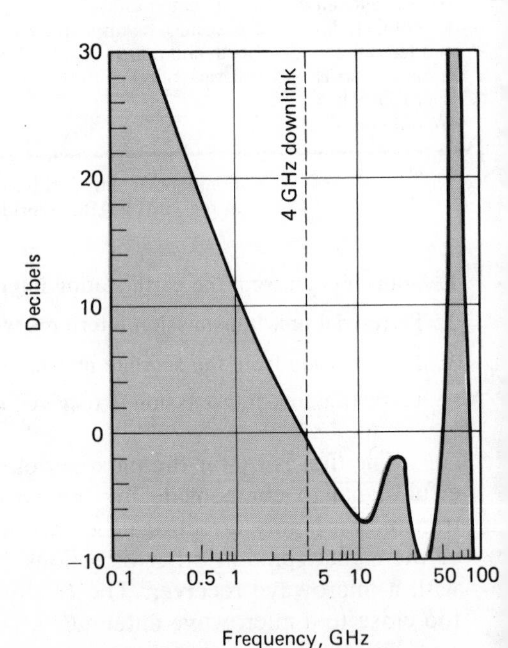

4. Composite effect of link loss, atmospheric attenuation and atmospheric noise

Figure 8.4 Charts drawn to compare how the relative effects of different types of loss and noise vary with frequency. All charts are drawn relative to the loss or noise power of a 4/6 GHz down-link. All charts are for a 15° elevation.

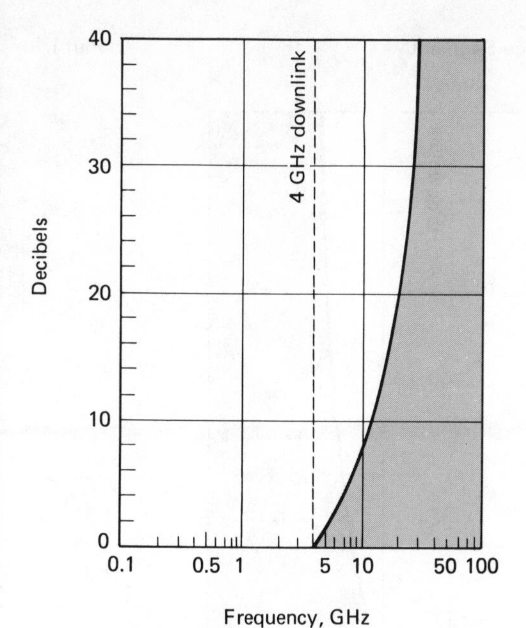

5. Relative attenuation caused by very heavy rain

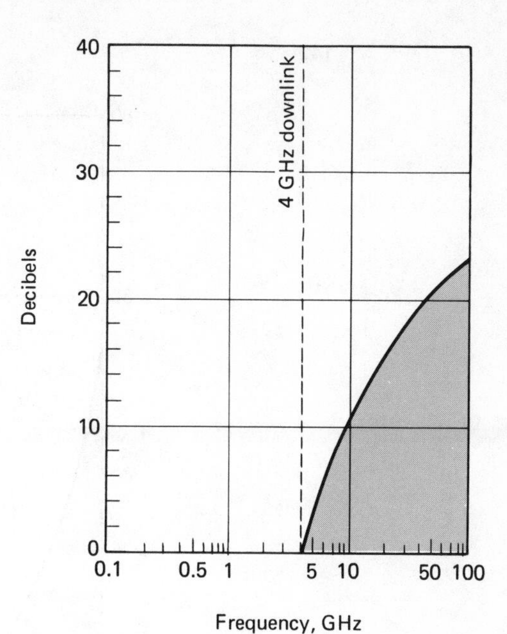

6. Relative power of noise from very heavy rain

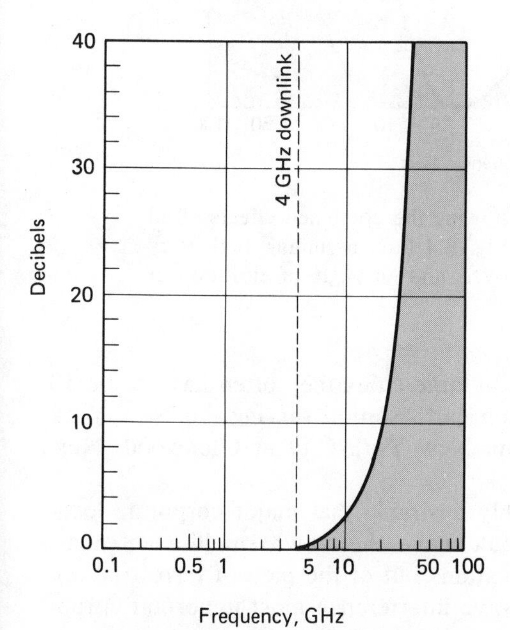

7. Relative attenuation caused by very thick fog or cloud cover

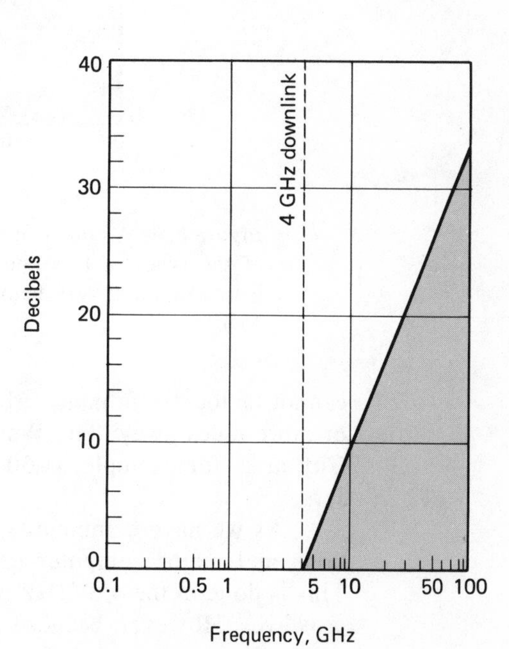

8. Relative noise power of very thick fog or cloud cover

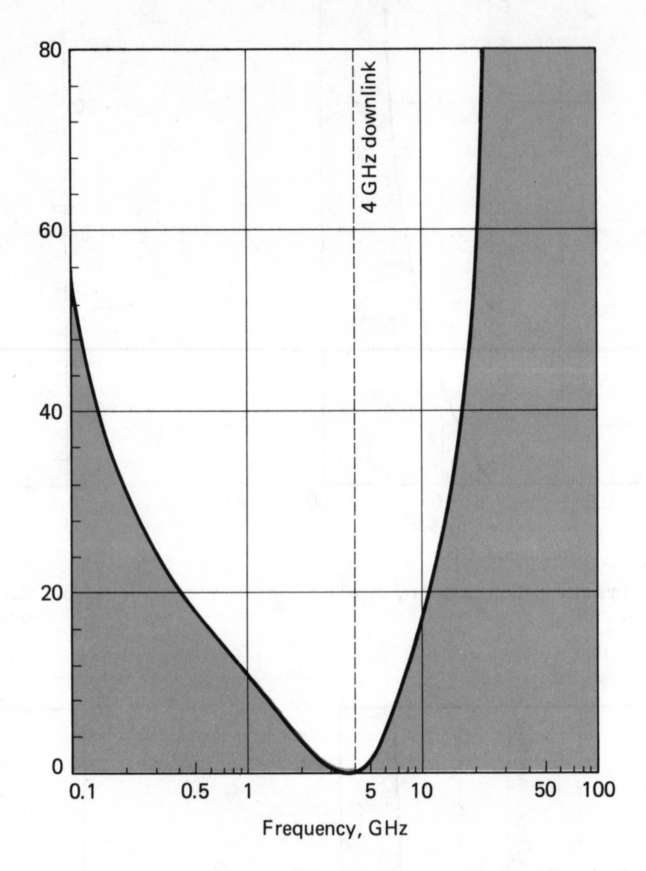

Figure 8.5 A composite diagram showing the combined effects of all of the types of loss and noise in Fig. 8.4 (i.e., assuming both very heavy rain and very heavy cloud cover, and an angle of elevation of 15°).

cannot be located in many urban areas. In large cities they often have to be 50 or more miles away. The Western Union earth station serving the New York City area, for example, is 50 miles from New York City at Glenwood, New Jersey.

As we have commented, it is highly desirable that major corporate locations and major computer centers should have their own satellite antennas. This is done in the 4/6-GHz band in locations out of the path of terrestrial microwaves. However, because of microwave interference most important corporate and government locations cannot have their own 4/6-GHz antenna and are many miles from the nearest possible antenna site. The cost of moving the wideband signals over common carrier facilities from the nearest antenna site is very high—often high enough to discourage satellite usage.

Because of the interference problem, one of the most important factors in

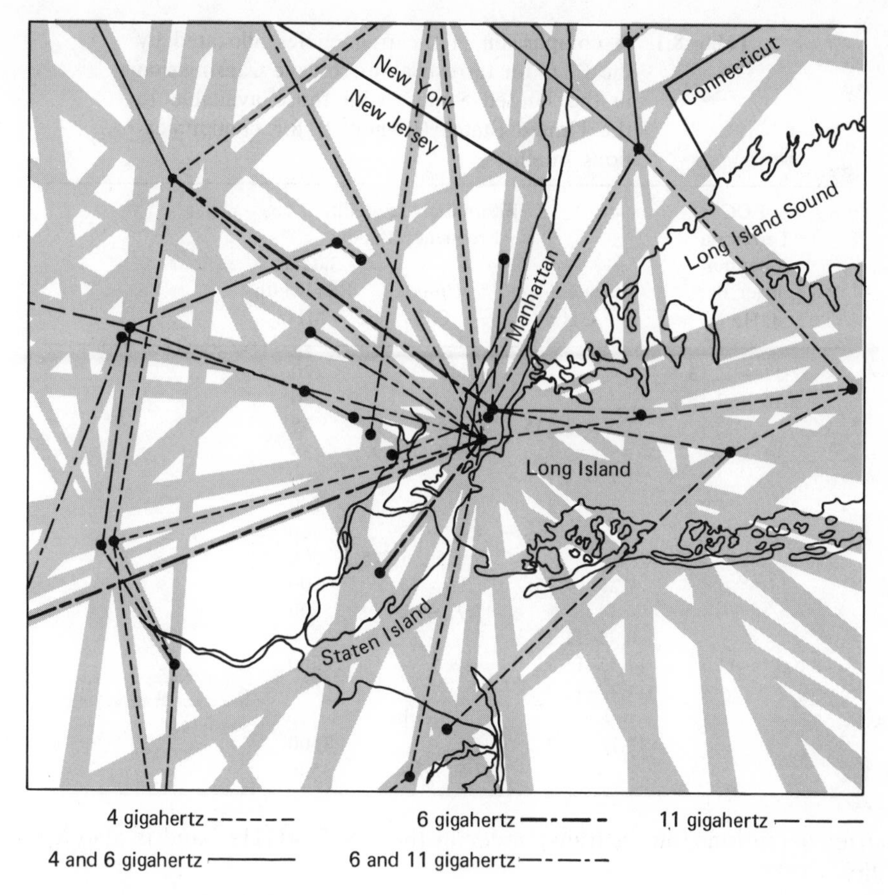

MICROWAVE INTERFERENCE

Figure 8.6 Criss-crossing microwave beams in New York. The microwave congestion is so great that 4/6-GHz earth stations have to be many miles from city centers. [1]

the selection of satellite frequencies for corporate use is that they should not be terrestrial common carrier frequencies.

Table 8.1 shows the frequencies allocated by the FCC for terrestrial microwave operation by the U.S. common carriers and compares them with the satellite bands allocated to the Western Hemisphere. It will be seen that several satellite bands do not clash with terrestrial microwaves. The 7/8-GHz band is not used by common carriers, but these satellite frequencies are allocated to government (mainly military) operations. The lowest 500-MHz bandwidth which does not conflict with terrestrial microwaves is that with a 14–14.5-GHz up-link and an 11.7–12.2-Ghz down-link. A 300-MHz bandwidth with 4.4–4.7 GHz up and 3.4–3.7 GHz down would avoid conflict with common

Table 8.1 A comparison of the frequencies allocated by the FCC for terrestrial microwave transmission in the United States, with those available to the United States (Region 2) for communications satellites

FCC Terrestrial Common Carrier Band GHz	International Satellite Frequency Bands			
	Down-link GHz	*Up-link* GHz	Satellite Bandwidth MHz	
2.11 - 2.13			20	
2.16 - 2.18			20	
	2.5 - 2.535		35	
		2.655- 2.69	35	
	3.4 - 3.7		300	
3.7 - 4.2	3.7 - 4.2		500	
		4.4 - 4.7	300	
5.925- 6.425		5.925- 6.425	500	
	7.25- 7.75		500	
		7.9 - 8.4	500	
	10.95–11.2			
10.7 -11.7	11.45–11.7		500	Frequencies of the next generation of satellites
	11.7 -12.2		500	Frequencies of the next generation of satellites
		14 -14.5	500	Frequencies of the next generation of satellites
	17.7 -21.2		3500	
		27.5 -31	3500	

carrier operations, and a transponder in the 2.5–2.6-GHz band is also a possibility.

12/14-GHz SATELLITES

Many satellites in the next decade will operate at 12/14 GHz. The SBS satellite system is designed for these frequencies, and in some countries television broadcasting satellites have been planned to use 12/14 GHz.

The 12/14-GHz band has a mixture of advantages and disadvantages over the traditional 4/6-GHz satellites.

The advantages of 12/14-GHz are as follows:

1. The band is generally not used for terrestrial common carrier links, so 12/14-GHz earth antennas can operate in city centers on the rooftops of buildings. Many major corporate locations will have their own antennas. In a crowded center there may be one earth station serving many local users and linked to them by short line-of-sight microwave millimeterwave radio hops.

2. The beam width from an earth antenna of a given size is less than half of that for a 4/6-GHz satellite. Therefore, about twice as many satellites could be used without interference, thus lessening potential congestion in the equatorial orbit. If a 12/14-GHz satellite is to serve the continental United States

(not including Alaska), it could be positioned so that no earth antenna need have an elevation less than 20°, and most would have an elevation more than 30°. As seen from Fig. 3.9, this results in much less absorption due to bad weather and atmosphere than with, say, 15° elevation.

3. A satellite antenna of a given weight can be made more directional. Multiple beams to or from the satellite could therefore be made to operate at the same frequency. The satellite could therefore have more transponders than a 4/6-GHz satellite without exceeding the 500-MHz bandwidth.

4. The antenna gain for a given size of satellite antenna is $(14.25/6.175)^2 = 5.33$ greater on the up-link and $(11.95/3.95)^2 = 9.15$ greater on the down-link than with a 4/6-GHz satellite. This gives a total improvement of 16.9 decibels. This improvement could be used either to compensate for the increased absorption and noise that occur in bad weather or to permit the earth-station antennas to be smaller and cheaper.

5. When the 4-GHz down-link is used there is a limit imposed on the radiated satellite power to prevent interference with terrestrial common carrier systems. No such limit is imposed on the 12-GHz down-link.

The disadvantage of the higher frequency is represented by the last four curves of Fig. 8.4 and the composite curve of Fig. 8.5. With very heavy rain, fog, or clouds, the received signal strength falls and the noise increases. These curves are drawn for a 15° angle of elevation and most 12/14-GHz links will be engineered so as to avoid such a low angle of elevation. With a 30° angle of elevation, the increased noise and absorption in bad weather very roughly balance the effect of increased antenna gain.

INTERFERENCE

The main argument for 12/14 GHz is to avoid interference with terrestrial systems. However, another form of interference may become troublesome at these frequencies in the future. The frequencies are ideal for satellites which broadcast television programs, and in order to make the television receivers as inexpensive as possible, the satellite should beam the 12-GHz signal to earth with as much power as possible. Careful constraints are then needed to prevent the broadcasting satellites from interfering with the two-way transmission satellites and vice versa.

Two-way transmission systems use relatively large earth antennas to transmit and receive over a fairly narrow cone. The larger the antennas, the smaller the beam angle, and the closer the satellites can be without mutual interference. The FCC specifies a minimum size for transmit/receive and receive-only antennas, and related to this a minimum spacing for satellites using the same frequencies. Satellite interference is treated as noise in the link calculations which are concerned with the noise-to-signal ratio. Calculations relating to the avoidance of interference are given in the World Administrative Radio Conference recommendations [2].

SBS estimates that 12/14-GHz satellites can be 3° apart in orbit when its

5-meter antennas are used. Somewhat closer spacing is possible, but 3° allow for some measure of orbital drift.

A broadcast satellite is likely to have much more power than satellites for telephone or data transmission. There is therefore more danger of it causing interference. If the receivers for the broadcast satellite are cheap enough for home use, they will have small antennas with little directional capability. There is therefore a danger that they will receive unwanted signals from other satellites.

It would be possible to reserve part of the 12/14-GHz band for broadcasting and part for the two-way operation. This is not a good idea, however, because it would limit the throughput of the 12/14-GHz satellite. A satellite which could use only half the band and hence had only half the capacity would cost almost as much as a full-capacity satellite, and hence the cost per channel would be almost twice as great.

Instead it is desirable to ensure that the broadcast satellite is sufficiently powerful that its receivers can discriminate its signals from those of other satellites, and that interference between satellites does not occur.

In the design of the SBS satellite system an examination was made of possible interference between an SBS satellite and a satellite similar to the Canadian CTS broadcast satellite [3]. In the calculations it was assumed that the satellites were not closer than 6° orbital spacing. The following table shows the ratio of the signal being received to the noise from an interfering satellite 6° away:

	Up-link		Down-link	
	Signal/ Interference (decibels)	*% of Total Noise Caused by Interference*	*Signal/ Interference (decibels)*	*% of Total Noise Caused by Interference*
CTS satellite interfering with SBS satellite	4.90	0.001	26.5	21.8
SBS satellite interfering with CTS satellite	28.9	6.4	48.5	0.07

The SBS satellite shining into the small receiving dishes of CTS has a negligible effect because the CTS satellite is so much more powerful. The more serious effect arises from the CTS satellite broadcasting into the customer antennas of SBS. SBS can control this effect by making their customer antennas larger and hence more directional. However, there is no need because a signal-to-interference ratio of 26.5 decibels is well within the design constraints. Up to 10 decibels would have been tolerable.

Although the above figures cause no problem, the interference could have been made much less by ensuring that the neighboring satellites transmitted beams of opposite polarization. Cross polarization does not eliminate interference completely because some depolarization is caused by rain. The interference when opposite polarizations are used is as follows:

	Up-link		Down-link	
	Signal/ Interference (decibels)	*% of Total Noise Caused by Interference*	*Signal/ Interference (decibels)*	*% of Total Noise Caused by Interference*
CTS satellite interfering with SBS satellite	59.0	0.0001	36.5	2.2
SBS satellite interfering with CTS satellite	38.9	0.6	58.5	0.007

Some countries may wish to reserve the 12-GHz down frequencies for broadcasting. Some may reserve them for two-way operation. Some may allocate certain frequencies within the 12-GHz band (500-MHz bandwidth) for two-way operation and some for broadcasting.

A satellite which broadcasts to one country will have an antenna shaped to direct the beam to that country. This shaping of the signal is not very precise, however, and some of one nation's signal may spill across the border to neighboring countries. A German television satellite could interfere with parts of France. An extremely powerful satellite which has been proposed in Russia could interfere with Finland.

Constraints on satellite broadcasting and allocation of orbit space are issues over which major arguments will probably ensue as more powerful satellites are launched and more countries want to take advantage of this potent technology.

REFERENCES

1. R. Bowers and F. Jeffrey, "Technology Assessment and Microwave Disorders," *Scientific American,* Feb. 1972.

2. *Final Acts of the World Administrative Radio Conference for Space Telecommunications,* International Telecommunications Union, Geneva, 1971.

3. *FCC Application of Satellite Business Systems,* Vol. II, Federal Communications Commission, Washington, D.C. 1976.

9 TRADE-OFFS IN SATELLITE DESIGN

In this chapter we summarize the major trade-offs in the design of the space segment. The design of the space segment is closely linked to the design of the ground segment, so a complete exploration of trade-offs ought to include the material we discuss in Part II.

Figure 9.1 summarizes the major trade-offs affecting the satellite design. There are many important aspects of its design not shown in Fig. 9.1, such as what type of power generation it has and whether it has three-axis stabilization, but these are secondary to the major design parameters shown in Fig. 9.1.

PRICE ELASTICITY

There are major economies of scale in satellite design. The cost per channel per year can be much less if the satellite is designed to handle a very large number of channels. The concept of the market for the satellite services is therefore very important. If a large quantity of satellite channels can be sold, for any purpose, then the channels are more likely to compete effectively with the terrestrial alternatives. The market for telecommunications has traditionally been one of high price elasticity. In other words, if channels are cheap, there are more customers for them. The reason for the price elasticity is that telecommunications users always have alternatives. You can send a letter, or mail a tape of data, instead of using telephone circuits. As the cost of long-distance telephone calls dropped, so the number of users increased greatly. The same was true with telegraphy, data transmission, and television transmission. It will be true with video telephones, video conferencing, and electronic mail.

ALL TYPES OF SIGNALS

Another step to achieving economies of scale is to ensure that the satellite can handle all types of signals. The marketing concept should not restrict itself

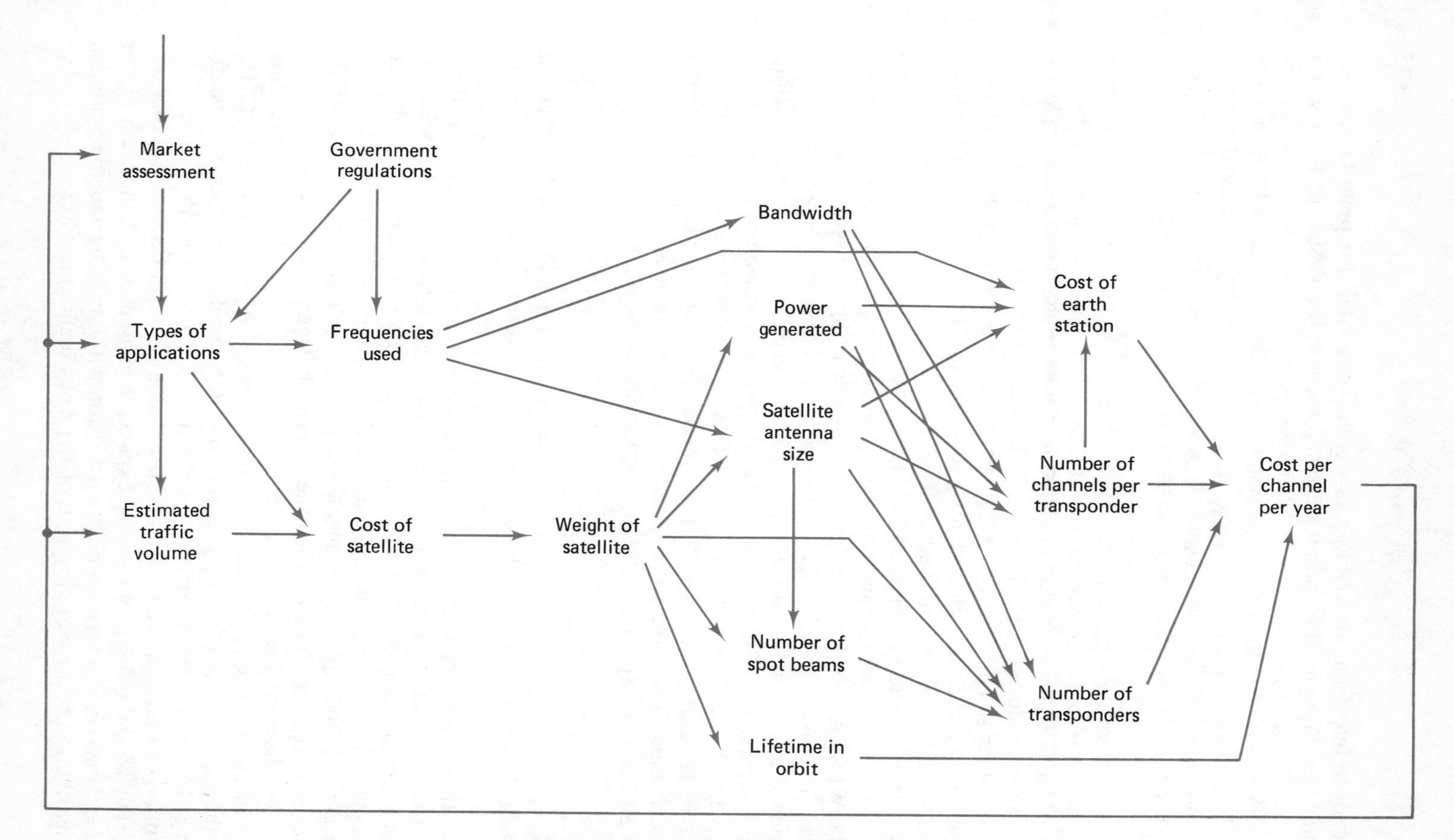

Figure 9.1 Major tradeoffs in satellite design.

to voice, or to television, or to data. As we shall see later, for point-to-point or multipoint transmission, this matter resides largely in the design of the ground equipment. There is, however, a difference between the ideal satellite for *broadcasting* and the ideal for multipoint transmission. The ATS-6 satellite (Fig. 4.17) is suitable for broadcasting. It has a large, highly directional antenna which can concentrate as much signal as possible on the area where the broadcast is received so that the receiving antennas can be as inexpensive as possible.

A design with a large antenna like ATS-6 would be suitable for common carrier operation, pointing narrow spot beams with high precision at a limited number of earth stations sited in areas of maximum traffic density. Fairchild Industries, who produced ATS-6 and made it perform so excellently, designed a variation of that satellite with 120 transponders. One transponder would have been capable of carrying more than 1800 voice channels using appropriate techniques (discussed later in the book).

A satellite could be designed for both broadcasting and multipoint transmission. As the Musak Corporation has demonstrated, today's domestic satellites like WESTAR are suitable for sound broadcasting to users with low-cost antennas. Television broadcasting to low-cost antennas needs about 400 times the bandwidth of sound broadcasting and hence needs a much larger satellite. It is more economical for the time being to transmit to medium-sized antennas connected to CATV systems or local television transmitters. Many hundreds of receive-only earth stations in North America are used for this form of television distribution. They use 10-meter and 4.5-meter antennas to receive signals via the 36 MHz transponders of domestic satellites (Fig. 9.2).

BREAK-EVEN DISTANCE

There is a special reason for high price elasticity in satellite channels. The investment cost for terrestrial channels is roughly proportional to their distance, whereas the cost of one-hop satellite channels is independent of distance. If the break-even cost between satellite channels and their terrestrial alternative occurs for transmissions of 3000 miles, then shorter transmissions will go by terrestrial links. If, however, the satellite costs are lower so that the break-even cost occurs at 300 miles, then much more traffic will go via satellite. If the break-even cost is at 30 miles, then the satellite business will be vast.

A low-distance break-even point implies that there will be a very large number of earth stations. If a satellite system has only half a dozen earth stations, then a major portion of the end-to-end transmission costs will be for trucking the signal across terrestrial links to and from the earth stations. The distribution of signals between the end users and the earth stations is expensive for all signals, and often prohibitively expensive for signals of bandwidth higher than telephone. In countries which permit competition in telecommunications, local distribution is the Achilles' heel of the new common carriers.

Figure 9.2 10-meter and 4.5 meter satellite earth stations are used in the U.S.A. to distribute CATV programs so that many towns can have services like *Home Box Office*. Satellites permit cable TV systems to be built in remote areas. Persons in isolated or rural towns can then enjoy first-run movies, sports events, and other CATV offerings. In the future they might be linked into interactive facilities. (*Photograph courtesy Scientific-Atlanta.*)

ANTENNAS

To avoid local distribution costs, the antennas need to be small enough to be on the premises of major users—at least as small as the antennas shown in Fig. 9.3. As discussed earlier, it is desirable that the frequencies used by such antennas should not be in the 4/6-GHz band because of the use of these frequencies for terrestrial microwaves (although the antennas in Fig. 9.3 all used 4/6 GHz).

Figure 9.3 Earth stations on corporate premises.

An objective of domestic satellite development should be to have a satellite antenna at most major corporate locations, if possible smaller antennas than these at Hughes Aircraft. The dish-shaped antennas here are covered with a plastic dome for weather protection.

An antenna in the parking lot at an IBM laboratory in Poughkeepsie, N.Y. This antenna transmits via the ANIK and WESTAR satellites. The use of SBS will permit smaller and cheaper corporate earth stations.

If a satellite is to serve a large number of earth stations—much larger than the number planned for WESTAR, for example—then the cost balance swings in favor of doing whatever is necessary in the satellite design to lower the earth-station cost. The satellite can become expensive in order to make the earth stations cheap. Frequencies should be used which allow users to have antennas at their own locations.

As the industry progresses toward satellite users having their own earth stations, a satellite will interconnect many hundreds and then thousands of earth stations. When this situation emerges there will be a powerful incentive to spend money on the satellite in order to save money on the earth stations.

WEIGHT

In an orbit 22,300 miles from earth, the factor which dominates all other design considerations is weight. It cost almost $7 million to place WESTAR I's 607 pounds in orbit. If a housewife's groceries from one supermarket visit were carried on that satellite, that would cost more than a quarter of a million dollars.

The cost per pound is less if the launch vehicle is larger. Figure 9.3 shows the approximate costs of launch vehicles available prior to the space shuttle and plots a trend showing economies of scale in launching. All the vehicles shown are small compared with the largest soviet rockets or with Saturn V, the moon-shot rocket. The costs of space launchings will drop substantially in the next decade with the general availability of the space shuttle vehicles, which can return to earth for refueling rather than being lost like today's rockets. The space shuttle is designed only to go into low earth orbit; a perigee kick motor will be used to take satellites from there to geosynchronous orbit (Fig. 9.4).

Satellite weight increased rapidly from the first geosynchronous satellite, SYNCOM II, in 1963, which weighed 78 pounds in orbit, to TACSAT I in 1969, which weighed 1617 pounds in orbit. The domestic satellites weighed much less (ANIK, 607 pounds; WESTAR, 654 pounds). INTELSAT IV A weighs about 1732 pounds. ATS-6 weighs 2990 pounds.

The U.S. domestic satellites are only about 40% of the weight of the intercontinental satellites of their era (both military and commercial). They can be smaller because they are designed to carry somewhat less traffic and because earth stations need not have the very low angles of elevation which cause excessive atmospheric absorption and noise. Making the intercontinental satellites much heavier would require assurance that the quantity of traffic and number of earth stations would justify the increased satellite cost.

Both the traffic volume and the expected number of earth stations are growing rapidly, and so heavier satellites will probably be justified before long, and the space shuttle will be available for launching them.

Greater satellite weight can be used for two major purposes (and many lesser ones). First, more power can be generated, and hence the output transmission power can be greater. Second, the antennas can be larger. Larger an-

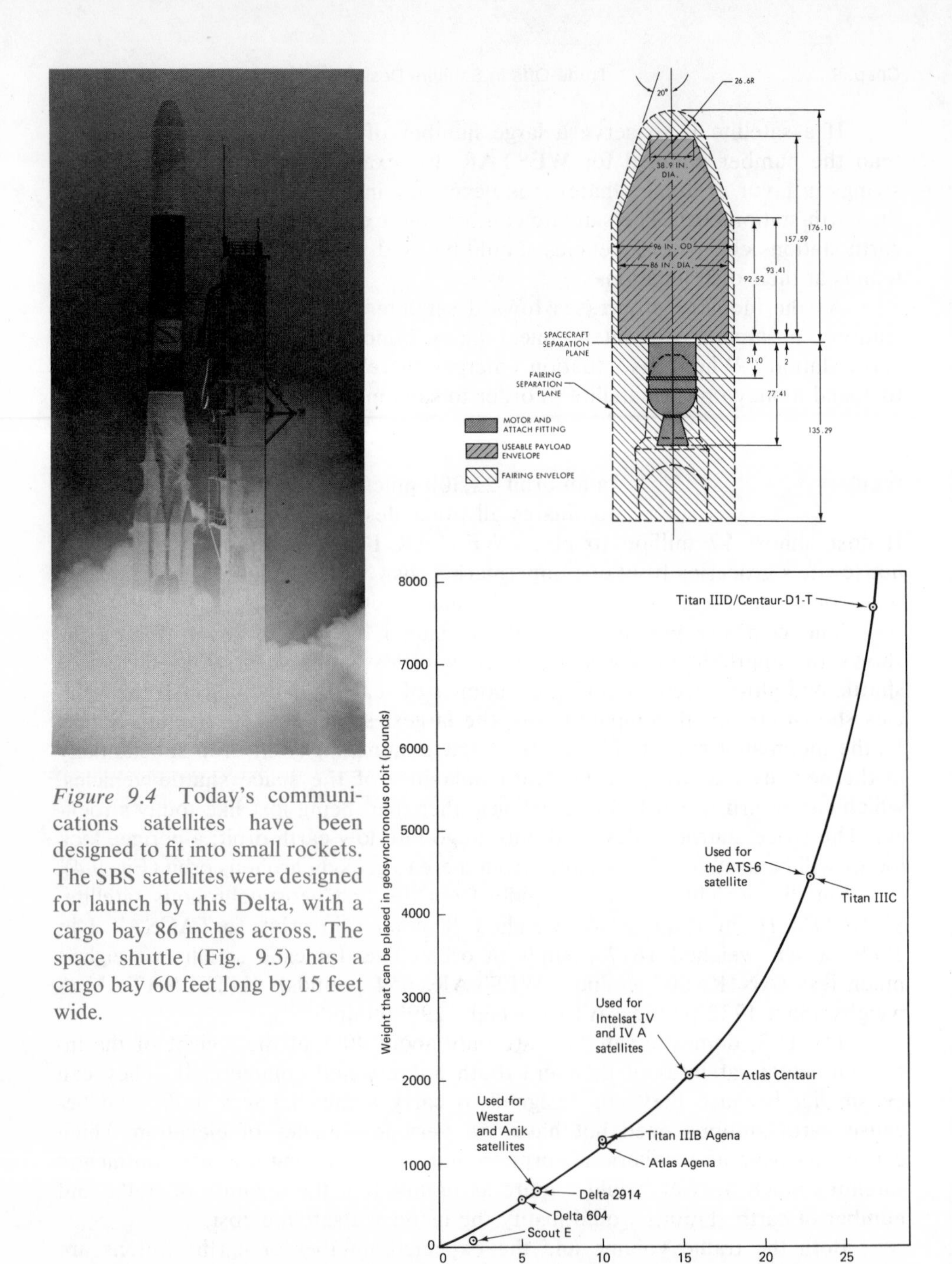

Figure 9.4 Today's communications satellites have been designed to fit into small rockets. The SBS satellites were designed for launch by this Delta, with a cargo bay 86 inches across. The space shuttle (Fig. 9.5) has a cargo bay 60 feet long by 15 feet wide.

tennas give a higher signal gain both when receiving and transmitting and permit the use of highly directional spot beams. The total gain for receiving and transmitting is proportional to the fourth power of the diameter of the antenna.

Multiple spot beams enable the same frequencies to be reused, thereby increasing the capacity of the satellite. If a stronger signal is relayed to earth, this can either permit more information to be sent over the same bandwidth than with today's satellites, or else permit the use of lower-cost earth stations. A third use of increased weight is to increase the number of transponders. Fourth, increased weight can enable the satellite to carry more propellant and hence have a longer life in orbit.

The uses of weight to increase satellite power and to increase antenna size are being demonstrated by two experimental satellites. The Canadian Technology Satellite (CTS) has a 1-kilowatt array of solar cells on two 28-foot solar sails (Fig. 4.6) and NASA's ATS-6 has its 30-foot antenna (Fig. 4.17). Both satellites unfold their bulky equipment in space, and both satellites need three-dimensional stabilization. Experimental designs have been done for satellites with *much* larger solar sails and *much* larger antennas. Great economies of scale would be achieved with larger satellites.

NUMBER OF CHANNELS

The number of channels a satellite can provide is related to the bandwidth available and to how the bandwidth is used. The available bandwidth is related to the frequency allocations. As shown in Table 8.1, the bandwidths of 4/6-GHz, 7/8-GHz, and 12/14-GHz satellites is 500 MHz. Bands lower in frequency than 4/6 GHz have a lower bandwidth. 20/30-GHz satellites can have a much higher bandwidth, 3500 MHz, but more development work is needed at these frequencies. The enormous bandwidth of 20/30 GHz is very appealing for future satellite designs, though these frequencies are much more severely attenuated by bad weather. NASA is developing electronics that would open up the still higher frequencies of 41/43 and 84/86 GHz. It is developing powerful (200 watt) space-qualified travelling wave tubes for these bands which seem suitable for television broadcasting direct to homes with very small antennas.

The number of channels may also be increased by improving the efficiency with which a transponder is used. We shall discuss this in Part II. It is possible to increase the number of channels per transponder if the transmitting power used with that transponder is increased. More power requires more weight, and so again the number of channels is related to satellite weight.

To increase the number of channels per satellite without spilling beyond the bandwidth limitation, differently polarized beams and separate spot beams can be made to carry the same frequencies. The reader might imagine searchlight beams from the satellite illuminating different parts of the ground and carrying the same frequencies (Fig. 2.9). Multiple spot beams need a large antenna with multiple feeds, and more transponders, which means greater satellite weight and power.

Figure 9.5 The space shuttle.

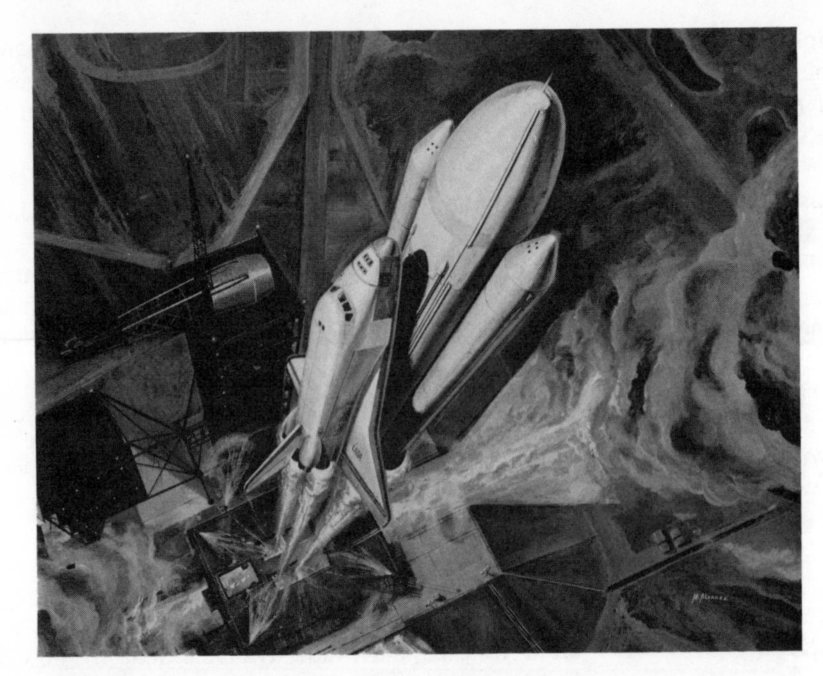

The space shuttle being launched with the help of two solid fuel booster rockets, and a large tank of liquid propellant.

The solid fuel booster rockets are jettisoned 27 miles above the earth. They parachute back and are recovered from the ocean. The large tank is carried almost to orbit, feeding the shuttle's main engines, and is then jettisoned.

The Interim Upper Stage being deployed. This is a vehicle which could take a payload from the shuttle orbit to geosynchronous orbit. Before it is available a perigee kick motor will be used to send communications satellites into geosynchronous orbit.

The shuttle will land on a conventional jet aircraft runway 2 miles long, at a speed of 200 miles per hour.

COST PER CHANNEL PER YEAR

The cost per channel per year is a measure of the efficiency of a satellite. The number of years the satellite will be operational is thus a major design consideration. As Fig. 1.4 shows, the cost per channel per year decreased dramatically during the first 10 years of communications satellite operation, and part of the reason was increasing satellite lifetime.

Most equipment for terrestrial circuits is designed to be operational for 40 years. It would be possible to design a satellite for a 40-year lifetime, but this would not make much sense because satellite technology is changing too fast. Satellites launched prior to 1968 rapidly became so obsolete that although most of them could still relay signals it was nevertheless uneconomical to operate them commercially. There can be little doubt that satellite design will change radically in the first decade of space shuttle operation, and it will probably become uneconomical to operate *today's* 4/6-GHz satellites. New generations of satellites will make white elephants of today's multimillion-dollar earth stations. Until the technology becomes more permanent a 7- to 10-year lifetime seems a reasonable objective.

To give a satellite a long lifetime there are three main requirements. It must have enough gas or fuel on board to maintain its orbital position. It must have enough solar cells to power the satellite even after many years of cell degradation (Fig. 5.1). Finally, it must be designed with sufficient reliability that it probably will function for the intended period.

END-OF-LIFE

It is interesting to speculate on the end-of-life of the satellites we are using today. Satellites do not normally die a sudden death. Catastrophic failure or destruction of the satellite is unlikely, although a vital component, such as the antenna despinning mechanism, may fail. A few of its many transponders may fail during its lifetime. Its solar cells will lose some of their power. Nevertheless, long after the planned end-of-life of the satellite there is likely to be enough power to operate several of the transponders. It may not be possible to charge up the storage batteries fully, and so the satellite *may* not be usable during eclipses. It seems likely, however, that many of the current satellites will each be capable of providing thousands of voice channels for a decade and perhaps several decades after their planned life.

At some time the last gas in the thrust subsystem will be used up. The satellite will slowly drift into an increasingly inclined orbit and will appear to move in increasingly large figures of eight in the sky. Unless it is at 101°W or 79°E it will also drift very slowly along the equatorial orbit to one of these points. Prior to running out of gas, the satellite may be given a push into a position in which its drift can be tolerated for a time. As the gas runs short, it may be conserved by giving the satellite less frequent adjustments. The end-of-life of a satellite may come when its drift becomes a nuisance and threatens to interfere with another satellite. It will then have its transponders switched off. Even that might not be the end; it may be switched on again if the other satellite fails or if it drifts past it.

Several of the early satellites are now "retired," including SYNCOM, ATS I, ATS III, EARLY BIRD, INTELSAT II, INTELSAT III, and TACSAT I. Nevertheless, they are not dead and some of them are sometimes used experimentally. A satellite which no longer gives adequate voice transmission can often be made to work well with bursts of data, switching its transmitters off between bursts to conserve power. Some satellites switch their transmitters on automatically only when they detect a carrier signal to be transmitted. These satellites automatically switch their transmitters on and off very rapidly when handling short bursts of data. When ATS-1 was no longer usable for telephone transmission the University of Hawaii was employing it quite satisfactorily for computer terminal connections (as will be discussed in Chapter 20), with a low duty cycle conserving its power. The University of Hawaii asked NASA if it could acquire the "dead" satellite. NASA encourages a wide variety of experimentation with its old satellites.

Perhaps by the late 1980s most domestic satellites will make efficient use of frequencies higher than 4/6 GHz, and there will be lots of semiretired satellites available for universities to experiment with. If regulatory bodies agreed, entrepreneurs might buy them cheaply and operate them as tramp satellites carrying electric mail or computer data.

ECONOMIES OF SCALE

As we have commented, there is great scope for future economies of scale in satellites. In the United States in 1980 about 15 million telephone calls will be taking place in a peak instant, and about 1 million of them will be long distance. The AT&T and GTE COMSTAR satellites can carry about 28,000 long-distance telephone calls, less than 3% of the 1980 total. If we consider all types of traffic, including new types such as electronic mail, and then consider television and radio broadcasting, it is clear that today's operations are only a tentative toe in the water of the satellite market. However, the forces of inertia are strong. Government regulations tend to protect the status quo. Communications managers in some corporations have grown up under the motherly wing of traditional common carriers and are not used to the idea that they can save their corporation vast sums of money by ingenious redesign of the corporate facilities.

When Western Union announced dramatically cheaper leased long-distance circuits, thanks to WESTAR, they did not sell most of them overnight. An immediate effect of WESTAR was a price war in long-distance circuits. The cost of leased long-distance voice, data, and television circuits dropped dramatically.

Probably the one place where the economies of scale could be most directly realized is in the Bell System.

TELEPHONE TRAFFIC

The part of the Bell System which handles long-distance transmission represents about 17% of the total cost. By 1980 this will probably have grown to $20 billion. This figure does not include the switching costs, and satellite circuits bypass much of the trunk switching.

By 1980 the volume of long-distance calls being made *simultaneously* on the Bell System in a peak period will have grown to almost a million. There will be about 20 billion long-distance telephone calls per year. A satellite with multiple spot beams, assuming no major breakthrough in launching capability, could be built to transmit a third of a million simultaneous telephone calls. As we commented earlier, a Fairchild version of the ATS-6 satellite could carry 120 transponders, each capable of transmitting about 1800 telephone calls. Six such satellites could handle the Bell System long-distance traffic. Such satellites should be buildable and launchable for less than $150 million each. (Both this and the other costs should be assumed in 1975 dollars.) The ground segment would cost substantially more than the space segment if the satellites were to give extensive coverage. Let us suppose that 500 cities have satellite antennas at, or near, their toll offices and that the total cost for each of these earth stations is $2 million. Certain earth stations which monitor and control the system would be more costly. The total costs, using these figures, are of the order of $2 billion. In addition, some relatively short terrestrial links would be needed to toll offices near the antennas. The entire long-distance transmission network could be built for much less than a tenth of the cost of terrestrial facilities. The ratio of costs for long-distance facilities will swing further in favor of satellites as space costs drop with the space shuttle and associated vehicles in the 1980s.

Today's terrestrial facilities *exist,* however, and are rapidly being added to. The traffic is growing at a rapid rate, with long-distance traffic growing much faster than local traffic. The volume of long-distance traffic will probably double in the next four years and double again in the following four years. A vast expenditure is therefore needed on *new* long-distance capacity. It makes sense that most of this new capacity should be in the form of satellite networks. There are, however, complex regulatory factors which could prevent this.

100 MILLION VOICE CIRCUITS

Not surprisingly, perhaps, one of the most ambitious satellite studies ever published was from AT&T. A Bell Laboratories report described how 50 satellites operating at 20/30 GHz could provide about 100 million voice channels or their equivalent [1]. The report was published when the enthusiasm for Picturephone was still strong, and if 220 million Americans could not be expected to make 100 million simultaneous telephone calls, Picturephone's gluttonous appetite for bandwidth still made the proposals appear interesting.

Two such satellites, rather than 50, could make a major impact on today's long-distance telecommunications costs. The report tactfully avoided giving cost figures, but the cost per channel of such transmission facilities must be a small fraction of that for the long-distance coaxial cable and microwave systems that Bell is building today.

In the Bell Laboratories study the satellites used frequency bands 4 GHz wide. Today 3.5 GHz would probably be used, following the WARC frequency allocations shown in Chapter 8. The bandwidth was divided into eight bands, each of which carried a pulse rate of 315 million pulses per second. Each pulse carried 2 bits, using four-level pulse code modulation (which we shall describe in Chapter 12). The satellite therefore had eight transponders for each 4-GHz beam, carrying eight channels of 630 million bits per second each.

Each satellite would be stabilized to $\pm 0.01°$ and would carry a 10-meter multibeam antenna. Such an antenna gives a high gain at 30 GHz (Fig. 6.1). Multiple antenna feeds would give multiple spot beams, each aimed at a specific earth station. Each beam would be able to carry the same frequency band without the beams interfering. In the maximum configuration there would be 50 beams aimed at 50 earth stations, and the satellite would carry 50 sets of 8 transponders. The total throughput of the satellite would thus be $50 \times 8 \times 630$ million $= 2.52 \times 10^{11}$ bits per second. It was estimated that such a satellite would require 7.2 kilowatts of power and weigh 11,250 pounds. A launch configuration using the space shuttle could perform the launch.

If the satellite had fewer than 50 spot beams, its weight and power requirements would be less. It could be anywhere from INTELSAT IV size upwards, as shown in Table 9.1.

The ground station antennas in the proposal would be designed to point at each satellite. If there were N channels per transponder, S satellites, and G ground stations, the total capacity of the system would be $8N \times S \times G$. In the proposal the 630-megabit throughput of the transponder would be used to carry 10,000 voice channels. A single satellite spot beam would carry $8 \times 10,000$

Table 9.1 Alternative satellites in the Bell Laboratories proposal [1]

	Number of Ground Stations Served			
	8	*16*	*32*	*50*
Weight of satellite (pounds)	1640	3280	6560	11,250
dc power (kW)	1.15	2.30	4.60	7.20
Transponders	64	128	196	400
Gigabits through satellite	40	80	160	252
One-way equivalent voice circuits through satellite (thousands)	640	1280	1960	4000
One-way equivalent TV channels through satellite	640	1280	1960	4000

voice channels. A satellite with 50 spot beams to 50 earth stations would carry $8 \times 50 \times 10{,}000 = 4$ million voice channels. The total system capacity would be $8 \times 10{,}000 \times 50 \times 50 = 200$ million channels of which not all would be utilized at any one time because of uneven and fluctuating load distribution.

An elaborate switching system would be a necessary component of such a network, as on the rest of the Bell System. The system must be designed so that the load will be reasonably well balanced among the ground stations. Many of the stations will be close to major metropolitan areas. Switching must be designed so that faulty circuits can be bypassed, as with today's terrestrial network. With a suitable switching matrix, a substantial number of failed repeaters in the satellites can be tolerated without inacceptable system degradation. The switching will make the system adaptable to changing traffic patterns and will permit ground stations affected by severe storms to be bypassed. Because of the transmission delay, the switching will be organized so that not more than one satellite link is switched into any circuit.

CHANNELS PER POUND

It is interesting to note that whereas the Western Union satellites carry up to 18 voice channels per pound weight of satellite, the AT&T proposal claims 390 voice channels per pound weight of satellite. It is unfair to compare the two figures because AT&T's proposal was advanced technology, which has not yet been built. Nevertheless, the fact that AT&T's figure is *so* much larger is, again, an indication of the economies of scale. A 120-transponder variant on ATS-6 using proven technology would carry about 120 channels per pound. AT&T uses a transponder with a capacity equivalent to 10 television channels, whereas Western Union uses a transponder equivalent to 1 television channel. Few organizations could afford a transponder of 10-television-channel capacity. The organizations that can afford it can have far more voice channels per pound weight of satellite and also each pound weight will cost much less.

GROUND SEGMENT TRADE-OFFS

It is one of the great ironies of the satellite story that the corporation which produced the first satellite to actively relay telephone signals, and the corporation most needing satellites, AT&T, was the one corporation prevented from developing the satellites it needed by government regulation of one form or another. AT&T has both the resources and the technological brilliance to develop satellites as advanced as the one just described. The thought of replacing much of AT&T long-distance technology with competitive satellites must be a hair-raising thought for AT&T top management.

Telecommunications in America would almost certainly have been better served in the space segment if AT&T had been allowed to go full steam ahead with satellites. However, this would probably have wiped out any potential competition and might have prevented (depending on the FCC) what now appear to be some of the most interesting satellite innovations—the developments in the ground segment permitting communications users to lease their own satellite antennas and build private corporate satellite networks.

Some of the most interesting trade-offs for the next few years apply to the ground facilities rather than the space segment. If no satellites of greater weight than today's could be launched, there would still be a dramatic drop in cost of satellite usage due to better ways of organizing the terrestrial segment. It is probably on the ground rather than in the sky that the most important innovations are now possible. They look nothing like the AT&T scheme just outlined. Left to their own devices, the last thing that traditional common carriers would want is users having their own antennas, bypassing much of the common carrier facilities.

In Part II we shall discuss the ground segment, and in Chapter 24 we shall summarize trade-offs in the ground segment.

REFERENCE

1. Leroy C. Tillotson, "A Model of a Domestic Satellite Communication System," *Bell System Tech. J.*, Dec. 1968.

PART II THE GROUND SEGMENT

10 MULTIPLEXING

A satellite has a bandwidth or digital throughput much larger than any one user can employ. Its capacity needs to be split up into many channels which different users can employ independently.

The term *multiplexing* has been used for a century in telecommunications to refer to any technique which permits more than one independent signal to share one physical facility. In a satellite a high level of multiplexing is needed so that many signals can share the bandwidth.

There are two aspects to sharing the satellite's capacity. The first is multiplexing. If the satellite merely provided a point-to-point link between two earth stations, multiplexing would be all that was needed. However, one of the most valuable aspects of satellites is that they can be shared by many antennas scattered across the earth. In addition to simple multiplexing a scheme must be devised which will permit many geographically dispersed earth stations to share the satellite. This is referred to as *multiple access*. The demand for channels will vary among the separate geographical locations from moment to moment, and therefore multiple access needs to be assigned in a time-varying fashion according to the demands of the moment. This dynamically variable allocation of channels is called *demand-assigned multiple access,* sometimes abbreviated *DAMA*.

In this chapter we shall discuss simple multiplexing, and later in the book, techniques for demand-assigned multiple access will be contrasted.

THREE METHODS OF MULTIPLEXING

For many decades, three types of multiplexing have been used in terrestrial telecommunications, and the same three are important with satellites:

1. Space-division multiplexing.

2. Frequency-division multiplexing.
3. Time-division multiplexing.

Space-division multiplexing means that more than one physical transmission path are grouped together. (The term was used long before "space" referred to extraterrestrial activities.) Wire-pair cables, for example, are constructed containing many hundreds of wire pairs.

Frequency-division and *time-division multiplexing* are alternative techniques for splitting up a single physical path. The information to be transmitted can be thought of as occupying a two-dimensional continuum of frequency and time, as illustrated on the left of Fig. 10.1. The quantity of information that can be carried is proportional to the period of time used and to the range of frequencies, the bandwidth, used. If the quantity of information required from one channel is less than that which the physical facilities can carry, then the space available can be divided up either into frequency slices or time slices, as in Fig. 10.1 (center and right).

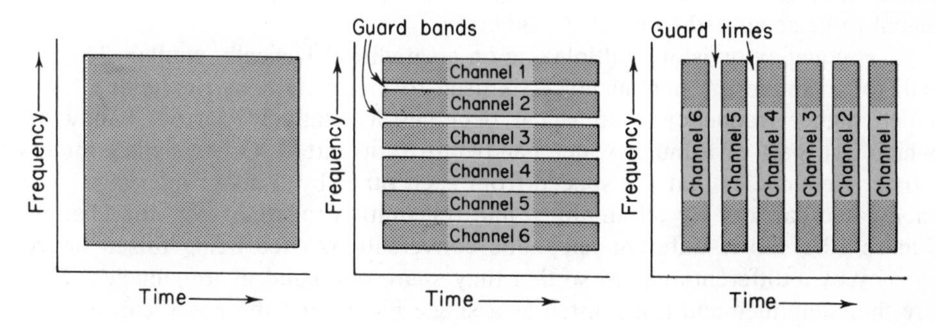

Figure 10.1 (Left) The space available for communication. (Center) Frequency-division multiplexing. (Right) Time-division multiplexing.

In either case the engineering limitations of the devices employed prevent the slices from being packed tightly together. With frequency-division multiplexing a guard band is needed between the frequencies used for separate channels, and with time-division multiplexing a time is needed to separate the time slices. Logs cannot be sawed without some sawdust. If the guard bands or guard times were made too small, the expense of the equipment would increase out of proportion to the advantage gained.

SPACE-DIVISION MULTIPLEXING

A satellite's capacity can be shared by channels using the same frequency band and time if it has directional antennas. One spot beam, for example, could illuminate an area several hundred miles across on the East Coast of the United States; another could illuminate the West Coast. The east and west coasts would then use the same frequency band at the same time in different transponders to transmit different signals. Future satellites may have multiple independent spot beams. Narrower beams can be used when the transmission frequencies are higher. A 20/30-GHz satellite, accurately stabilized, with an antenna as large as ATS-6, could conceivably relay 50 spot beams to North America.

The capacity of each spot beam, however, needs to be subdivided, with other types of multiplexing.

FREQUENCY-DIVISION MULTIPLEXING

A familiar example of frequency multiplexing is radio broadcasting. The signals received by a domestic radio set contain many different programs traveling together but occupying different frequencies on the radio bandwidth. The speech and sounds from each radio station *modulate* a carrier of a frequency allocated to that station; i.e., the frequency is modified in accordance with the information to be sent and hence "carries" the information. We shall discuss modulation in Chapter 11. The tuning circuits in the radio set allow one such signal to be separated from all the others.

Frequency-division multiplexing on a satellite is basically similar. The signal, for example the human voice, is used to modulate a carrier which has a much higher frequency. The signal occupies a relatively narrow bandwidth which is a part of a much wider bandwidth transmitted. Other signals modify carrier frequencies that are spaced from each other by a given interval. These modulated carriers are all amplified and transmitted together over the channel. Figure 10.2 shows different telephone conversations each being raised in frequency by a different amount so that they share one band of frequencies. They are then amplified and transmitted as a single block, and the block can be used in further modulation processes.

Figure 10.3 illustrates the principle of the equipment used for frequency-division multiplexing. Here 12 signals each needing a bandwidth of not more than 4 kHz are combined so that they can be sent together over one physical channel.

At the sending end there are 12 modulators and at the receiving end 12 demodulators. For simplicity Fig. 10.3 illustrates only one-way transmission. The signals pass through 12 low-pass filters to remove any high-frequency components and are then used to modulate 12 separate carrier signals each 4 kHz apart. The frequencies resulting from each modulation process must be restricted to their own band. If they spill over into the bands occupied by other signals, the signals will not be separated correctly at the receiving end. The modulation process produces components spread over a frequency range wider than that of the original signal. Therefore, the outputs of each of the 12 modulators must be filtered again to stop them from interfering with each other. Band-pass filters are used to restrict each signal to the allocated 4 kHz band shown.

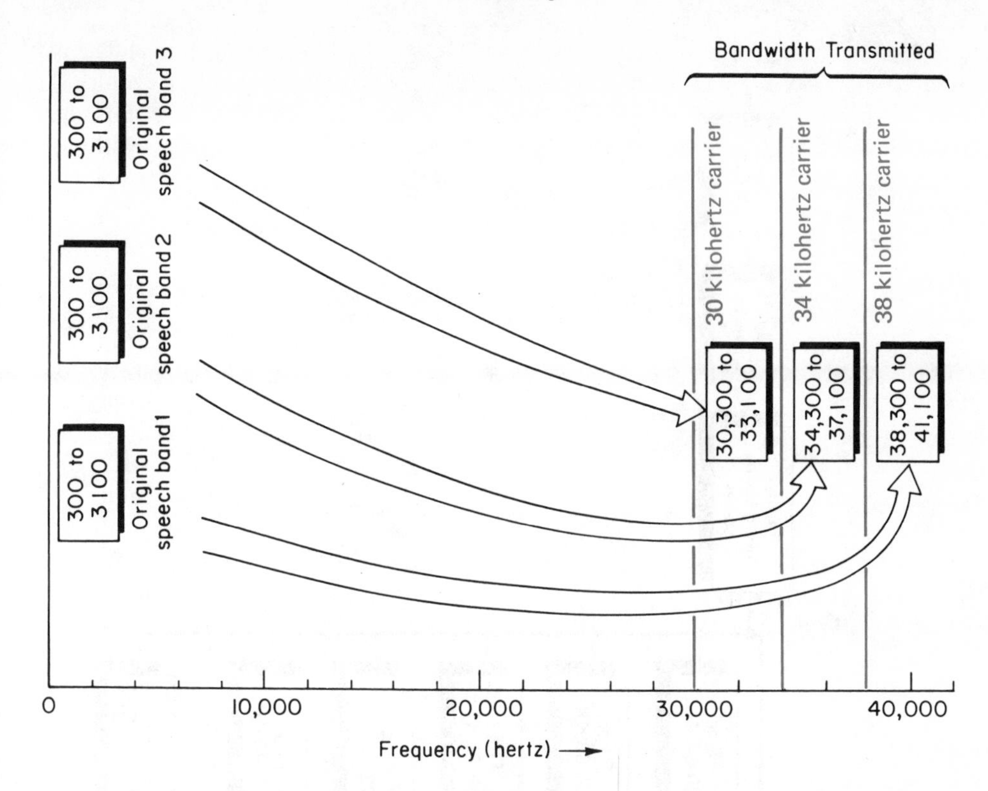

Figure 10.2 Telephone multiplexing. Each speech channel is raised in frequency to fit into a given 4 KHz bandwidth which is assigned to it.

When the signal is received a converse process takes place. Twelve band-pass filters let through the frequencies of one signal only, as shown. These then pass into 12 demodulating circuits, and the original signal is recovered.

GROUPS

The block of 12 voice channels that is formed is referred to as a *channel group* in American terminology or as a *primary group* or simply *group* in international CCITT† terminology. Its bandwidth is much lower than that of the satellite, so additional multiplexing steps take place.

Five channel groups are multiplexed together in a similar fashion to form a *supergroup*. Similarly, a *master group* is then formed from 10 supergroups in North America or 5 or 15 supergroups in other countries. An American master group thus contains 600 voice channels; a CCITT master group can contain 300 or 900 voice channels.

†CCITT (Comité Consultatif International Télégraphique et Téléphonique) is the international organization which establishes standards for telephony, telegraphy, and data transmission.

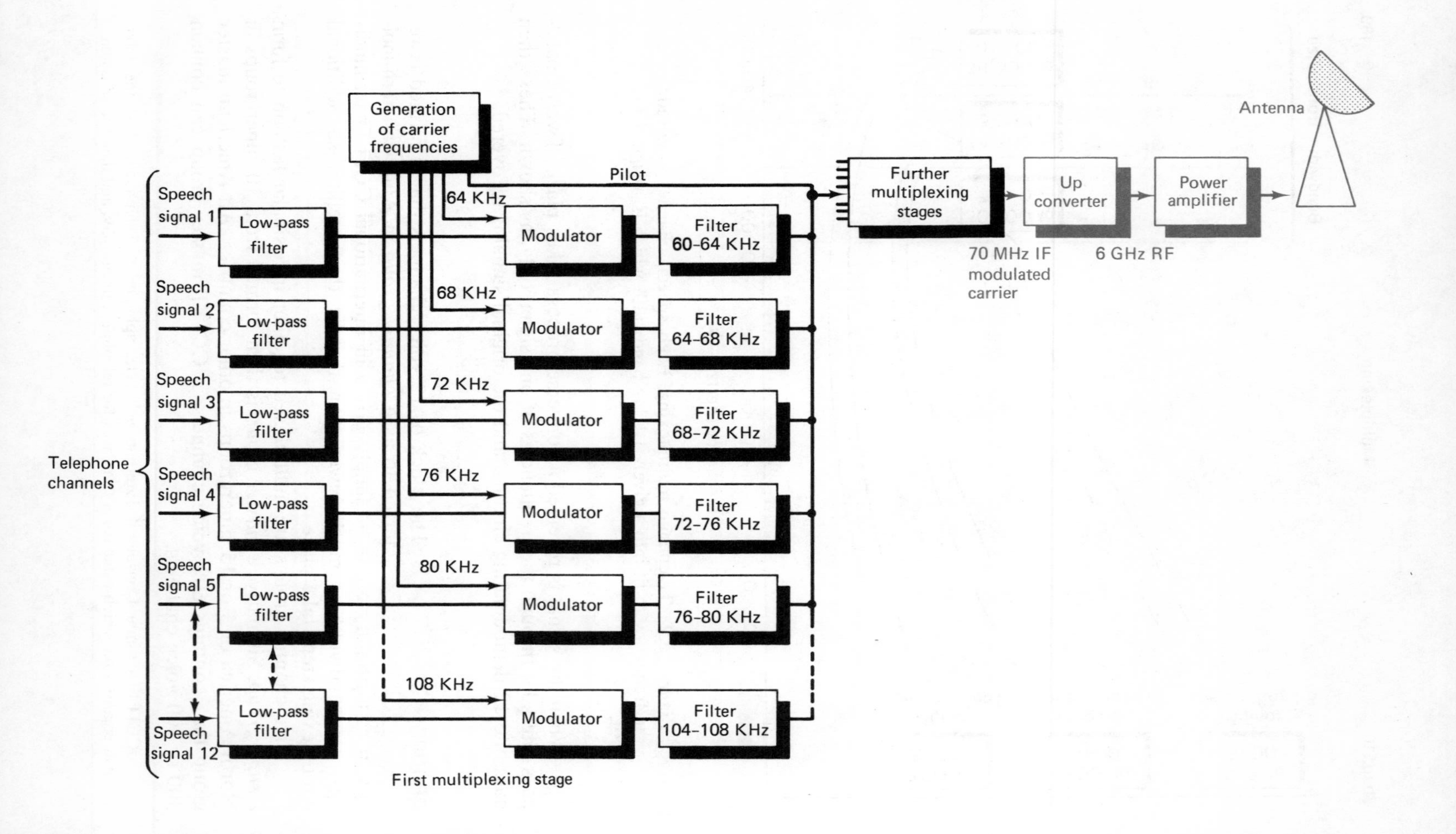
Telephone channels
Speech signal 1
Speech signal 2
Speech signal 3
Speech signal 4
Speech signal 5
Speech signal 12
Low-pass filter
Generation of carrier frequencies
Pilot
64 KHz
68 KHz
72 KHz
76 KHz
80 KHz
108 KHz
Modulator
Filter 60–64 KHz
Filter 64–68 KHz
Filter 68–72 KHz
Filter 72–76 KHz
Filter 76–80 KHz
Filter 104–108 KHz
First multiplexing stage
Further multiplexing stages
70 MHz IF modulated carrier
Up converter
6 GHz RF
Power amplifier
Antenna

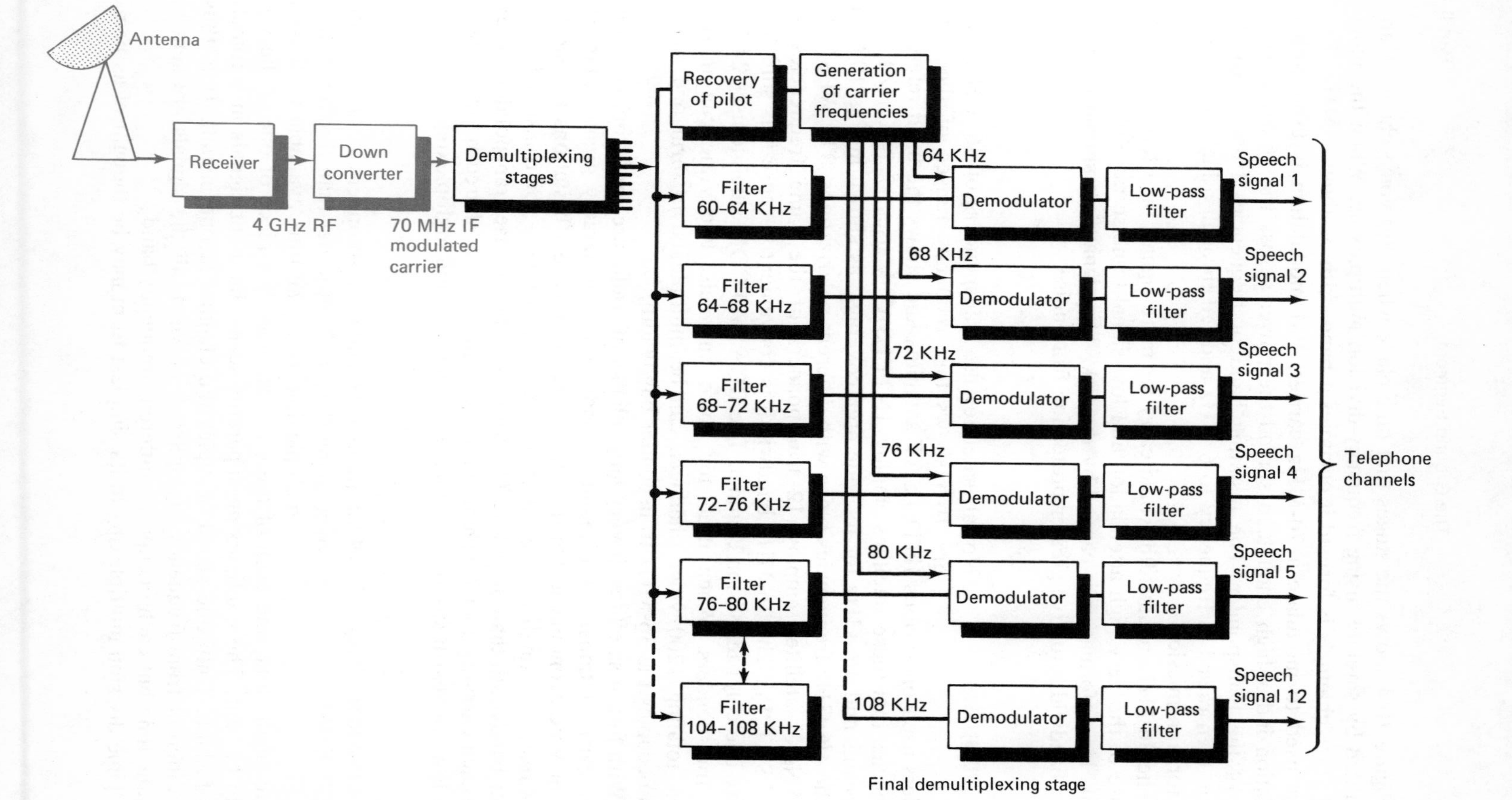

Figure 10.3 Frequency division multiplexing of telephone channels.

Figure 10.4 shows the stages that take place when 900 voice channels are transmitted by satellite, using frequency-division multiplexing. When the block of 900 voice channels is formed it occupies a bandwidth of about 4 MHz. This band is used to modulate a 70-MHz carrier. FM modulation is used with a modulation index high enough to spread the power across the 36-MHz bandwidth of the transponder. The resulting band of frequencies from 52 to 88 MHz is then raised in frequency to 6 GHz and fed through a high-power amplifier for transmission.

The 36-MHz bandwidth would carry far more signals on a terrestrial link. Because of the very high attenuation and low signal-to-noise ratio of the satellite link, less information is carried by the link. The signals are spread across a greater bandwidth to give greater protection from noise.

TRANSPONDERS

Not all satellites use transponders of 36 MHz. Figure 10.5 shows the frequency ranges of the transponders used on various INTELSAT satellites and shows the fixed reference frequencies that these satellites employ. INTELSAT V uses 12/14 GHz for its spot beams and 4/6 GHz for its wide area coverage, as shown in Fig. 10.6. It uses the 4/6 GHz frequencies twice with different polarizations. The WESTAR and ANIK satellites employ 12 transponders of the same frequencies as INTELSAT IV. The 36-MHz bandwidth is convenient because it is sufficient to relay one color television channel. The RCA SATCOM satellite uses 24 36-MHz transponders. When used for voice channels, two American master groups, totaling 1200 voice channels, can be multiplexed to form one signal which occupies the 36-MHz transponder bandwidth.

With future satellites, if very large blocks of traffic are sent from common carrier earth stations, a transponder bandwidth considerably greatly than 36 MHz may be economical. On the other hand, if most earth stations are small, with a relatively small number of channels, there is a case for transponders of smaller bandwidth than 36 MHz. The SBS satellite will use transponders of 54 MHz each carrying a 41 Mbps digital bit stream which is shared by many relatively low-cost earth stations, by means of time-division multiplexing.

TIME-DIVISION MULTIPLEXING

All of today's satellite transponders could be employed using *time-division* rather than *frequency-division* multiplexing. Here the time available is divided up into small slots, and each of these is occupied by a piece of one of the signals to be sent. The multiplexing apparatus scans the input signals in a round-robin fashion. Only one signal occupies the channel at one instant. It is thus quite different from frequency multiplexing in which all the signals are sent at the same time but each occupies a different frequency band.

Time-division multiplexing in its simplest form may be thought of as being

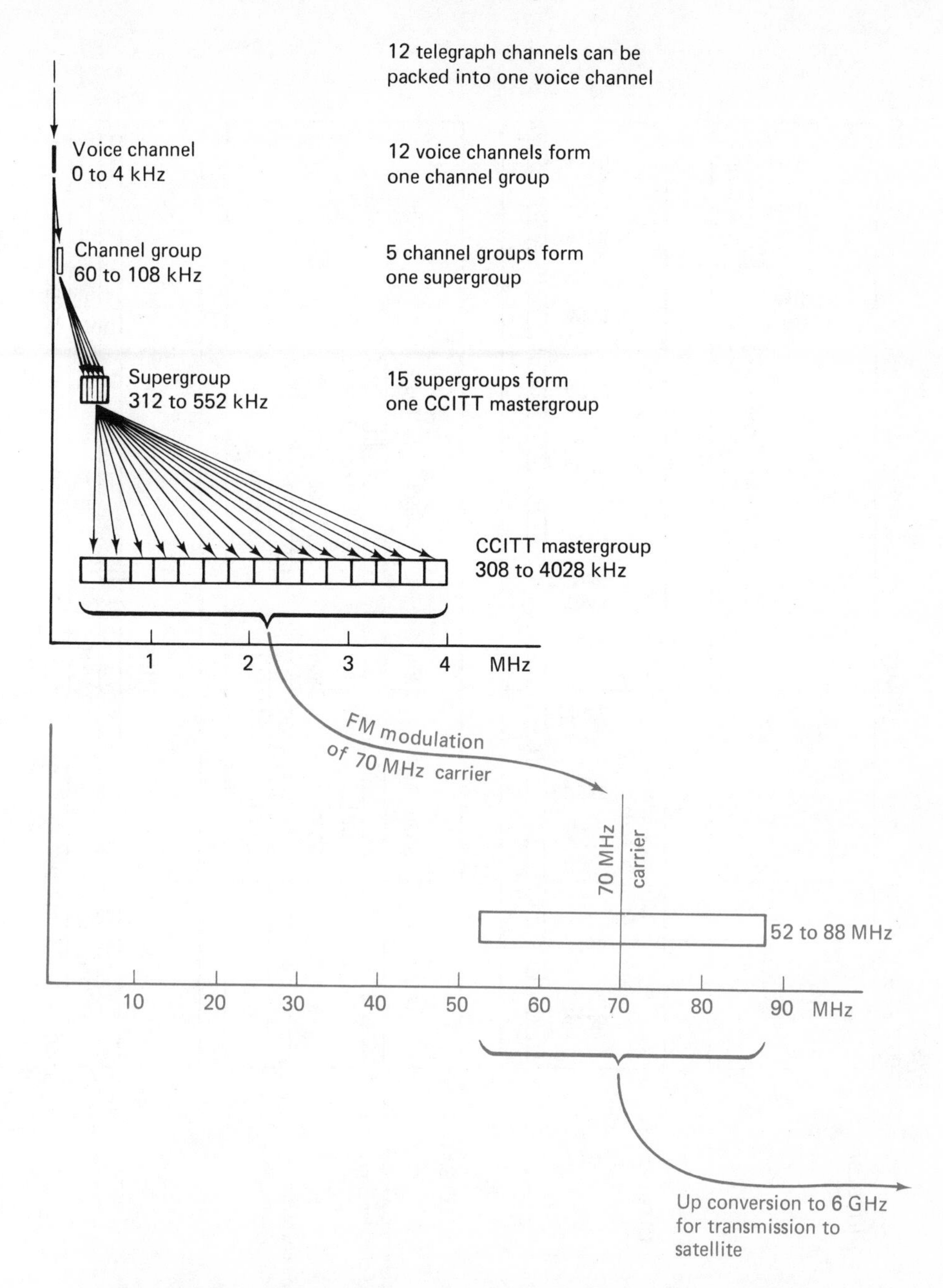

Figure 10.4 Multiplexing stages when frequency-division multiplexing is used on a transponder carrying a total of 900 voice channels. Each of the 12 transponders of INTELSAT IV can carry such a block at the frequencies shown in Fig. 10.5.

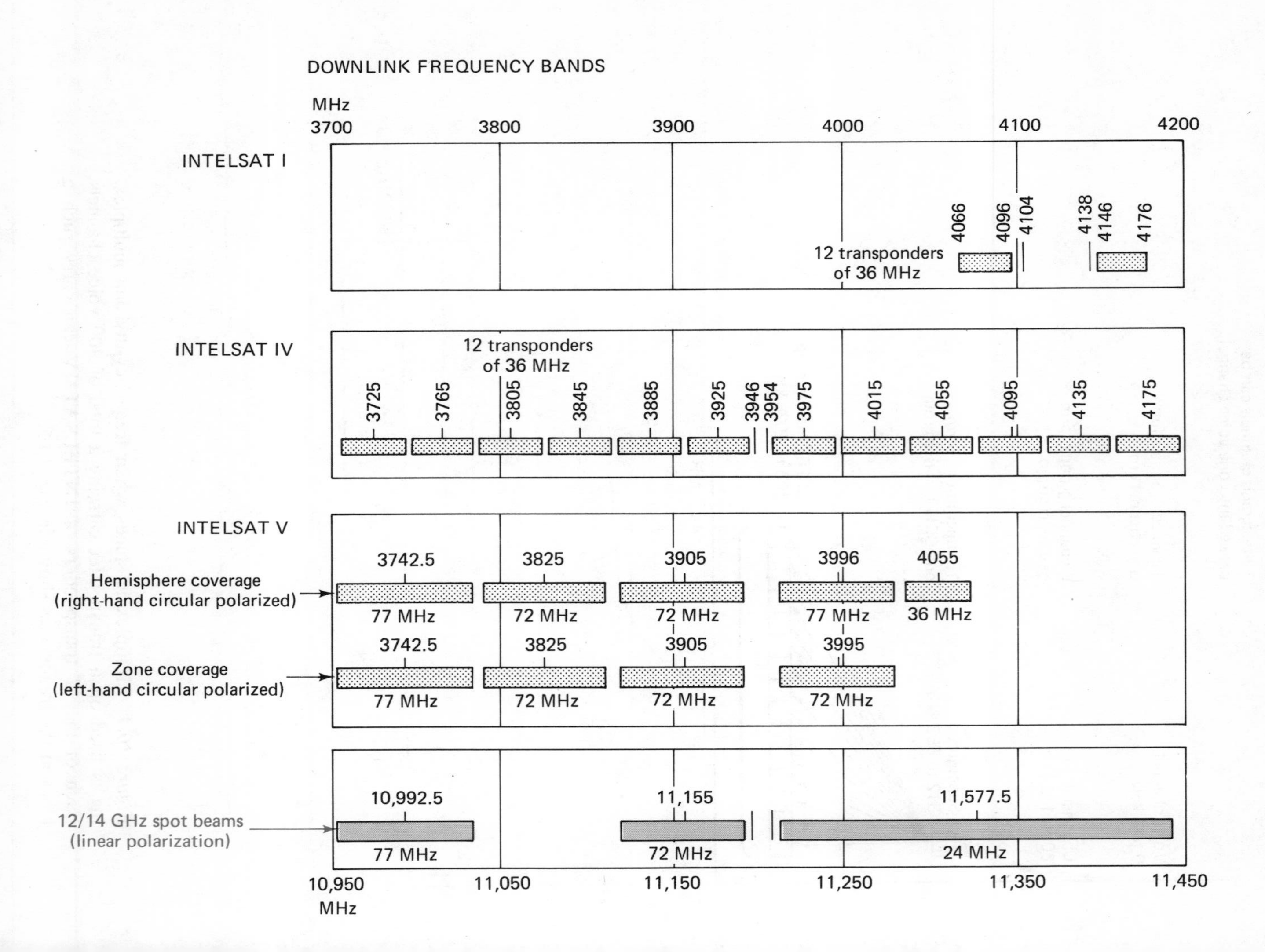
DOWNLINK FREQUENCY BANDS
MHz
3700
3800
3900
4000
4100
4200
INTELSAT I
12 transponders of 36 MHz
4066
4096
4104
4138
4146
4176
INTELSAT IV
12 transponders of 36 MHz
3725
3765
3805
3845
3885
3925
3946
3954
3975
4015
4055
4095
4135
4175
INTELSAT V
Hemisphere coverage (right-hand circular polarized)
3742.5
77 MHz
3825
72 MHz
3905
72 MHz
3996
77 MHz
4055
36 MHz
Zone coverage (left-hand circular polarized)
3742.5
77 MHz
3825
72 MHz
3905
72 MHz
3995
72 MHz
12/14 GHz spot beams (linear polarization)
10,992.5
77 MHz
11,155
72 MHz
11,577.5
24 MHz
10,950 MHz
11,050
11,150
11,250
11,350
11,450

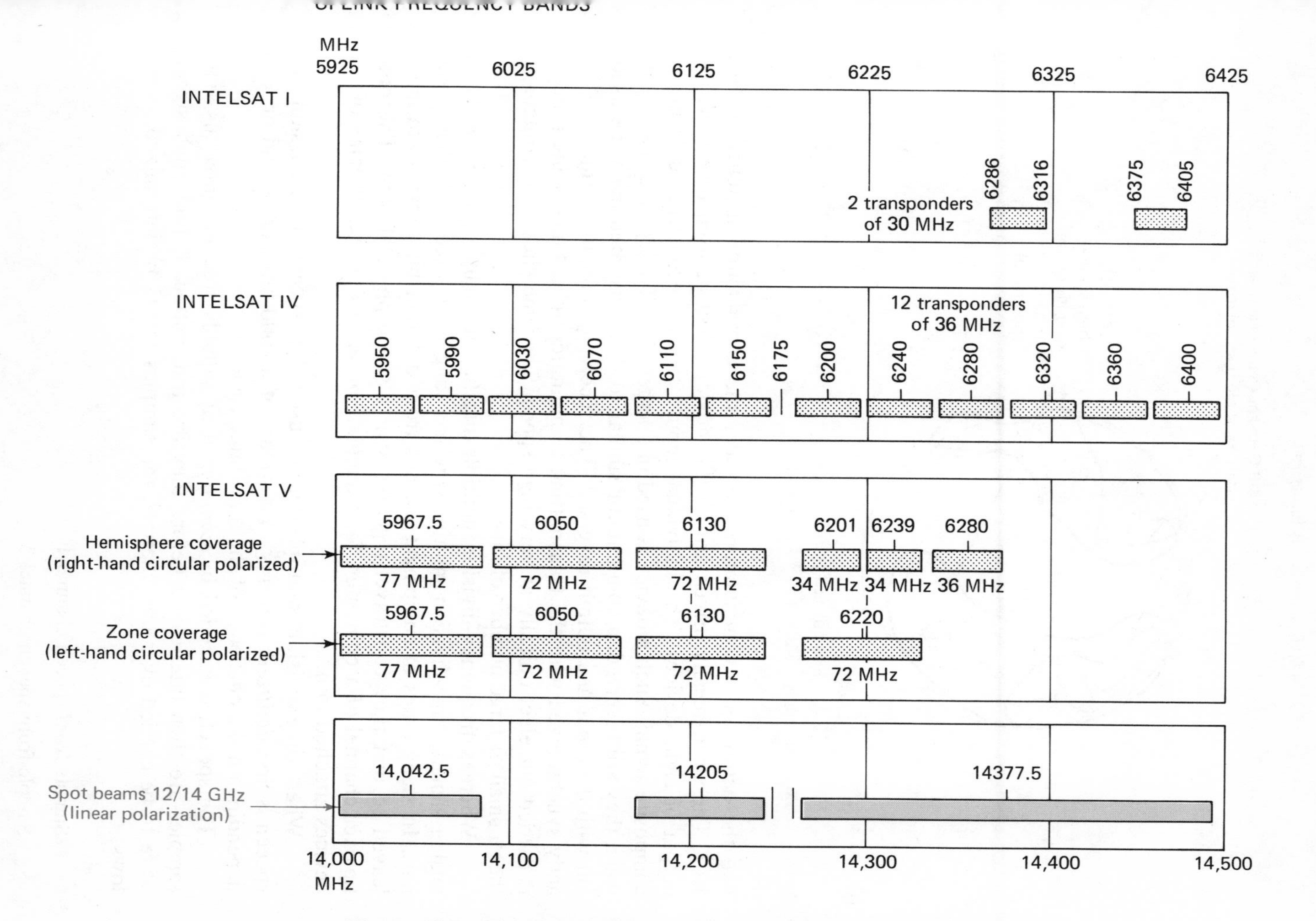

Figure 10.5 The frequencies of the transponders in INTELSAT I, IV, and V. INTELSAT uses the 4/6 GHz bank twice with different polarizations, once for the hemisphere beam (Fig. 10.6) and once for the zone coverage beam. It also uses 12/14 GHz for spot beams.

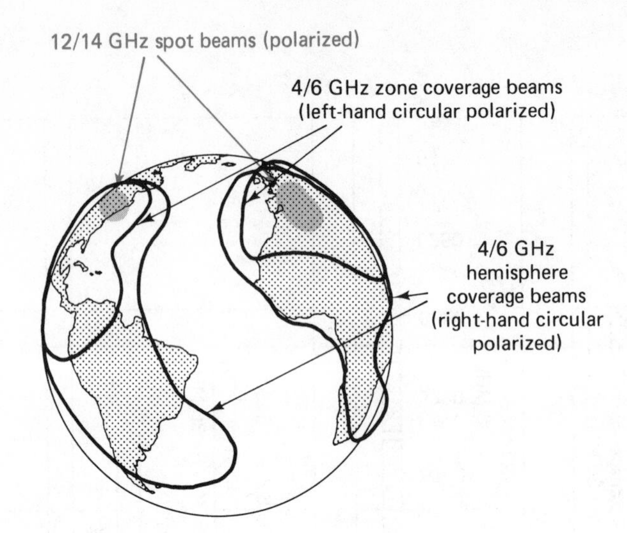

Figure 10.6 The three types of beams used in INTELSAT V, using the frequencies shown in Fig. 10.5.

like the action of a commutator. Consider the commutator sketched in Fig. 10.7. The mechanically driven arm of this device might be used to sample the output of eight devices. The individual inputs can be reconstructed from the composite signal. Such a device is used in telemetering. To separate the signals when they are received, a commutator similar to that illustrated might be used, but with the input and output reversed. The receiving commutator must be exactly synchronized with the transmitting commutator. This time-slicing process is carried out electronically at very high speed. The time slices can contain either a small or large number of bits.

Whereas frequency-division multiplexing fits naturally into the world of analog signals, time-division multiplexing has long been used with digital signals. In many types of equipment the bits from different bit streams are interleaved so that they can travel together over a single physical path. The multiplexer channel of a computer, for example, interleaves the bit streams of the devices attached to it.

When time-division multiplexing is used for speech transmission the speech is first *digitized.* It forms a bit stream of 64,000 bits per second or less, depending on the efficiency of the digitizing process.

The speech is encoded by *sampling* it at suitably frequent intervals. The samples are then interleaved before they are transmitted. If four such signals were to be carried over one channel, the samples would be intermixed as follows:

Sample from speech channel 1

Sample from speech channel 2

Sample from speech channel 3
Sample from speech channel 4
Sample from speech channel 1
Sample from speech channel 2
Sample from speech channel 3
⋮

The transmission must be accurately synchronized so that the receiving equipment knows which sample is which.

Many streams of samples are interleaved to form a stream which is often many millions of bits per second. This high-speed digital signal is superimposed upon a frequency which the satellite transponder can handle.

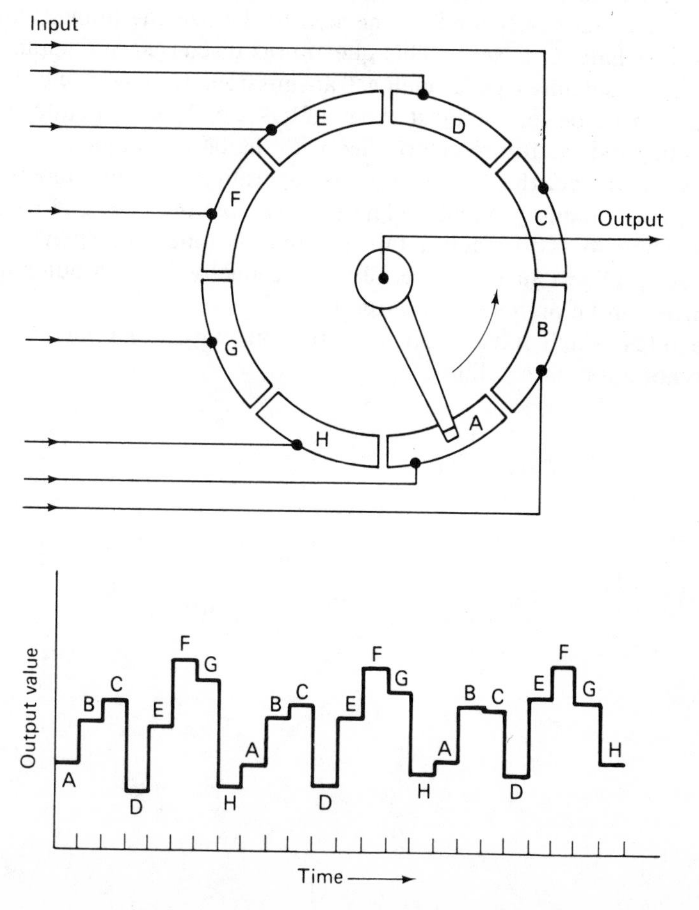

Figure 10.7 Time-division multiplexing time-slices the signal like a commutator. It is done in very high speed digital circuitry.

Just as there are standard *groups* in frequency-division multiplexing, so there are standard bit streams in time-division multiplexing. The standard specifies the bit rate and the composition of the bit stream. We shall discuss it further in Chapter 12.

ANALOG OR DIGITAL

The choice between frequency-division or time-division multiplexing is really a choice between whether the signal should be handled in an *analog* or a *digital* fashion. The satellite link, like radio and cable links on earth, can be made to carry either an analog or digital signal. The question to be answered is, Which is the most efficient?

The efficiency of digital signaling depends on two factors. First, how many bits per second can be relayed by the satellite? This depends on the design of the modulation equipment, which we shall discuss in Chapter 11. Second, how many bits per second are needed to digitize the human voice, music, television, or whatever is sent? This question is discussed in Chapter 12.

The early satellites used analog transmission. It is now clear that digital techniques can be better for most types of system. Much satellite transmission is now digital, and the future clearly lies with digital techniques.

As we shall see, the swing to digital operation has very interesting implications for the computer industry. The relative cost of voice and data transmission is swinging so as to make data transmission appear inexpensive compared to voice. A satellite can have a prodigious capacity for computer data if only the channels can be organized appropriately.

The rules which have been accepted on earth—one might say groundrules—do not apply to satellites.

11 MODULATION

Modulation is a process which permits a signal to be changed so that it can travel over a given transmission bandwidth.

When we wish to transmit music, or the human voice, or data signals, it often happens that their frequency spectra are unsuited to the medium we must use for transmission. Take the case of radio broadcasting, for example. You wish to listen to a violin concerto on the radio so you tune your radio to the station on which it is playing—96.3 MHz. This is the frequency, the midpoint of the frequency band at which this communication system is then operating. Waves of this frequency travel through the atmosphere and are picked up by your radio set. However, the violins are not being played at this frequency. You would not hear them if they were. The sound of the violins is in the range 30 to 20,000 cycles. This sound must therefore be used to modulate the waves of very much higher frequency which the medium uses for transmission. The same is true in the transmission of data. The higher-frequency signal must therefore be made to "carry" the lower frequency signal. The higher-frequency signal is said to be *modulated* by the information that it is to carry.

Satellite transponders operate at very high frequencies—the frequencies shown in Fig. 10.5. The frequencies of the radio link (called RF frequencies) are made to carry a lower IF (intermediate-frequency) signal centered at 70 MHz. The IF signal is in turn modulated by the information that is to be transmitted. Many signals may be multiplexed together and then as a group used to modulate the IF carrier, as shown in Fig. 10.3.

MODEMS FOR SATELLITES

If we wish to send data over a telephone circuit, a satellite transponder, or any other analog transmission facility, we must use a *carrier frequency* somewhere near the center of the range of frequencies available. Data are then superimposed upon the carrier in much the same way that the violin concerto was superimposed upon the broadcasting frequency of 96.3 MHz. The carrier selected is *modulated* with the data to be sent. After transmission a *demodulation* process is required to recover the data.

The electronic circuit which does the modulation and demodulation is called a *modem* (a combination of the first letters of "modulation" and "demodulation"). Most modems are designed for telephone lines and transmit between 1200 and 9600 bits per second over the telephone bandwidth. High-speed modems are available for transmission over satellite transponders. Modems used by Western Union to provide digital channels via the WESTAR satellites transmit 50 million bits per second.

The modulation process results in signal energy being scattered across the usable range of frequencies. In the case of a telephone line the resulting energy must lie between about 300 and 3000 Hz. In the case of a 70-MHz IF carrier for a transponder bandwidth of 36 MHz, the energy must lie between 52 and 88 MHz. Much ingenuity has been expended on improving the performance of modems for terrestrial telephone lines. A similar degree of ingenuity is now being expended on satellite modems. The modem must match the characteristics of the signal being sent to the characteristics of the transmission medium.

THREE TYPES OF MODULATION

When we employ a *sine-wave carrier* to convey data, it has three parameters which we could modulate: its amplitude, its frequency, and its phase. There are thus three basic types of modulation in use: *amplitude modulation, frequency modulation,* and *phase modulation,* referred to as AM, FM, and PM. Each of these methods is in common use today. The sine-wave carrier may be represented by

$$a_c = A_c \sin(2\pi f_c t + \theta_c) \tag{11.1}$$

where a_c = instantaneous value of carrier voltage at time t
A_c = maximum amplitude of carrier voltage
f_c = carrier frequency
Θ_c = phase

The values of A_c, f_c, or θ_c may be varied to make the wave carry information.

This is illustrated in Fig. 11.1. A sinusoidal carrier wave of, say, 1500 Hz, a midfrequency of a voice channel band, is modulated to carry the information bits 0 1 0 0 0 1 0 1 1 0 0. In this simplified diagram the channel is being operated inefficiently because *far* more bits could be packed into the carrier oscillations shown. The tightness of this packing determines the speed of operation.

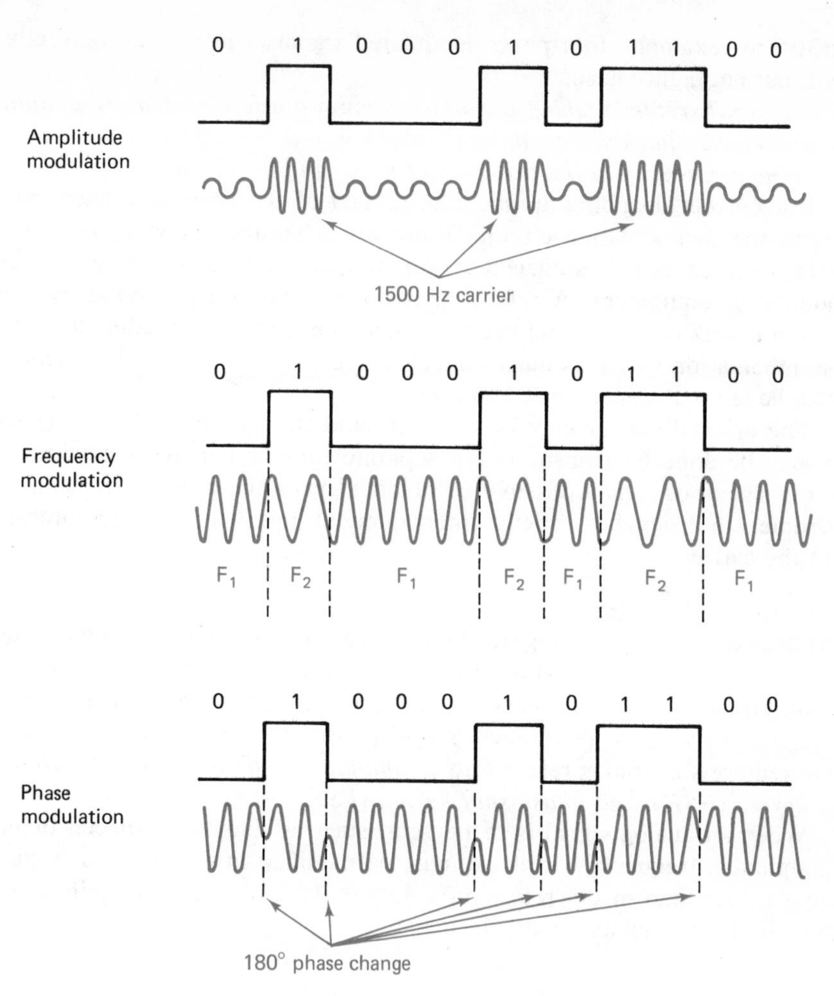

Figure 11.1 The three basic methods of modulating a sine wave carrier with a binary data stream. The amplitude, the frequency, or the phase can be modified to carry the data.

Furthermore, here we modulate the carrier by placing it in one of two possible states in each case. With amplitude modulation we could send several different amplitudes, as in the case illustrated in Fig. 11.9. Similarly, with frequency modulation, we could use several frequencies rather than just the F_1 and F_2 shown. With phase modulation Fig. 11.9 illustrates only 180° phase changes; we could use phase changes which are multiples of 90°, giving four possible states, or 45° giving eight, and so on. Increasing the number of states of the carrier that are used increases the complexity of the decoding or demodulation circuits and considerably increases the susceptibility of the transmission to noise and distortion. If distortion can change the carrier phase by $\pm 30°$, for example, four-phase modulated signals can still be correctly detected, but not eight-phase.

We are therefore seeking a workable compromise between the quantity of data that can be packed into the transmission and the ability of the modem to decode them correctly in the presence of noise and distortion.

For correct decoding an accurate replica of the original carrier must be given to the demodulating circuit. There are a number of ways of obtaining this. In some cases it is sufficient to generate the replica independently in the demodulating equipment. A reference frequency may be generated by a high-precision quartz oscillator and used for decoding frequency modulation. It is in phase modulation that it is most difficult to obtain a reference. The demodulator can have no absolute sense of phase.

The original carrier may be reconstructed from information in the signal. This may be done by transmitting a separate tone of narrow bandwidth along with the signal, or it may possibly be obtained from the modulated signal itself. Sometimes the signal is briefly interrupted at intervals to give information about the carrier.

ANALOG OR DIGITAL MODULATION

Figure 11.1 represents a relatively simple modulation process because the signal to be carried is a binary on-off signal. Discrete changes in amplitude, frequency, and phase are used. The three processes illustrated are sometimes referred to as *amplitude shift keying (ASK), frequency shift keying (FSK),* and *phase shift keying (PSK).*

When an analog signal is to be sent, such as telephone speech or music, the amplitude, frequency, or phase must be modified in a continuously variable manner rather than in discrete jumps. Figure 11.2 illustrates amplitude modulation with a continuously variable signal.

SIDE BANDS

Each of the three methods of modulating a sine-wave carrier results in bands of energy at frequencies above and below the original carrier frequency. These are referred to as *side bands.* They can be illustrated simply if we discuss what happens when a sine-wave component of the information is carried. Boxes 11.1, 11.2, and 11.3 contain equations for a carrier of frequency f_c being modulated by a sine wave of frequency f_m.

Amplitude modulation results in two side bands, one at a higher frequency than the carrier, $f_c + f_m$, and one at a lower frequency, $f_c - f_m$. The upper and lower side bands are mirror images of one another and hence contain the same information. In addition to the side bands a component is produced at the original carrier frequency, f_c. This component contains no information but is of greater amplitude than the side bands.

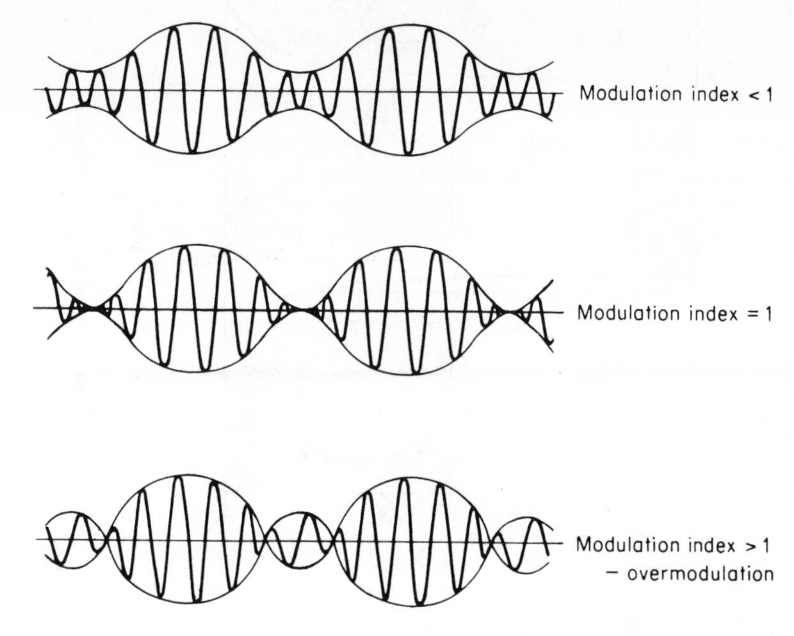

Figure 11.2 Amplitude modulation by a continuously varying signal, showing undermodulation, overmodulation, and ideal modulation with a modulation index = 1.

In reality the modulating signal will consist not of one sine wave but many, spread across a given bandwidth. Any signal that is to be sent can be represented by a series of sine waves using Fourier analysis.

In telephone transmission, a carrier at say 60,000 Hz may be modulated with speech filling a band of 300 to 3300 Hz. The result after amplitude modulation is shown in Fig. 11.3. The lower side band occupies 56,700 to 59,700 Hz and the upper side band 60,300 to 63,300 Hz. In telephone practice a band-pass filter removes all but one side band. Many such modulation operations occur, with different carriers spaced 4 kHz apart. The resulting side bands travel together as one signal. This constitutes the multiplexing operation illustrated in Figs. 10.1 and 10.3.

The amplitude of the carrier, A_c, should approximately equal the amplitude of signal carried, A_m. The ratio A_m/A_c is called the *modulation index*. If the modulation index is greater than 1, the resultant wave would have an envelope with more peaks than the modulating wave, and the original signal would not be recovered without distortion. This is called overmodulation and is shown in the bottom diagram in Fig. 11.2. The top diagram of Fig. 11.2 shows undermodulation—a modulation index less than 1. Undermodulation is inefficient,

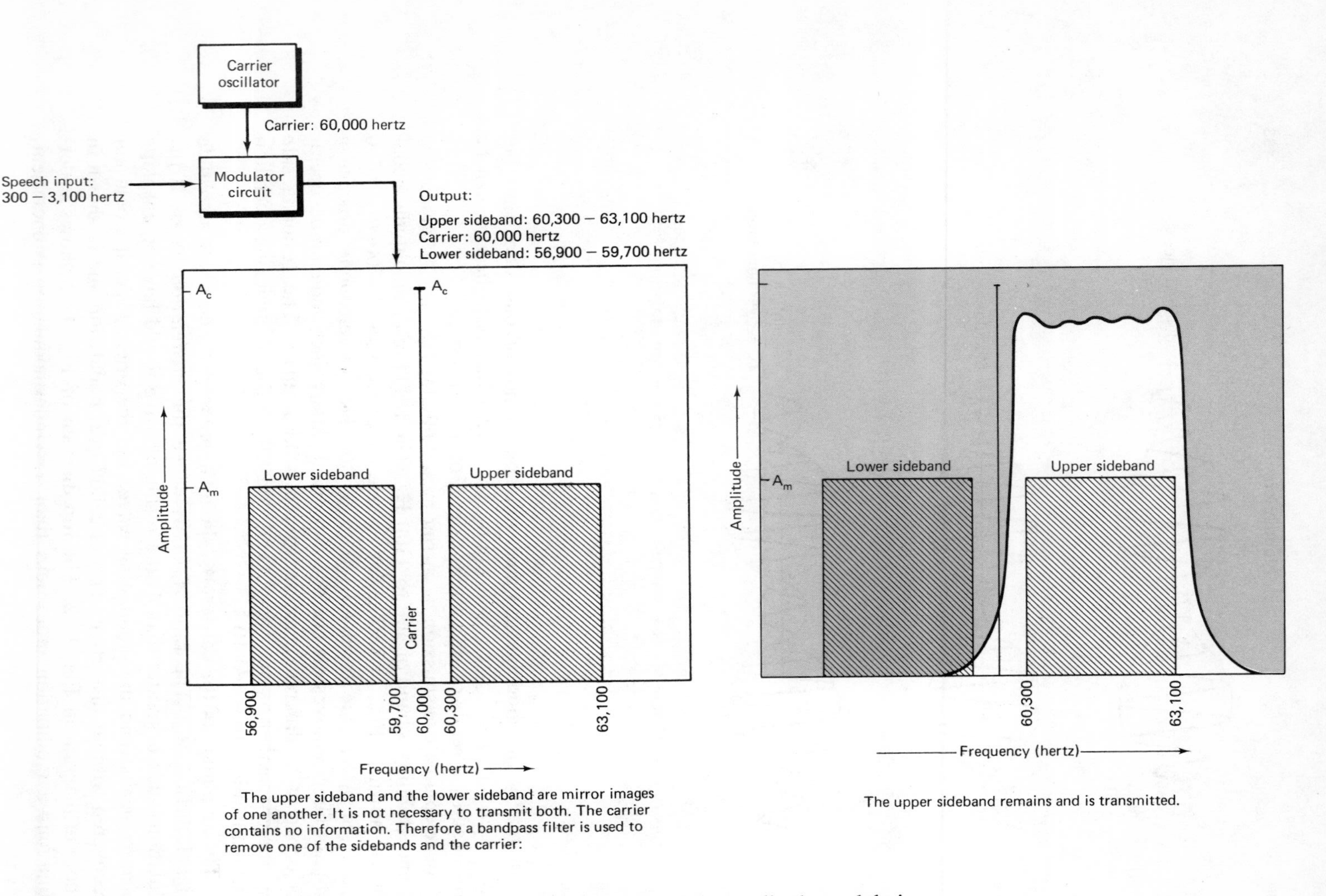

Figure 11.3 Upper and lower sidebands in amplitude modulation.

though many of the component frequencies of signal to be transmitted are of amplitude lower than the carrier.

The power needed for transmission is proportional to the square of the signal amplitude. The power needed to transmit the carrier is proportional to A_c^2 and that to transmit the side band is proportional to $(A_m/2)^2$. For a modulation index of 1 the power needed to transmit the carrier is four times that needed to transmit the side band. For signal components less than the carrier amplitude the ratio is higher.

It is common, therefore, to find suppressed carrier amplitude modulation in which the carrier has been removed by a filter and only the side bands transmitted, and also single-side-band amplitude modulation in which one side band has been removed. It can be shown that the latter improves the signal-to-noise ratio over full-amplitude modulation by 4 decibels or more. It also halves the bandwidth required.

DETECTION WITH AM

When the modulated waveform reaches the demodulator a *detection* process must convert it back to the original signal. For amplitude modulation there are two main types of detection: *synchronous* (also called *coherent,* or *homodyne*) detection and *envelope* detection.

Synchronous detection involves the use of a locally produced source of carrier which has the same frequency and phase as that bringing the received signal. The transmission is multiplied by this carrier, and this enables the signal to be extracted. Some additional components appear with it consisting of side bands centered around $2f_c$, $4f_c$, $6f_c$, and so on. These are filtered off with a low-pass filter.

In synchronous detection, to produce the reference wave of the same frequency and phase as the carrier, it is normally necessary to transmit some information with the signal for this purpose. Such a wave can be extracted from the carrier, and so the carrier may not be completely suppressed. It is usually partly suppressed because of its relatively high energy content. We now have a form of transmission in which one side band is suppressed and only enough carrier is sent to give a reference frequency and phase for synchronous detection.

Envelope detection involves rectifying and smoothing the signal so as to obtain its envelope, as illustrated in Fig. 11.4. Envelope detection does not require the reference wave to be produced and so is considerably less expensive, as producing it is one of the main problems of synchronous detection. Envelope detection does, however, need both side bands and full-amplitude carrier. Suppressed side-band amplitude modulation would lead to an envelope shape which differs from that of the original signal.

BOX 11.1 Amplitude modulation

The sine-wave carrier is represented by $a_c = A_c \sin 2\pi f_c t$.

A sine-wave component of the signal to be carried is represented by $a_m = A_m \sin 2\pi f_m t$.

Amplitude modulation results in A_c being changed as follows:

$$
\begin{aligned}
a_{mc} &= (A_c + a_m) \sin 2\pi f_c t \\
&= (A_c + A_m \sin 2\pi f_m t) \sin 2\pi f_c t \\
&= A_c \sin 2\pi f_c t + A_m (\sin 2\pi f_m t) \sin 2\pi f_c t \\
&= A_c \sin 2\pi f_c t + \frac{A_m}{2} \cos 2\pi(f_c - f_m)t - \frac{A_m}{2} \cos 2\pi(f_c + f_m)t \\
&= A_c \sin 2\pi f_c t + \frac{A_m}{2} \sin \left[2\pi(f - f_m)t + \frac{\pi}{2} \right] \\
&\quad + \frac{A_m}{2} \sin \left[\pi(f_c + f_m)t - \frac{\pi}{2} \right]
\end{aligned}
$$

This contains the three components, the carrier at frequency f_c, which contains no information, and the two side bands at frequencies $f_c - f_m$ and $f_c + f_m$, which do contain information because their amplitude is proportional to A_m. Thus,

$$
\begin{aligned}
\text{Carrier:} \quad & A_c \sin 2\pi f_c t \\
\text{Lower side band:} \quad & \frac{A_m}{2} \sin \left[2\pi(f_c - f_m)\, t + \frac{\pi}{2} \right] \\
\text{Upper side band:} \quad & \frac{A_m}{2} \sin \left[2\pi(f_c + f_m)\, t - \frac{\pi}{2} \right]
\end{aligned}
$$

A_m/A_c is referred to as the *modulation factor* or *modulation index*.

In choice of detection method we thus have a compromise between speed and cost. Envelope detection needs twice the bandwidth of synchronous detection because both side bands must be transmitted. Single sideband transmission is more expensive but gives twice the number of channels.

FREQUENCY MODULATION

When frequency modulation was developed it was used to replace amplitude modulation where better performance in the presence of impulse noise was needed. The signal is transmitted at constant amplitude and so is resistant to

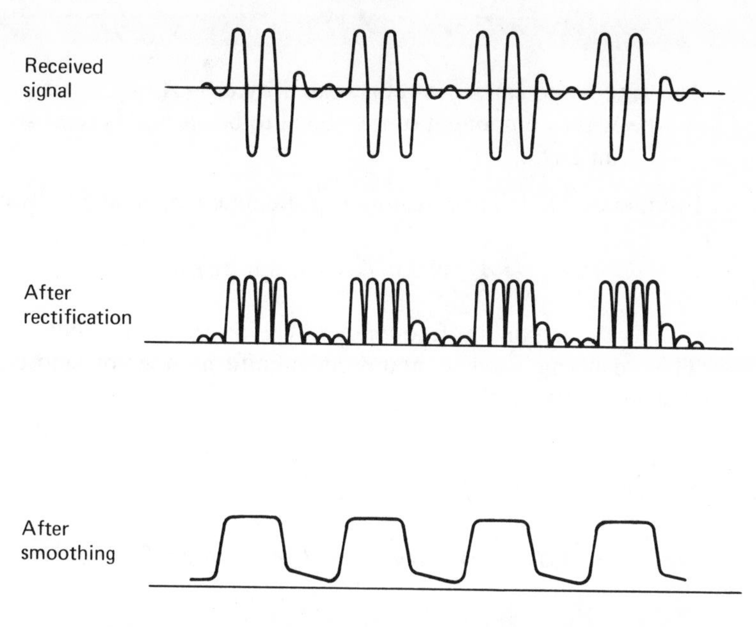

Figure 11.4 Envelope detection of an AM signal.

changes in amplitude caused by noise. However, a larger bandwidth is needed. A satellite channel has a large bandwidth but severe noise, so frequency modulation is commonly used.

Frequency modulation gives an important tradeoff between use of bandwidth and use of power. It can permit the use of very low cost earth stations for signals which are only a fraction of the transponder bandwidth.

As shown in Box 11.2, frequency modulation produces not two side bands, like amplitude modulation, but an infinite number. Unlike amplitude modulation the *modulation index,* $\Delta f_c/f_m$, is variable over a wide range. By varying the modulation index, a signal of large or small bandwidth can be adjusted to fit the available channel bandwidth. The same information can be concentrated into a narrow range of frequencies when $\Delta f_c/f_m$ is small, and spread over a wide range when $\Delta f_c/f_m$ is large.

Figure 11.6 shows the modulation of an IF carrier of 70 MHz with a sine wave of frequency 1 MHz by five different circuits. The top one uses amplitude modulation, and the two side bands can be seen on either side of the non-information-carrying component at 70 MHz. The other four use frequency modulation with different modulation indices. When the modulation index is low the energy tends to be clustered around the carrier frequency—though not so much as with amplitude modulation. When the modulation index is high the energy can be spread over a wide range of frequencies.

In practice the signal transmitted is not composed of one sine wave but of many or of a continuous band of frequencies. Looking at the bottom diagram of Fig. 11.6, the reader may imagine a band of frequencies up to 1 MHz being

BOX 11.2 Frequency modulation

The sine-wave carrier is represented by $a_c = A_c \sin 2\pi f_c t$.

A sine-wave component of the signal to be carried is represented by $a_m = A_m \sin 2\pi f_m t$.

Frequency modulation results in f_c being changed as follows:

$$a_{mc} = A_c \sin 2\pi(f_c + \Delta f_c \sin 2\pi f_m t)t$$

;

where Δf_c is the maximum frequency deviation that can occur.

This resulting wave contains an infinite number of side bands spaced at intervals f_m; thus,

$$\begin{aligned}
a_{mc} = {} & A_c J_0\left(\frac{\Delta f_c}{f_m}\right) \sin 2\pi f_c t \\
& + A_c J_1\left(\frac{\Delta f_c}{f_m}\right) [\sin 2\pi(f_c + f_m)t - \sin 2\pi(f_c - f_m)t] \\
& + A_c J_2\left(\frac{\Delta f_c}{f_m}\right) [\sin 2\pi(f_c + 2f_m)t + \sin 2\pi(f_c - 2f_m)t] \\
& + A_c J_3\left(\frac{\Delta f_c}{f_m}\right) [\sin 2\pi(f_c + 3f_m)t - \sin 2\pi(f_c - 3f_m)t] \\
& + A_c J_4\left(\frac{\Delta f_c}{f_m}\right) [\sin 2\pi(f_c + 4f_m)t + \sin 2\pi(f_c - 4f_m)t] \\
& + \ldots \text{ etc.}
\end{aligned}$$

where $J_0(\Delta f_c/f_m)$, etc., are Bessel functions.

The Bessel functions giving the relative amplitudes of these spectral components are shown in Fig. 11.5.

Figure 11.6 plots the above spectrum for a carrier of frequency 70 MHz being modulated by a sine wave of frequency 1 Mz.

The ratio $\frac{\Delta f_c}{f_m}$ is referred to as the *modulation index.*

smeared across the bandwidth of a satellite transponder—from 52 to 88 MHz.

There is a trade-off between the use of power and the use of bandwidth. The signal at the bottom of Fig. 11.6 is of less power than the one at the top but requires more bandwidth. On terrestrial transmission links plenty of power is available, and so the signal is packed tightly into a small bandwidth, thereby

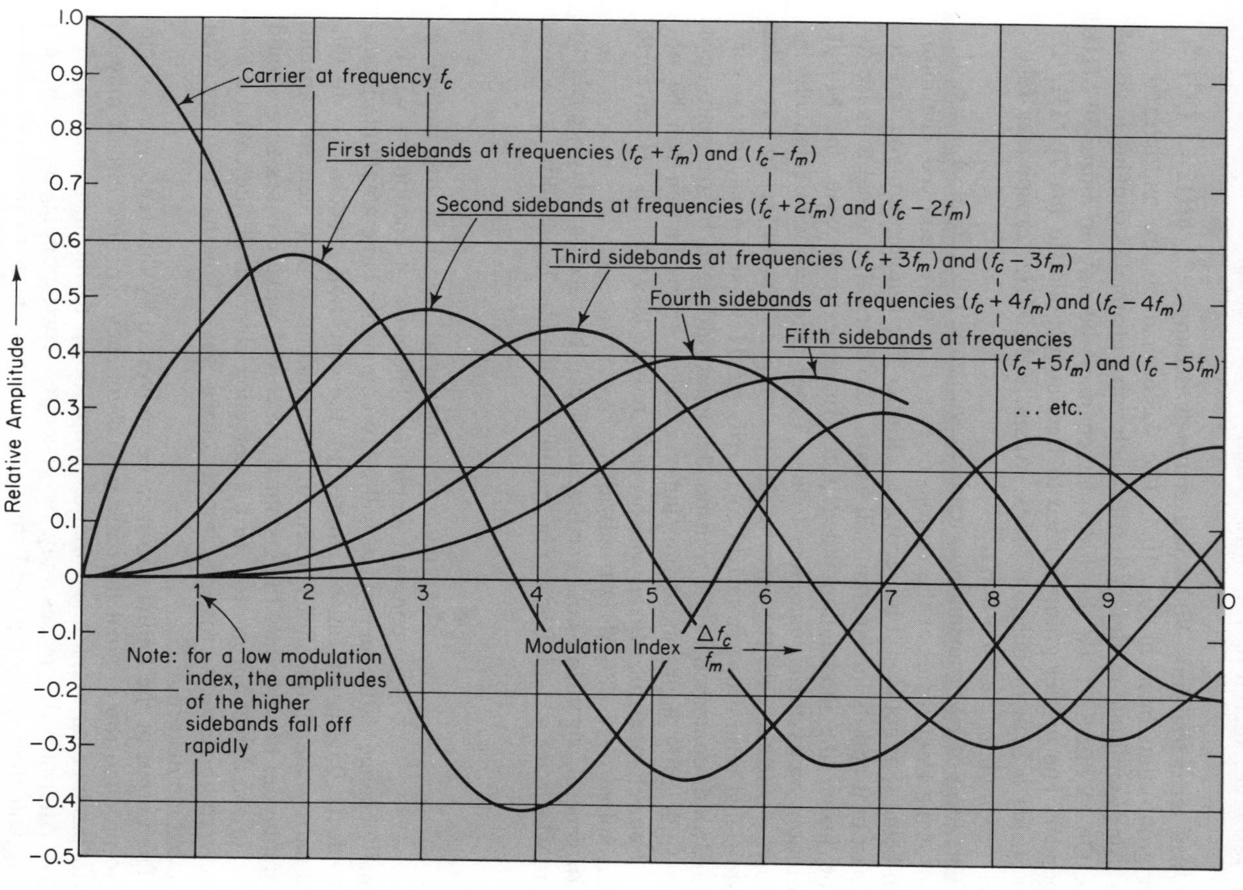

Figure 11.5 Spectral components of a carrier of frequency f_c, frequency modulated by a sine wave of frequency f_m.

increasing the total information-carrying capacity. On satellite links power is a scarce resource; therefore FM with a fairly high modulation index may be used.

On a terrestrial link, 900 voice channels occupy 3.72 MHz—a CCITT master group, illustrated in Fig. 10.4. The voice channels are multiplexed together using amplitude modulation, with each voice channel occupying a slot of 4 kHz. On INTELSAT IV the 900 voice channels occupy an entire 36-MHz transponder. The master group is used to frequency-modulate the 70-MHz IF carrier, using a high modulation index. Again, a 36-MHz transponder relays one television channel of 4.6 MHz.

This form of FM transmission tends to withstand the effects of noise. If a portion of the frequencies at the bottom of Fig. 11.6 were removed, the information being sent could still be recovered. Rather like a hologram, you can destroy part of it and its information still exists. The worse the signal-to-noise ratio, the higher the modulation index used. In an extreme case, the Muzak Corporation wants their customers to use 3-foot antennas with relatively inexpensive electronics. As the antenna size is dropped from 40 feet to 3 feet the noise-to-signal ratio drops from bad to horrifying. However, music has a very low bandwidth compared to the transponder bandwidth, and by transmitting it with a sufficiently high modulation index excellent reproduction can be obtained. Four channels of Muzak which would require less than 80 kHz on earth are transmitted over a 36-MHz bandwidth.

In general, small low-cost earth stations can be used to relay signals by satellite which are a small fraction of the full transponder bandwidth.

FM DETECTION

The frequency-modulated signal is transmitted at constant amplitude. The noise it encounters will occasionally change its frequency but will more commonly have amplitude modulation effects. The latter can be ignored by the detection process. To do this only a narrow amplitude slice is used for the detection. This is centered around zero amplitude. Ideally, only the instant the received wave crosses zero should be used in the detection process. In the detection circuit a device called a *limiter* converts these zero crossings into a square wave. This has then removed any amplitude distortion.

The output of the *limiter* can then be converted by different types of circuits to produce the original bit pattern. A frequency-sensitive circuit may be used to produce an amplitude variation proportional to the instantaneous frequency. Alternatively, pulses may be generated corresponding to each zero crossing and these pulses passed through a low-pass filter to produce a wave with an amplitude variation equivalent to the bit pattern transmitted. This is illustrated in Fig. 11.7.

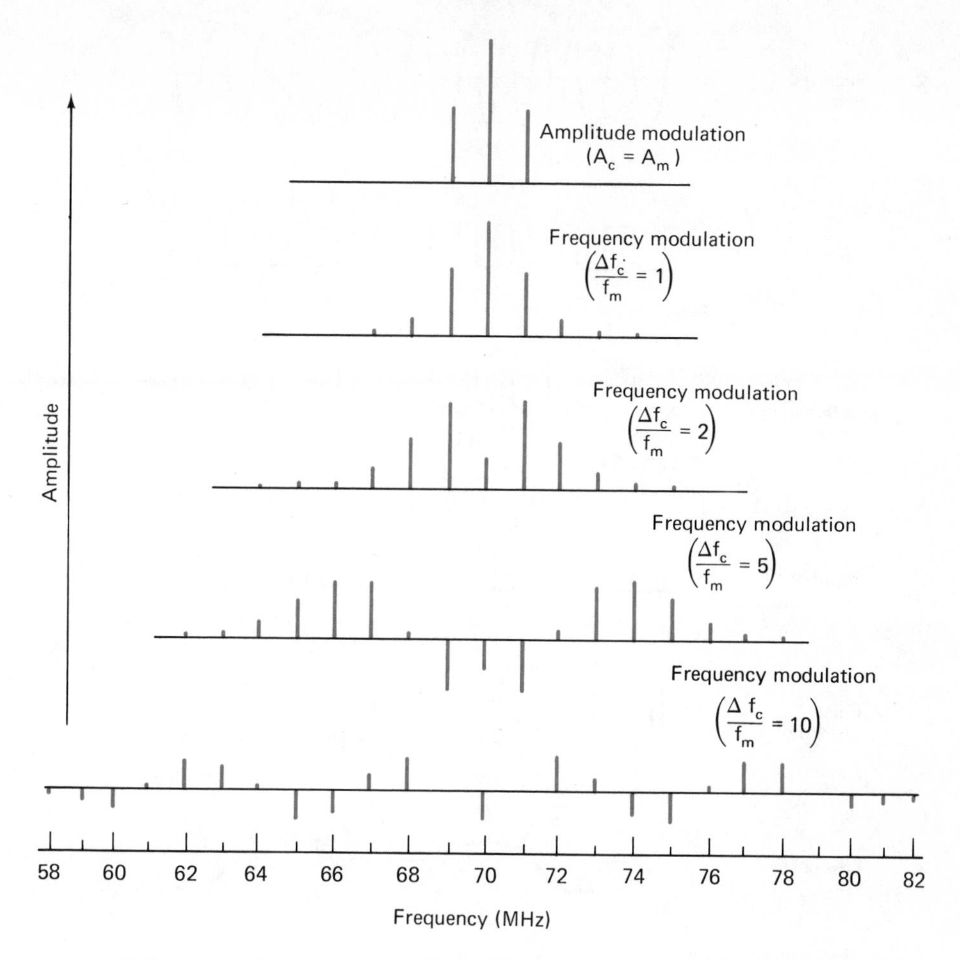

Figure 11.6 Spectra resulting from the modulation of a carrier of frequency 70 MHz with a sine wave of frequency 1 MHz. Frequency modulation can spread the energy across an entire transponder bandwidth.

PHASE MODULATION Phase modulation is commonly used for data transmission, mainly because it tends to be resistant to the effects of noise. The phase of the carrier is varied in accordance with the data to be sent. A sudden phase change of +180° cannot be differentiated from a change of −180°. Therefore the maximum range over which the phase can be varied is ±180°. As small changes in phase cannot be transmitted and detected with accuracy, phase modulation is not normally used for the transmission of

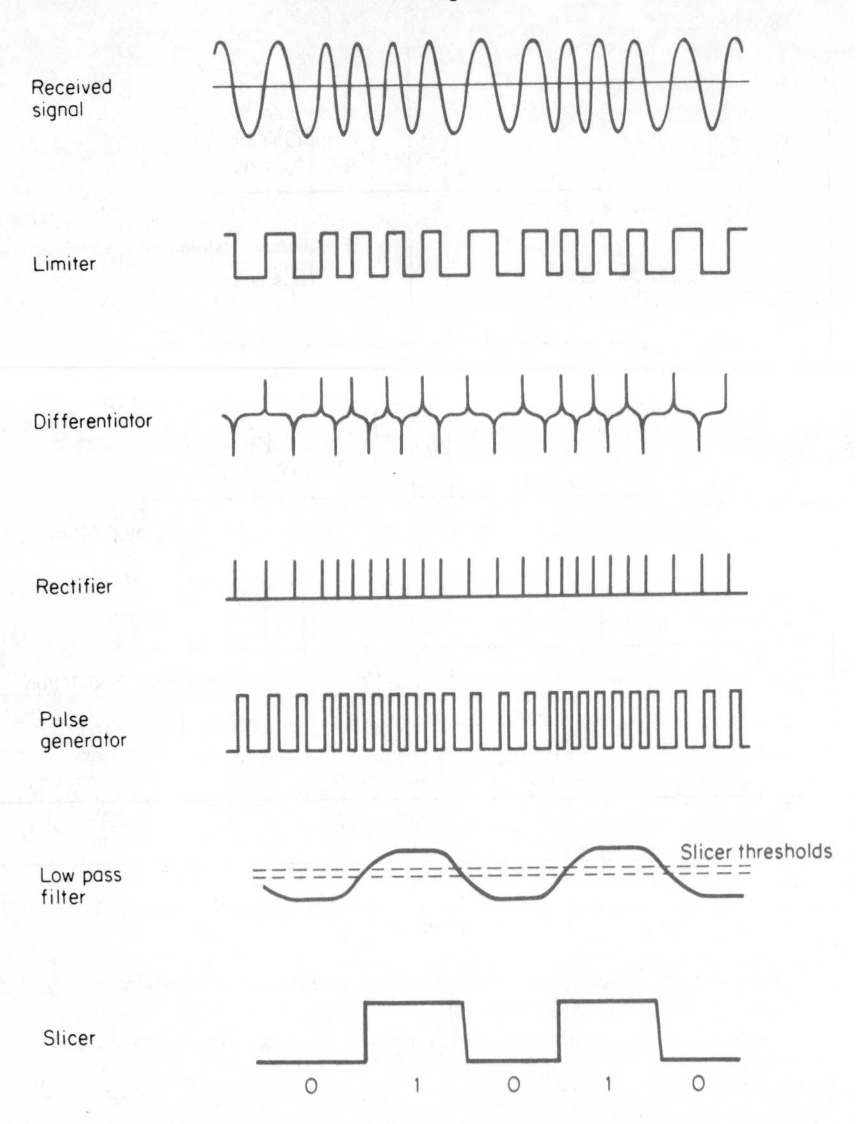

Figure 11.7 Detection of an FM signal.

analog signals, for which frequency and amplitude modulation are commonly used. The small range of variations can be used, however, to code the 2 bits of binary transmission, or 4 bits, or 8 bits, when multiple-level codes are used. Generally the highest data rates that can be transmitted over a given channel are obtained with phase modulation. Modems are in use which transmit 60 million bits per second over a 36-MHz transponder.

PM DETECTION

There are two basically different methods of detection in phase-modulated systems: *fixed-reference detection* and *differential detection.*

The receiver has no absolute sense of phase. It is therefore necessary either to use the signal in some way to generate information about the phases at the source or else to manage without it and operate by examining the *changes* in phase that occur.

The former approach needs a *fixed reference* giving the source phase. There are a number of ingenious methods of obtaining the reference phase from the carrier frequency. Alternatively, a separate tone may be sent: a very narrow band outside the data band, harmonically related to the carrier frequency so that it may contain information about the phase of the latter. Again, a phase reference may be sent in bursts at intervals in the transmission. As we shall see, satellite signals are often organized into bursts, separately sent, and checked. Each of these must be preceded by a signal which enables the carrier reference to be recovered.

Differential detection does not attempt to generate a fixed-reference phase at the receiver. Instead, the data are coded by means of *changes* in phase. Thus in two-phase transmission a 1 bit may be coded as a +90° change in the phase of the signal and a 0 bit as a −90° change. The detector now merely looks for changes in phase and does not need a reference-phase signal. There is no need to have the coding start at any specific phase. If the phase of the signal slips or drifts because of interference, the system recovers without aid.

To carry out the detection, the signal received is delayed one symbol interval and compared with the signal then being received. This comparison indicates the phase change that has occurred between the symbol intervals. The phase change detected is then converted into bits.

MULTIPLE-LEVEL DATA TRANSMISSION

In Figure 11.1, one of two states is encoded at each instant in time. The signal is represented by one of two amplitude levels, frequencies, or phase changes. More than 1 bit can be represented at one instant, using multiple amplitudes, frequencies, or phase changes. In Figure 11.9, 2 bits are represented at one instant. To do so one needs four alternative states that can be detected. Figure 11.9 shows four possible amplitudes and four possible phase changes. Three bits could be represented with 8 states or N bits with 2^n states.

In Figure 11.9 the bits are grouped into *dibits* 00, 01, 10, or 11. This doubles the possible transmission speed but makes the transmission more susceptible to noise or distortion. In general, the larger the number of states repre-

BOX 11.3 Phase modulation

The sine-wave carrier is represented by $a_c = A_c \sin(2\pi f_c t + \theta_c)$. A sine-wave component of the signal to be carried is represented by $a_m = A_m \sin 2\pi f_m t$.

Phase modulation results in θ_c being changed as follows:

$$a_{cm} = A_c \sin(2\pi f_c t + \Delta\theta_m \sin 2f_m t)$$

where $\Delta\theta_m$ is the maximum change in phase and is here called the *modulation index*.

The instantaneous frequency of the wave is $(1/2\pi) \times$ (the rate at which its angle is changing at that instant), in this case

$$\frac{1}{2\pi} \times \frac{d}{dt}(2\pi f_c t + \Delta\theta_m \sin 2\pi f_m t) = f_c + f_m \,\Delta\theta_m \cos 2\pi f_m t$$

Thus the instantaneous frequency is $f_c + f_m \,\Delta\theta_m \cos 2\pi f_m t$. This is equivalent to frequency modulation of the carrier frequency f_m by a wave of frequency f_m.

Δf, the maximum frequency deviation, is $f_m \,\Delta\theta_m$.

Phase modulation is thus equivalent to frequency modulation

sented at one instant, the higher the transmission speed but the lower the margin for error in the detection equipment.

Quaternary phase shift keying, using four possible phase changes to represent 2 bits, is commonly employed in the modems for satellite data transmission.

A useful way to quote the effectiveness of a given digital modulation technique over a given channel is to quote $\left(\frac{\text{transmission rate in bits per second}}{\text{channel bandwidth in hertz}}\right)$. This ratio is sometimes referred to as *the bandwidth expansion factor.*

In Chapter 8, we quoted Shannon's law,

$$c = w \log_2\left(\frac{P_r}{P_n} + 1\right) \tag{8.1}$$

with a modulation index $(f_m \Delta\theta_m/f_m) = \Delta\theta_m$. (This holds only when the modulation is sinusoidal.)

Thus again the resulting wave will contain an infinite number of side bands spaced at intervals equal to the modulating frequency, i.e., side bands at frequencies $f_c \pm f_m$, $f_c \pm 2f_m$, $f_c \pm 3f_m$, and so on.

It can be shown that the spectrum is as follows:

$$\begin{aligned} A_{cm} = {} & A_c J_0(\Delta\theta_m) \sin 2\pi f_c t \\ & + A_c J_1(\Delta\theta_m)[\sin 2\pi(f_c + f_m)t - \sin 2\pi(f_c - f_m)t] \\ & + A_c J_2(\Delta\theta_m)[\sin 2\pi(f_c + 2f_m)t + \sin 2\pi(f_c - 2f_m)t] \\ & + A_c J_3(\Delta\theta_m)[\sin 2\pi(f_c + 3f_m)t + \sin 2\pi(f_c - 3f_m)t], \\ & + \ldots \text{etc.} \end{aligned}$$

where $J_0(\Delta\theta_m)$, etc., are Bessel functions. These are illustrated in Fig. 11.8.

The spectra resulting from phase-modulating a carrier with a sine wave would thus be similar to those for frequency modulation in Fig. 11.6. Because of the greater modulation index possible with frequency modulation, it is possible to spread the data content over a greater bandwidth than with phase modulation.

where c = theoretical maximum channel capacity in bits
w = channel bandwidth
P_r = received power
P_n = received noise

In other words,

$$R \leqslant W \log_2 \left(\frac{P_r}{P_n} + 1\right)$$

where R is the transmission rate in bits per second.

$$P_r = E_b R$$

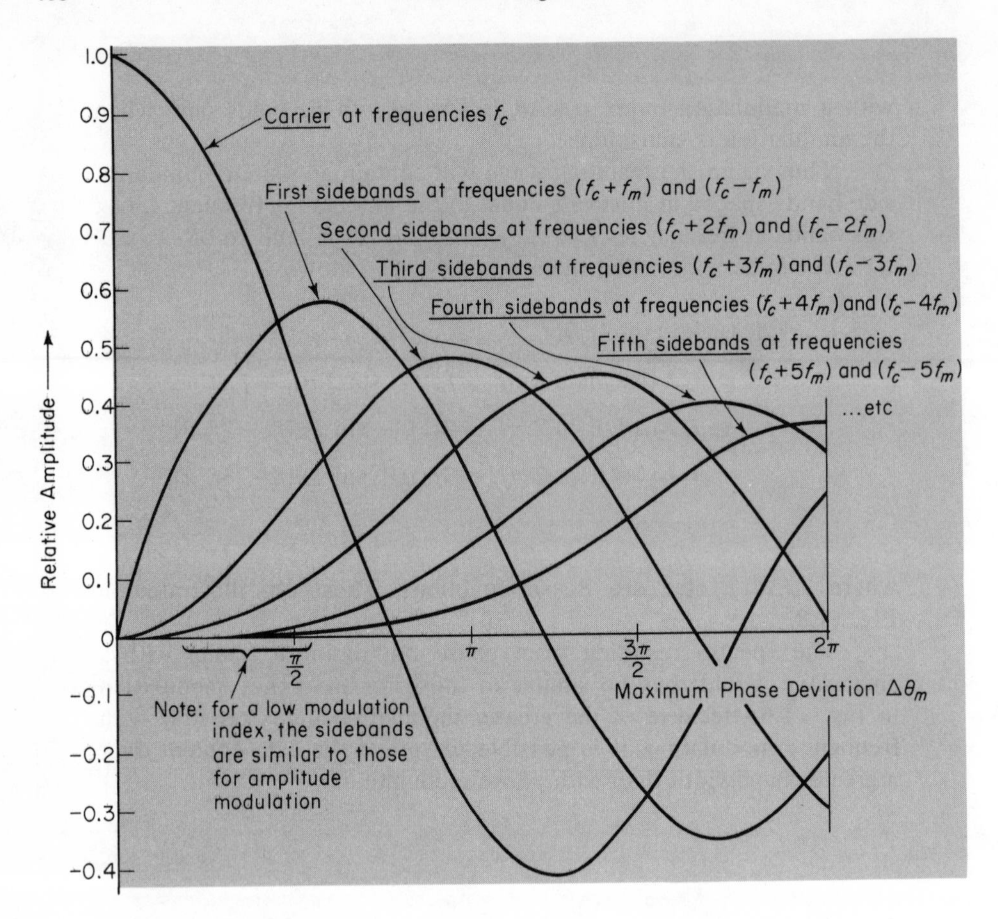

Figure 11.8 Spectral components of a carrier of frequency f_c, phase modulated by a sine wave of frequency f_m.

where E_b is the received energy per bit.

$$P_n = N_0 W$$

where N_0 is the noise density. Hence,

$$R \leqslant W \log_2 \left(\frac{E_b R}{N_0 W} + 1\right)$$

$$2^{(R/W)} \leqslant \frac{E_b R}{N_0 W} + 1$$

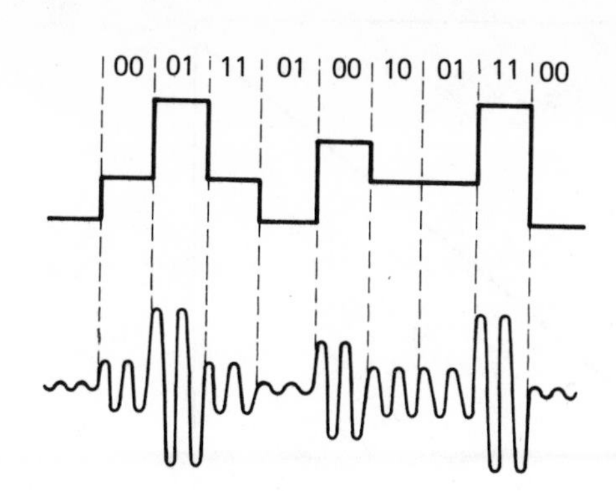

Quaternary amplitude modulation

Bit pair	Phase change
00	−135°
01	− 45°
10	+ 45°
11	+135°

Quaternary phase shift keying (QPSK)

Figure 11.9 At one instant in time more than one bit can be encoded.

$$\frac{E_b}{N_0} \geqslant \left(\frac{2^{(R/W)} \mp 1}{\text{R/W}}\right) \qquad (11.2)$$

This equation gives the minimum possible energy per bit per noise density that is required to reliably communicate with a bandwidth expansion factor R/W.

$$\frac{E_b}{N_0} \text{ (in decibels)} \geqslant 10 \log_{10} \left(\frac{2^{(R/W)} - 1}{\text{R/W}}\right) \qquad (11.3)$$

This is plotted in Fig. 11.10.

The probability that a bit will be received in error decreases as $\frac{E_b}{N_0}$ increases. Figure 11.11 shows the bit error probabilities for different types of modulation.

As we shall indicate in Chapter 17, satellite links are operated with a much worse value of $\frac{E_b}{N_0}$ than terrestrial links. *Consequently error-correcting codes are used in conjunction with satellite modems.* The improvement in bit errors that results from such codes is shown in Fig. 17.10.

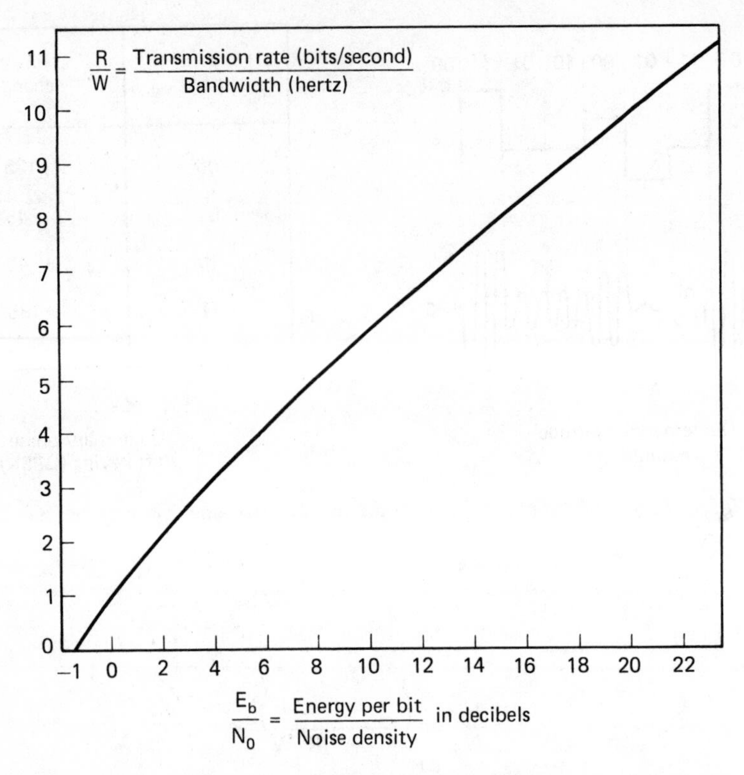

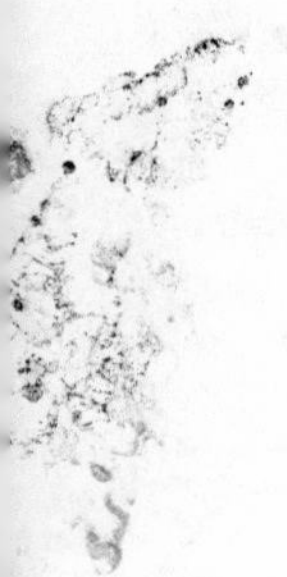

Figure 11.10 Maximum theoretical bandwidth expansion factor, R/W, plotted as a function of E_b/N_o, from Shannon's Law.

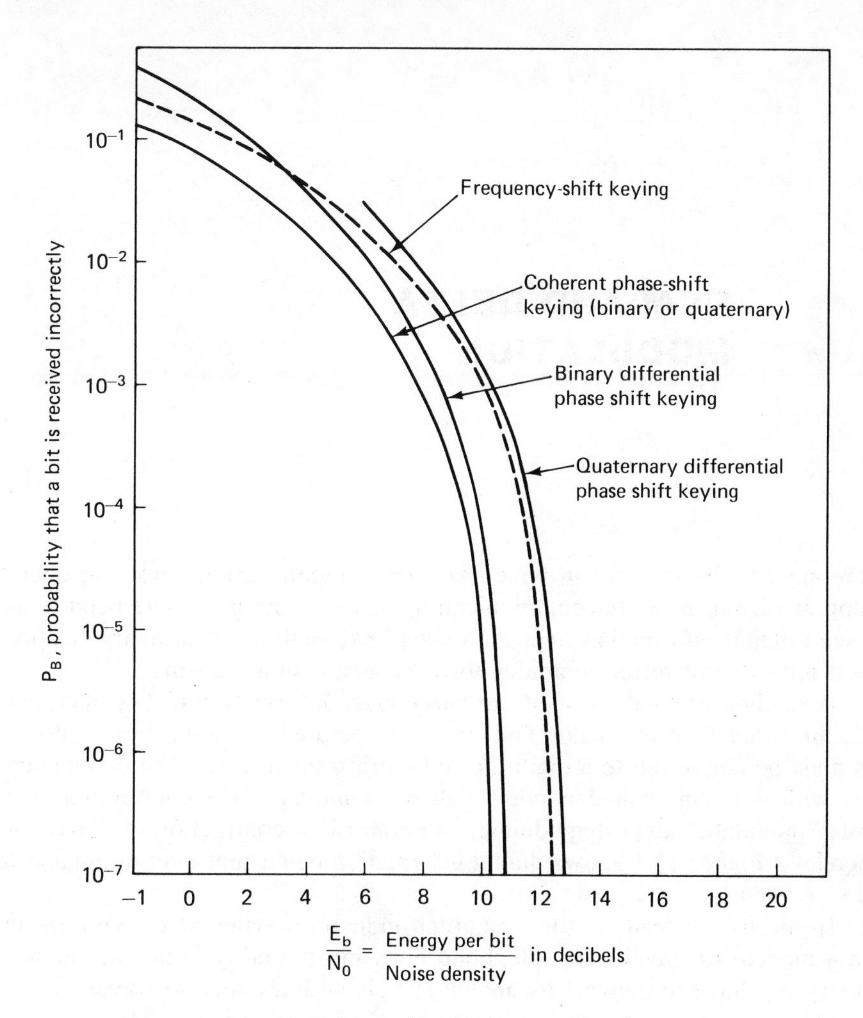

Figure 11.11 Variation in bit error probability with E_b/N_o.

12 PCM AND DELTA MODULATION

There are two basic ways in which any telecommunications link can operate: analog or digital. Most telecommunications links of the past have been analog. To send digital information over an analog link, such as an ordinary telephone line, it must be converted to analog form by means of a modem.

A satellite link can be made to carry more information if it is operated in a *digital* rather than an analog fashion. If it operated digitally, then analog signals must be converted to a digital form before transmission. The conversion is done with a circuit called a *codec*. (Just as *modem* is a contraction of the words "modulate" and "demodulate," so *codec* is a contraction of "code" and "decode".) Figure 12.1 shows digital information being sent over an analog line and vice versa.

Ironically, instead of the computer industry having to convert its data with a modem to travel over telephone lines in an analog form, the telephone industry will have to convert its analog signals with a codec to travel over digital lines.

Much ingenuity has been expended on making modems efficient so that they transmit a high bit rate. Now, much ingenuity is being spent on codecs, so that analog signals such as telephone speech or television can be converted into a small bit rate.

PCM

To convert an analog signal such as speech into a pulse train, a circuit may sample it at periodic intervals. The simplest form of sampling produces pulses the amplitude of which is proportional to the amplitude of the signal at the sampling instant (see Fig. 12.2). This process is called *pulse amplitude modulation* or PAM.

Compare the PAM illustration in Fig. 12.2 with that for amplitude modulation of a sine-wave carrier in Fig. 11.2. Envelope detection can be used for

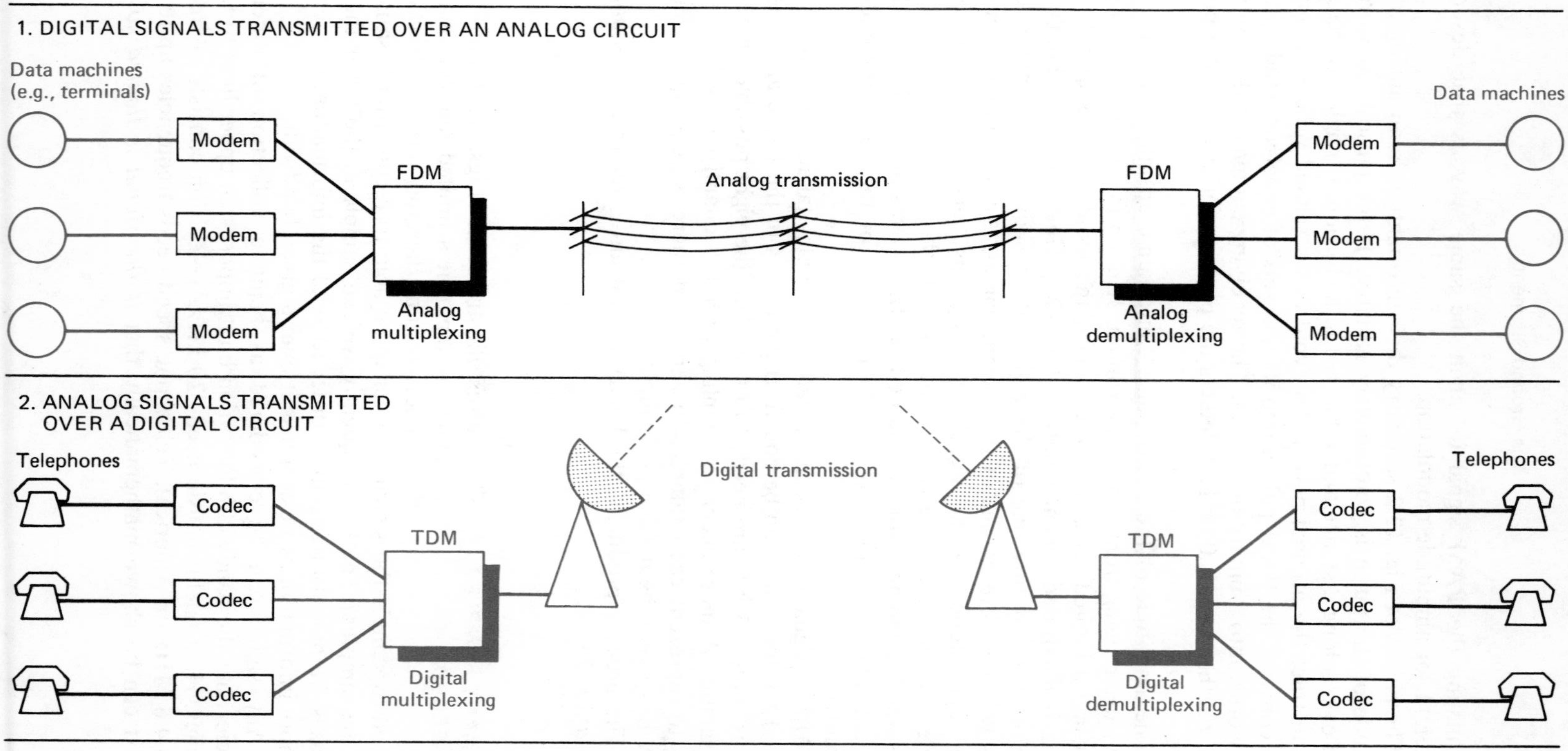

Figure 12.1 Modems and codecs.

demodulating the PAM signal in much the same way as that described in Chapter 11 for amplitude modulation.

The pulses of Fig. 12.2 still carry their information in an analog form; the amplitude of the pulse is continuously variable. If the pulse train were transmitted over a long distance and subjected to distortion, it might not be possible to reconstruct the original pulses. To avoid this a second process is employed which converts the PAM pulses into unique sets of equal-amplitude pulses—in other words, into a binary bit stream. The receiving equipment then only has to detect whether a bit is 0 or 1; it detects the presence or absence of a pulse, not its size.

The amplitude of the PAM pulse can assume an infinite number of possible values ranging from zero to maximum. It is normal with pulse modulation to transmit a limited set of discrete values; the input signal is *quantized.* This process is illustrated schematically in Fig. 12.3. Here the signal amplitude can be represented by any one of the eight values shown. The amplitude of the pulses will therefore be one of these eight values. An inaccuracy is introduced in the reproduction of the signal by doing this, analogous to the error introduced by rounding value in a computation. If there were more representable values, the "rounding error" would be less. In systems in actual use today, 128 pulse amplitudes are used, or 127 to be exact for the zero amplitude is not transmitted.

After a signal has been quantized and samples taken at specific points, as in Fig. 12.3, the result can be coded. If the pulses in the figure are coded in binary, as shown, 3 bits are needed to represent the eight possible amplitudes of each sample. A more accurate sampling with 128 quantized levels would need 7 bits to represent each sample. In general, if there were N quantized levels, $\log_2 N$ bits would be needed per sample.

The process producing the binary pulse train is referred to as *pulse code modulation,* PCM.

HOW MANY SAMPLES ARE NEEDED?

The pulses illustrated in Figs. 12.2 and 12.3 are sampling the input at a limited number of points in time. The question therefore arises, How often do we need to sample the signal in order to be able to reconstruct it satisfactorily from the samples? The less frequently we can sample it, the lower the number of pulses we have to transmit in order to send the information, or, conversely, the more information we can transmit over a given bandwidth.

Any signal can be considered as being a collection of different frequencies, but the bandwidth limitation on it imposes an upper limit to these frequencies. High-fidelity music needs 20 kHz. Television consists of frequencies up to 4.6 MHz (in America). Telephone speed needs frequencies up to 3 kHz.

It can be shown mathematically that *if the signal is limited so that the*

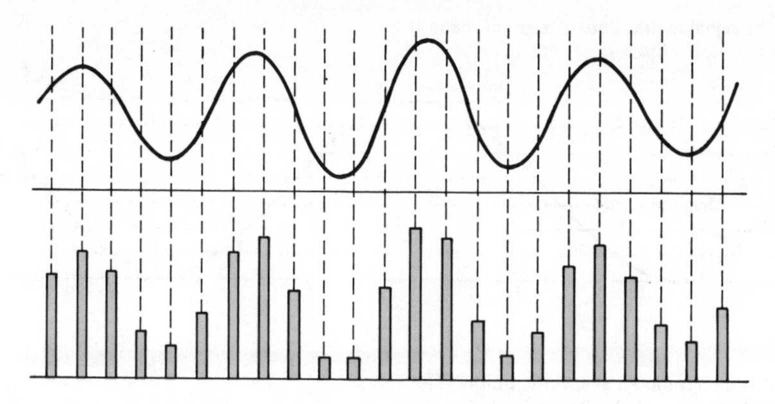

Figure 12.2 Pulse amplitude modulation (APM).

highest frequency it contains is W Hz, then a pulse train of 2W pulses per second is sufficient to carry it and allow it to be completely reconstructed.

The human voice, therefore, if limited to frequencies below 4000 Hz, can be carried by a pulse train of 8000 PAM pulses per second. The original voice sounds, below 4000 Hz, can then be *completely* reconstructed.

Similarly, 40,000 samples per second could carry monaural hi-fi music and allow complete reproduction. (If samples themselves were digitized, as with PCM, the reproduction would not be quite perfect because of the quantizing error.)

Table 12.1 shows the bandwidth needed for four types of signals for human perception, plus the digital bit rate used or planned for their transmission with PCM.

In telephone transmission, the frequency range encoded in PCM is somewhat less than 200 to 3500 Hz. 8000 samples per second are used. Each sample is digitized using 7 bits so that $2^7 = 128$ different volume levels can be

Table 12.1 Bandwidths and equivalent PCM bit rate for typical signals

Type of Signal	Analog Bandwidth Used (kHz)	Number of Bits Per Sample	Digital Bit Rate Used or Needed (1000 bps)
Telephone voice	4	7	$4 \times 2 \times 7 = 56$
High-fidelity music	20	10	$20 \times 2 \times 10 = 400$
Picturephone	1000	3	$1000 \times 2 \times 3 = 6000$
Color television	4600	10	$4600 \times 2 \times 10 = 92{,}000$

Note: More complex forms of digitization achieve much lower bit rates.

1 The signal is first "quantized" or made to occupy a discrete set of values

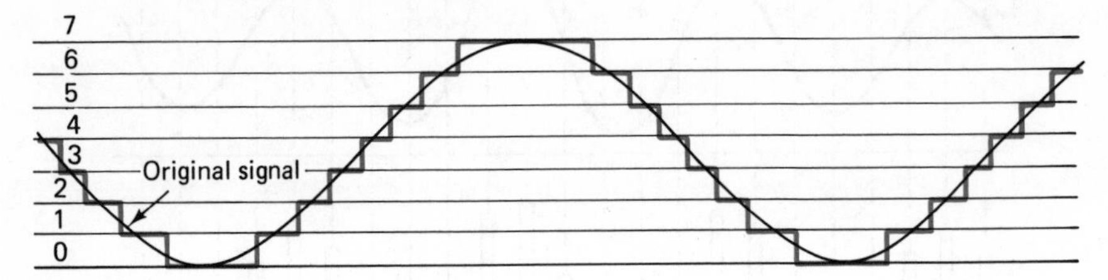

2 It is then sampled at specific points. The PAM signal that results can be coded for pulse code transmission

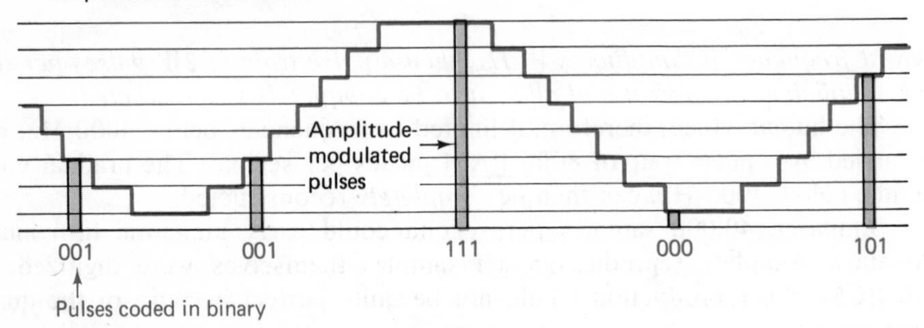

3 The coded pulse is transmitted in a binary form

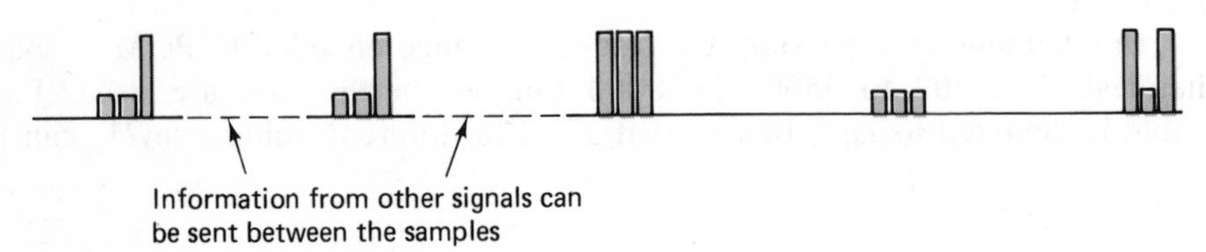

Figure 12.3 Pulse code modulation (PCM).

distinguished. This gives $7 \times 8000 = 56{,}000$ bits per second. High-fidelity music with five times this frequency range would need five times as many samples per second and, to achieve subtle reproduction, more bits per sample.

THE T1 CARRIER

Both terrestrial and space transmission facilities carry bit streams much greater than the 56,000 bits per second necessary for PCM speech. Hence time-division multiplexing is used to interleave the samples.

The most widely used PCM transmission system at present is the *Bell System T1 carrier.* This system was designed to make more efficient use of the vast quantity of terrestrial wire-pair circuits that exist. The wire pairs are fitted with digital repeaters (replacing their previous analog repeaters and load coils). The repeaters, every 6000 feet on the line, enable the line to transmit 1.544 million bits per second. Into this bit stream 24 voice channels are encoded, using PCM and time-division multiplexing.

The T1 carrier is used for short-haul transmission over distances of up to 50 miles. It has been highly successful, and more than 70 million voice channel miles of it are in operation. Most readers of this book have talked over a digital telephone line without knowing it. The data rate of 1.544 million bits per second has become a standard in North America and is used in some satellite systems for carrying speech.

The Bell T1 PCM System multiplexes together 24 voice channels using 7 bits for coding each sample. The system is designed to transmit voice frequencies up to 4 kHz, and therefore 8000 samples per second are needed; 8000 frames per second travel down the line. Each frame, then, takes 125 microseconds. A frame is illustrated in Fig. 12.4. It contains 8 bits for each channel. The eighth forms a bit stream for each speech channel which contains network signaling and routing information, for example, to establish a connection and to terminate a call. There are a total of 193 bits in each frame, giving $193 \times 8000 = 1.544$ million bits per second.

The last bit in the frame, the 193rd bit, is used for establishing and maintaining synchronization. The sequence of these 193 bits from separate frames is examined by the logic of the receiving terminal. If this sequence does not follow a given coded pattern, then the terminal detects that synchronization has been lost. If synchronization does slip, then the bits examined will in fact be bits from the channels—probably speech bits—and will not exhibit the required pattern. The synchronization pattern must therefore be chosen so that it is unlikely that it will occur by chance. An alternating bit pattern, 0 1 0 1 0 1 . . . never occurs for long in any bit position. Such a pattern would imply a 4-kHz component in the signal, and the input filters used would not pass this. Therefore the 193rd bit transmitted is made alternately a 1 and a 0. The receiving terminal inspects it to ensure that this 1 0 1 0 1 0 . . . pattern is present. If it is not, then it examines the other bit positions that are 193 bits apart until a 1 0 1 0 1 0 . . . pattern is found. It then assumes that these are the framing pulses.

This ingenious scheme works very well with speech transmission. If synchronization is lost, the framing circuit takes 0.4 to 6 milliseconds to detect the fact. The time required to reframe will be about 50 milliseconds at worst if all the other 192 positions are examined, but normally the time will be much less, depending on how far out of synchronization it is. This is quite acceptable on a speech channel. It is more of a nuisance when data are sent over the channel and would necessitate the retransmission of blocks of data. Retransmission is

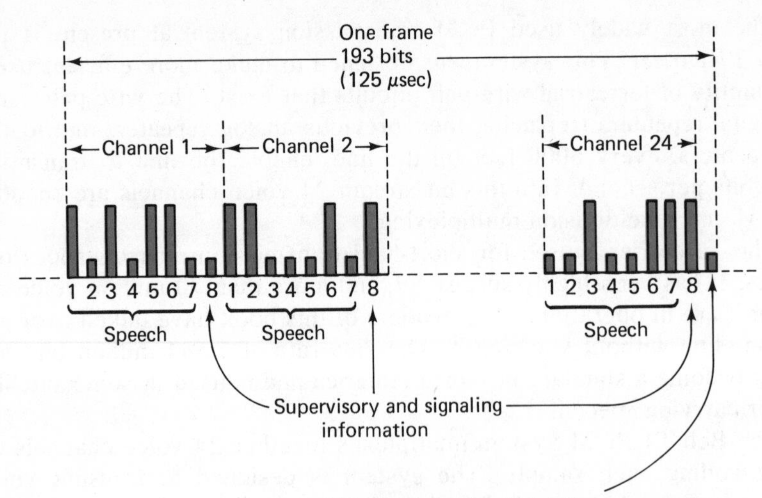

Figure 12.4 The bit structure of a North American T1 transmission link operating at 1.544 million bits per second. The above frame is repeated 8000 times per second, thus giving 8000 samples per second on each channel plus an 8000 bps bit stream for control signaling. The CCITT Recommendation for 1.544 bps PCM is slightly different [2].

required on most data transmission, however, as a means of correcting errors that are caused by noise on the line and detected with error-detecting codes.

INTERNATIONAL STANDARDS

International standards exist for PCM transmission. These standards are important because they allow connection between different networks and allow different manufacturers to make compatible equipment. International standards are established by *the CCITT (Comité Consultatif International Télégraphique et Téléphonique)* in Geneva.

The CCITT has made two recommendations for PCM transmission, one for transmission at the T1 carrier speed of 1.544 million bits per second (bps) [1] and one for transmission at 2.045 million bps [2], which can be achieved over most telephone wire pairs.

As is often the case, the CCITT recommendation for 1.544 million-bps transmission is slightly different from the North American standard. Conversion is necessary on international satellite channels. The CCITT scheme employs a 193-bit frame with 8 bits per channel as in Fig. 12.4, but the frame alignment bit is the first bit, not the 193rd bit as in Fig. 12.4, and it carries a different synchronization pattern.

If separate signaling is provided for each channel, *two* signaling bit streams are derived from the eighth bits, and only every sixth frame contains signaling bits. This gives a smaller bit rate for signaling but leaves *8 bits per channel* in five-sixths of the frames for carrying speech or information.

Figure 12.5 shows the CCITT 2.048 million-bps recommendation, which most of the world outside North America is starting to use for PCM transmission. In this, 16 frames of 256 bits each form a multiframe. There are 32 8-bit time slots in each frame, giving 30 speech channels of 64,000 bps each, plus one synchronization and alarm channel and one signaling channel which is submultiplexed to give four 500-bps signaling channels for each speech channel.

DIFFERENTIAL PCM

Although PCM is by far the most widely used technique for digital telephony, there are other methods which require fewer bits per second. To encode speech in fewer bits, more complex encoding methods are needed. On satellite links the cost of increased coding complexity is worthwhile.

Sampling the signal 8000 times per second can reproduce any frequency up to 4000 Hz. However, most of the energy in speech is at frequencies well below 4000 Hz; most of it is below 1000 Hz. The signal does not change too fast. Consequently, rather than encoding the absolute value of each sample, the *difference* in value between a sample and the previous sample may be encoded. This technique is called *differential pulse code modulation (DPCM)*. Fewer bits are needed per sample.

In practice, differential PCM encodes the difference between the amplitude of the current sample and a *predicted* amplitude, estimated from past samples. A DPCM circuit typically employs the last three speech samples to make a guess of what the next sample will be. The error in the guess can typically be encoded in 5 bits rather than the 7 bits of conventional PCM. Hence $5 \times 8000 = 40{,}000$ bps are needed instead of the 56,000 of conventional PCM.

A lower bit rate could be achieved if the guess were better. Elaborate schemes have been devised for improving the prediction based on measured characteristics of the speech. Different voices, for example, have a different pitch—the fundamental frequency of vibration of the vocal chords. The shape of the voice waveform tends to repeat itself at the pitch-frequency interval. The waveform which tends to repeat at this interval depends on the shape and position of the palate, tongue, and other speech articulators. A delay circuit with a delay equal to the above interval is used in some predictive circuits. The delay and other parameters which affect the prediction are updated continually. This form of encoding is known as *adaptive predictive encoding*, or *differential PCM with an adaptive predictor*. Bit rates of 20,000 have been used with experimental adaptive circuits.

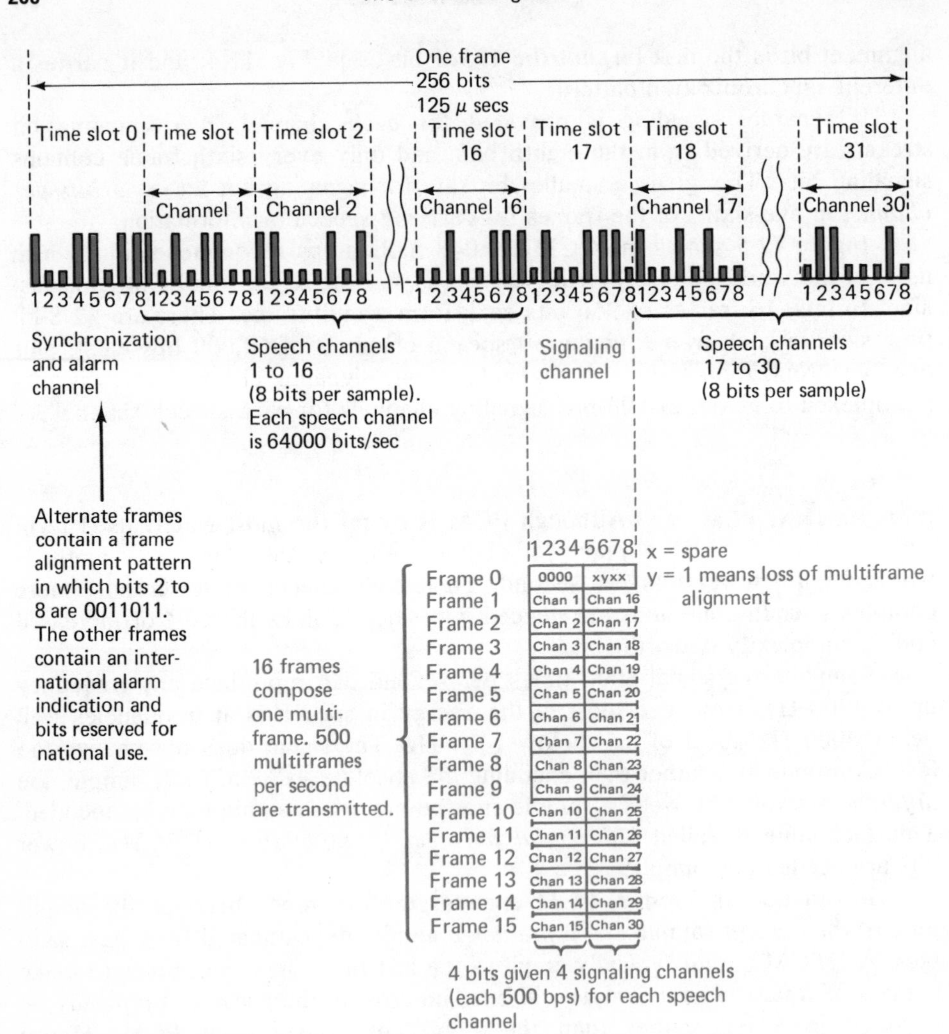

Figure 12.5 CCITT Recommendation for the structure of PCM channels for transmission at 2.048 million bits/sec. [*3*]. 30 speech channels of 64,000 bps are derived, each with a signaling channel of 500 bps.

DELTA MODULATION A technique called *delta modulation* also encodes signal *differences* but uses only 1 bit for each sample. The encoding indicates whether the waveform amplitude increases or decreases at the sampling instant. This binary sampling is illustrated in Fig. 12.6.

The number of pulses needed for this form of encoding depends on the rate of change of the signal amplitude. If the peak amplitudes are of low fre-

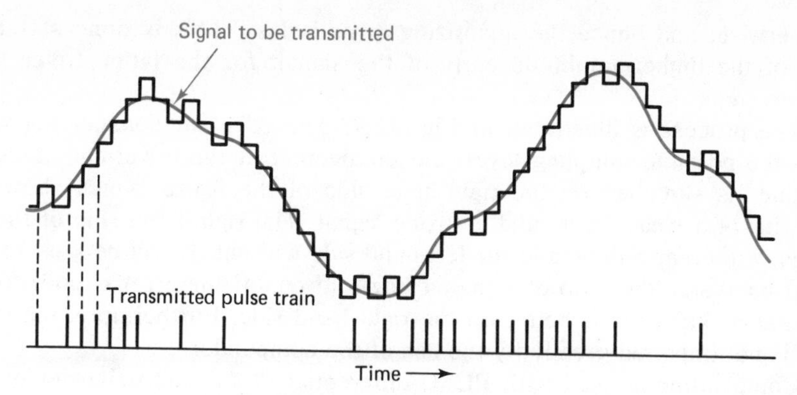

Figure 12.6 Delta modulation.

quency and if the high-frequency components are of low amplitude, fewer bits will be needed for the encoding than if all frequency components are of the same amplitude. This is the case with speech. It is also the case with television and videophone signals. In a 1-MHz Picturephone signal, most of the energy is concentrated below 50 kHz.

Overloading with this type of modulation will come not from too great a signal amplitude but from too great a rate of change. The encoding on the right-hand side of Fig. 12.6 is barely keeping up with the signal change.

It is inefficient to use only 1 bit per sample, and another variation on this scheme uses 3 or 4 bits per sample, which permits 8 and 16 gradations of signal change to be recorded each sample time, and then uses fewer samples. The levels are not linearly spaced.

Just as differential PCM can be made *adaptive,* so adaptive techniques can be used for delta modulation. Relatively inexpensive delta modulation circuits are in use which encode speech with good quality into 24,000 bps.

COMPANDORS

Quantizing noise is heard as raspy quality in the speech received. The quality of the speech can be improved, without increasing the requisite bit rate, by increasing the number of sampling levels for those amplitude values which occur most often and decreasing the number for those values which occur least often. In other words, there is a variable spacing between the sampling levels, with the levels closer together where they are used the most often.

Small amplitude values of speech occur more often than large ones. A scheme called *companding* therefore increases the spacing of the quantizing levels for the stronger signal values.

A *compandor* is a device that, in effect, compresses the higher-amplitude parts of a signal before modulation and expands them back to normal again after demodulation. Preferential treatment is therefore given to the weaker parts of a signal. The weaker signals traverse more quantum steps than they would do otherwise, and hence the quantizing error is less. This is done at the expense of the higher-amplitude parts of the signal, for the latter cover fewer quantum steps.

The process is illustrated in Fig. 12.7. The effect of companding which moves the possible sampling levels closer together at the lower-amplitude signal values is sketched on the right-hand side of the figure, which shows the quantizing of a weak signal and a strong signal. The right-hand side of the diagram is with companding, and the left-hand side without. It will be seen that on the left-hand side the ratio of signal strength to quantizing error is poor for the weak signal. The ratio is better on the right-hand side. Furthermore, the strong signal is not impaired greatly by the use of the compandor.

Companding is used with PCM, differential PCM, and delta modulation. CCITT standards exist for the spacing of the quantizing levels of compandors [3].

TASI

A technique has long been used on suboceanic cables for doubling their capacity by continually reassigning the channels to telephone speakers. It is called *TASI (time-assignment speech interpolation)*.

On a link carrying a conversation, both parties do not normally speak at once, and for a small proportion of the total connection time (usually about

Figure 12.7 With a compandor the quantization of the weak signal gives more separate values, and therefore a better ratio of signal to quantizing noise. Standard sets of companding rules are specified in the CCITT Recommendation G.711 [1].

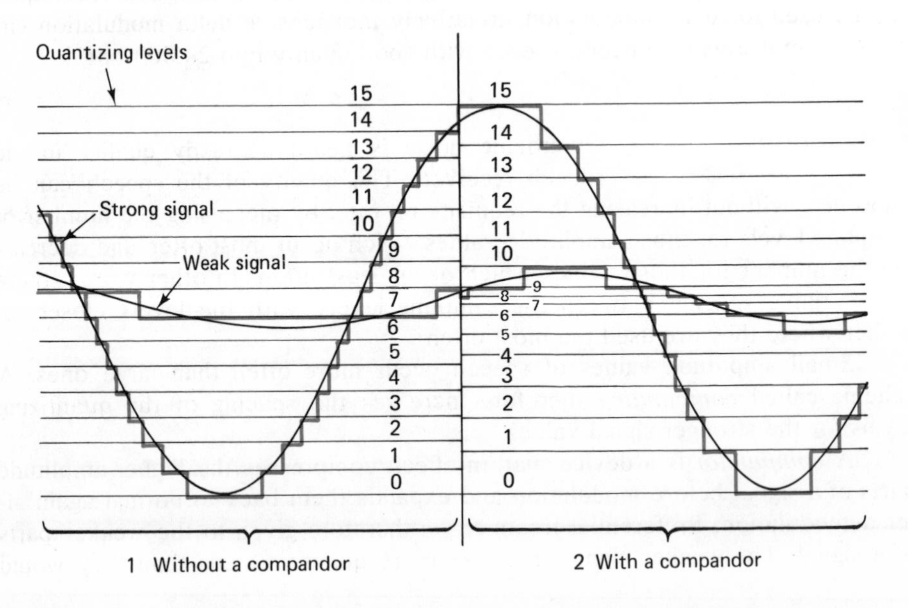

10%) nobody is speaking. The long-distance line is normally a "four-wire" circuit, meaning that there is a transmission path in both directions, so the path in one direction is in use on average only about 45% of the total time. In other words, for 100 talkers, only about 45 on average will be speaking simultaneously. There is a spread about this average, and so it is necessary to provide more than 45 one-way channels. However, if a large number of voice channels are carried together, statistics indicate that the ratio of talkers to one-way channels required becomes close to 1/0.45.

The TASI equipment is designed to detect a user's speech and assign him a channel in milliseconds after he begins to speak. An almost undetectable amount of his first syllable is lost. He retains the channel until he stops speaking. It is then taken away from him so that it can be allocated to another speaker if necessary.

DSI

With digitized speech, a digital form of TASI is used, called DSI (digital speech interpolation). DSI is much faster, and more efficient, than its analog precursor.

Satellite circuits can handle more telephone calls than suboceanic cables, and they are handled digitally. DSI is effective on such circuits. As soon as a speaker pauses, the subchannel he used is made available for dynamic allocation to other speakers. The high-speed circuits detect speech almost immediately when it begins and allocate a subchannel to it. If there is no free subchannel at that instant, the speaker will usually have to wait only milliseconds before one of the other speakers pauses and *his* subchannel is reallocated. He will not detect the wait.

DSI has been found to increase the capacity of satellite channels by factors of 2.2 and higher, without degrading speech quality.

Some packet-switching systems have transmitted telephone speech in 1000-bit packets, and have found that this technique can increase channel capacity for speech by a factor of 3 or higher.

DIGITAL ECHO SUPPRESSION

It is extremely important that echoes be removed when telephone calls are transmitted by satellite. Because of the long propagation times, they have a more serious effect on speakers than with terrestrial channels. When speech is processed digitally the removal of echoes becomes a function of the speech-processing circuitry and can be done cleanly and efficiently.

DIGITAL USE OF A TRANSPONDER

Today's modems applied to the 36-MHz bandwidth of a typical transponder transmit 60 million bps. A telephone channel using today's delta modulation equipment requires 24,000 bits per channel in both directions. DSI more than doubles the number of telephone calls transmitted.

As we shall see later, some of the channels are used for control purposes. Even so, the transponder can handle over 2000 telephone calls. An INTELSAT IV transponder (of the same bandwidth) operated in an analog fashion handles a CCITT master group of 900 channels. WESTAR handles two American master groups totaling 1200 channels.

Much of the future of satellites lies in digital techniques.

REFERENCES

1. "CCITT Recommendation A.733" (on PCM multiplex equipment operating at 1544 KB/S), *CCITT Green Book,* Vol. III, Line Transmission, International Telecommunications Union, Geneva, 1973.
2. "CCITT Recommendation A.732" (on PCM multiplex equipment operating at 2048 KB/S), *CCITT Green Book,* Vol. III, Line Transmission, International Telecommunications Union, Geneva, 1973.
3. "CCITT Recommendation A.711" (on compandors), *CCITT Green Book,* Vol. III, Line Transmission, International Telecommunications Union, Geneva, 1973.
4. J. S. Mayo, "A Bipolar Repeater for Pulse Code Signals," *Bell System Tech. J.,* Jan. 1962.
5. J. S. Mayo, "Experimental 224 MG/S PCM Terminals," *Bell System Tech. J.,* Nov. 1965.
6. I. Dorros, J. M. Sipress, and F. D. Waldhauer, "An Experimental 224 MG/S Digital Repeatered Line," *Bell System Tech. J.,* Sept. 1966.
7. C. G. Davies, "An Experimental Pulse Code Modulation System for Short-Haul Trunks," *Bell System Tech. J.,* Jan. 1962.
8. M. R. Aaron, "PCM Transmission in the Exchange Plant," *Bell System Tech. J.,* Jan. 1962.
9. R. H. Shennum and J. R. Gray, "Performance Limitation of a Practical PCM Terminal," *Bell System Tech. J.,* Jan. 1962.
10. H. Mann, H. M. Straube, and C. P. Villars, "A Companded Coder for an Experimental PCM Terminal," *Bell System Tech. J.,* Jan. 1962.
11. K. E. Fultz and D. B. Penick, "The T1 Carrier System," *Bell System Tech. J.,* Sept. 1965.
12. J. F. Travis and R. E. Yeager, "Wideband Data on T1 Carrier," *Bell System Tech. J.,* Oct. 1965.

13 USES OF DIGITAL CHANNELS

Because satellite channels will be digital, with telephone traffic being carried in digitized form, they will be especially cost-effective for data transmission. The relative cost of data transmission and telephone transmission swings in favor of data. Data are transmitted on terrestrial telephone lines at between 1200 and 9600 bps. In space, a telephone channel will be equivalent to about 32,000 bps or so (depending on the codec design). The 60 million bps of existing satellite modems represent a vast amount of data traffic, which could be carried on any of many transponders.

Furthermore, telephone traffic has to be transmitted in *real time;* i.e., when a person speaks his speech must be transmitted almost immediately. To achieve a good grade of service, i.e., low probability of a caller encountering a *busy* signal, there have to be idle channels ready for immediate use. No real-time systems achieve 100% utilization of their facilities. Probability calculations determine how much idle capacity is needed to provide a given grade of service.

The transmission capacity must be designed for the peak telephone traffic. The traffic during the peak hour of the day is several times higher than the average traffic. The peak hour of the peak day is substantially higher than that of the average day. The unshaded part of Fig. 13.1 represents idle capacity. On a typical corporate network the channels are idle 80% of the total time. Such is the nature of real-time traffic.

Much of the information we transfer is not real-time. We are happy if it is delivered an hour, or a day, later. We write letters, leave messages, send telegrams, order catalogues, transmit batches of computer data, and request books from libraries. This information transmission has two important characteristics. First, it can wait until channels are not occupied with telephone or other real-time traffic. Second, it can be interrupted in the middle of transmission pro-

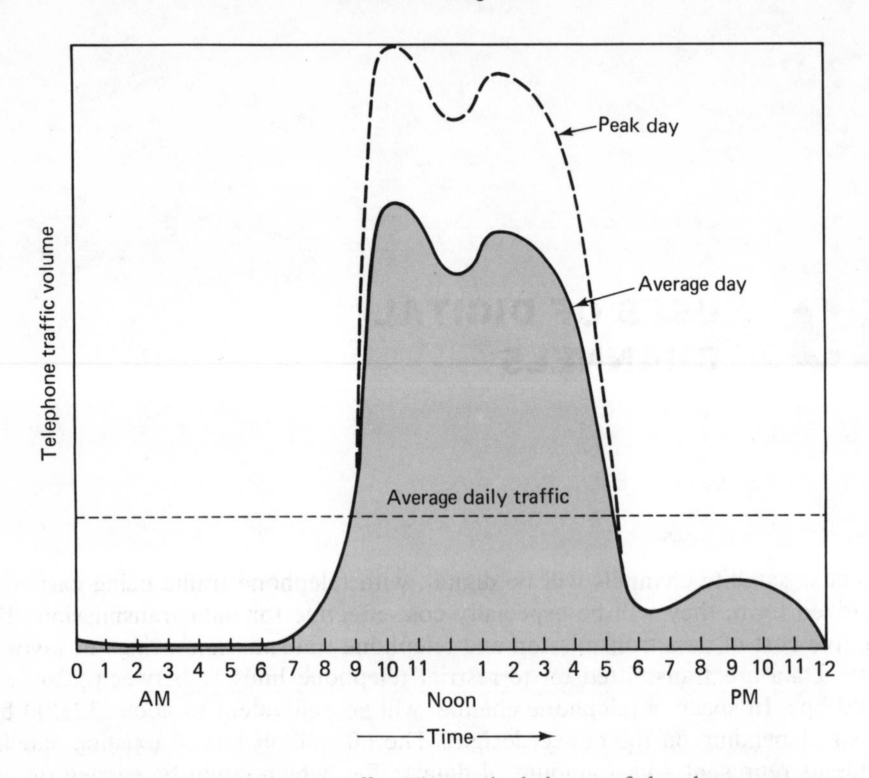

Figure 13.1 Telephone traffic. The unshaded part of the chart represents idle channel capacity. A *stutter port* can transmit non-real-time traffic in the idle capacity. It keeps trying to transmit non-real-time traffic at every instant when there is no real-time traffic.

vided that the control mechanisms and buffers are such that no information is lost.

The most efficient way to utilize communications channels is to organize them so that real-time and non-real-time traffic can be intermixed and so that real-time traffic has absolute priority over non-real-time traffic. Non-real-time traffic should never delay the real-time traffic more than a small fraction of a second – 50 milliseconds, say. Most telephone networks today do not carry interruptable non-real-time traffic, and consequently 75% to 85% of their total daily capacity is unutilized. When digital equipment is used for multiplexing voice signals a mechanism of that equipment can handle the intermixing of real-time and non-real-time traffic. This should be done on satellite channels. If the mechanism exists for doing it, a remarkably large capacity for non-real-time traffic is available. It is worth looking closely at society's non-real-time uses of information.

If a satellite were designed to carry a peak of 100,000 voice calls, with one voice call equivalent to 32,000 bps, the total capacity of the satellite would be $100{,}000 \times 32{,}000 \times 60 \times 24 \times 365 = 1.68 \times 10^{15}$ bits per year. If 80% of this capacity were unutilized by telephone traffic, as would probably be the case, then 1.35×10^{15} bits per year would be available for non-real-time traffic.

How can this prodigious capacity be used?

DIGITIZED MESSAGES

It is possible to convert any type of message into a digital form for transmission. Different messages require different numbers of bits. Box 13.1 gives some examples of satellite message lengths when converted to bits and compressed ready for transmission or storage. The figures are conservative in that a smaller number of bits can be used in some of the items if complex encoding techniques are employed. Telephone speech, for example, has been successfully encoded into far fewer than 32,000 bps with a level of distortion acceptable for one-way messages. A vocoder is a device which synthesizes speech from encoded parameters describing the speech-production mechanism and the sounds it is making. PCM sound can be synthesized from a much smaller bit stream. Vocoder telephone speech, listed in Box 13.1, can be designed to make the spoken words intelligible but not necessarily make the speaker's voice recognizable. Using vocoder techniques, spoken messages can be transmitted in one-tenth of the number of bits of speech using PCM or delta modulation.

Much more compaction can be achieved with still less natural sounding speech if the messages are composed of prerecorded words. A prerecorded vocabulary might have up to 1000 words, each addressed with 10 bits, thus permitting a 30-word message to be sent with 300 bits plus addressing and error-detection bits. After transmission, the 300 bits would trigger a spoken message from a voice response unit (many of which are in use today). A code book of words commonly used in business was employed for a similar purpose in the early days of telegraphy.

Varying degrees of compaction can be achieved with other types of message delivery. There is a trade-off between message length and encoding complexity.

A digital transmission system with the high bit rate of a satellite can be given the capability to transmit any of the message types in Box 13.1, the non-real-time messages fitting in gaps between the real-time traffic. The signals will be interweaved in the bit stream with whatever is the most appropriate form of multiplexing.

PRIORITIES

It is necessary to have some form of priority structure in the system. At its simplest there could be two priorities: *real-time* and *non-real-time*. There are, however, different degrees of urgency in the non-real-time traffic, so several priority levels may be used to help ensure a fast delivery of messages which require it. The system

BOX 13.1 Typical numbers of bits needed for different types of messages.

Message Type	Bits
1. A high-quality color photograph	2 million
2. A newspaper-quality photograph	100,000
3. A color television frame	1 million
4. A Picturephone frame	100,000
5. A brief telephone voice message	1 million
6. A vocoder telephone voice message	100,000
7. A voice message of code-book words	400
8. A document page in facsimile form	200,000
9. A document page in computer code	10,000
10. A typical interoffice memo	3,000
11. A typical telegram	2,000
12. A typical flip chart	1,000
13. A typical computer input transaction	500
14. A typical electronic fund transfer	500
15. A typical airline reservation	200
16. A coded request for a library document	200
17. A fire or burglar alarm signal	40

organization may be designed to permit the following categories of end-to-end delivery time:

1. Almost immediate (as with telephone speech).
2. A few seconds (as with interactive use of computers).
3. Several minutes.
4. Several hours.
5. Delivery the following morning.

When more than one message is waiting for transmission at any point, the higher-priority messages will be sent first. The fact that much of the traffic is not in the highest-priority category will make it possible to achieve a substantially higher line utilization than on a network which guarantees real-time transmission for all messages. In addition the facilities will be well utilized at night, when on other systems they would be largely idle.

A mechanism called a *shutter port* is used in some corporate satellite systems to transmit electronic mail and non-time-critical traffic during gaps in the telephone and real-time traffic.

TRAFFIC GROWTH

If transmission systems come into existence for transmitting non-real-time information at a low cost, there can be a major growth of such traffic, and the growth will probably incorporate traffic which is not sent electronically today.

Table 13.1 is a forecast from a NASA study showing demand trends for data record transmission. In addition to data records, there are other types of traffic of very high volume, for example, mail.

ELECTRONIC MAIL

The total cost of mail delivery is gigantic, especially in North America. Americans are not only the most communicative people by telephone; they also receive the most mail. More mail is sent in New York City than in the whole of Russia.

Some types of mail could be sent and delivered by electronic means, and where the volumes are high this could be done at a fraction of the cost of manual delivery. To send a handwritten letter electronically it is fed into a facsimile machine, transmitted, and received by another facsimile machine, which produces a copy. Most of today's facsimile machines transmit an analog signal over telephone lines. They can be designed to transmit a digital signal, and some digital facsimile machines are in use.

Table 13.2 breaks down the U.S. mail by type. The asterisks indicate which mail could be sent by electronic means and hence potentially by satellite channels. A single asterisk refers to mail which could be delivered electronically to the end user. It is assumed that individual households can neither send nor receive electronic mail. They have neither the equipment nor the desire to change their mail-sending habits. At some time in the future electronic mail will reach into consumers' homes, but we shall assume that for the time being only businesses and government will use it. When government and businesses send mail to households this mail could be delivered to the local post offices already sorted for delivery. All local post offices could have a receive-only satellite antenna on the roof (like the Musak antenna) and a high-speed facsimile printer. Advertising letters and promotional materials have not been included as potential electronic mail because they may contain glossy or high-quality reproductions. Some advertising letters could be sent by facsimile machines. Newspapers and magazines have not been included, although there has been much discussion of customized news sheets being electronically delivered to homes.

On this basis, 22.7% of all mail is potentially deliverable to end users via satellites, and 22.8% is potentially deliverable to post offices. In 1980 this will be a total of about 50 billion pieces of mail per year. If only a tenth of this were realized, it would pay for a large satellite system. The practicality of satellite mail would be aided by the fact that three-quarters of all U.S. mail originates in only 75 cities, and only 20.2% of all mail originates from individuals—the rest is from business and government.

Table 13.1 Demand trends for transmission of records (from a study about satellite uses commissioned by NASA [1])

		1950	1960	1970	1980	1990
Stolen vehicle information transfer	Cases/yr × 10^3	160	320	820	1950	4600
Facsimile transmission of "mug shots," finger prints, and court records	Cases/yr × 10^6	2	4	7	13	25
Stolen property information transfer	Cases/yr × 10^3	430	880	1700	3500	7000
Motor vehicle registration	Items/yr × 10^6	49	74	110	164	245
Driver's license renewal	Items/yr × 10^6	38	48	60	75	90
Remote library browsing	Accesses/yr × 10^6	0	0	Neg.	5	20
Remote title and abstract searches	Searches/yr × 10^6	0	0	Neg.	8	20
Interlibrary loans	Books/yr × 10^6	–	–	Neg.	40	100
Remote medical diagnosis	Cases/yr × 10^6	0	0	20	60	200
Remote medical browsing	Accesses/yr × 10^6	0	0	20	60	200
Electrocardiogram analysis	Cases/yr × 10^6	0	Neg.	20	60	200
Patent searches	Searches/yr × 10^6	6	6	6.5	7	7
Checks and credit transactions	Trans/yr × 10^9	11	25	56	135	340
Stock exchange quotations	Trans/yr × 10^9	0	0	1	2	4
Stock transfers	Trans/yr × 10^6	290	580	1200	2500	4900
Airline reservations	Pass/yr × 10^6	19	62	193	500	1400
Auto rental reservations	Reserv/yr × 10^6	0	Neg.	10	20	40
Hotel/motel reservations	Reserv/yr × 10^6	–	–	25	50	100
Entertainment reservations	Reserv/yr × 10^6	–	–	100	140	200
National Crime Information Center	Trans/yr × 10^6	0	0	6	20	70
National legal information center	Trans/yr × 10^6	0	0	Neg.	5	30

The 50 billion pieces of digitizable mail would require on average approximately 200,000 bits each to encode; some would need more than this; many would require less because alphanumeric encoding rather than facsimile would be used. The annual total would be roughly 50 billion × 200,000 = 10^{15} bits. This is less than the bits left over from a satellite handling 100,000 voice calls, discussed above. In other words, *the required channel capacity could be an unused by-product of a telephone satellite.*

ELECTRONIC FUND TRANSFER

Non-real-time digital transmission capacity will be required in the future for the transfer of funds between banks and other institutions. The Federal Reserve Board has made it clear that electronic fund transfer is essential for

Table 13.2 The composition of the U.S. mail, with an indication of which mail is potentially deliverable by satellite or other digital channels

Type of Mail		Percentage
Individual households to:		
Business	5.8	
Individual households	14.0	
Government	0.4	
TOTAL		20.2
Government to:		
Business*	1.8	
Individual households**	3.8	
Government*	0.6	
TOTAL		6.2
Business to business:		
To suppliers*	3.9	
Intracompany*	1.4	
To stockholders*	0.7	
To customers: order acknowledgement*	0.2	
bills*	6.7	
product distribution	1.3	
promotional materials	5.4	
Other*	6.2	
TOTAL	25.8	
Business to households:		
Letters:		
Bills*	10.1	
Transactions**	1.2	
Advertising	12.6	
Other**	4.5	
TOTAL LETTERS	28.5	
Postcards:		
Bills**	0.7	
Advertising**	2.1	
Other**	0.4	
TOTAL POSTCARDS	3.2	
Newspapers and magazines	13.6	
Parcels	1.3	
TOTAL BUSINESS TO HOUSEHOLDS	46.7	
Business to government:*	1.2	
TOTAL BUSINESS		73.6

*: Potentially deliverable by satellite to the end user (22.7%).
**: Potentially deliverable by satellite, sorted, to a post office (22.8%).
} (Shown by the red bands)

America, if only to halt the growing burden of paperwork such as check processing.

The international electronic fund transfer network, SWIFT, links banks in many countries. A few hundred participating banks will exchange money on the SWIFT network. It is forecast that SWIFT will handle 65 million financial transactions a year by 1980.

About 30 billion checks per year, representing $20 trillion per year, are written in the United States. An electronic fund transfer network could speed up the clearing time for checks by at least one day on average and probably more. This represents a float of $\frac{\$20 \text{ trillion}}{365} = \$54.8$ billion savable by electronic check transfer. At 8% interest this gives a saving of $4.38 billion per year.

If one check requires 500 bits for transmission, the total capacity needed is 30 billion $\times$ 500 = 1.5×10^{13} bits per year—a little more than 1% of the spare capacity in a 100,000-voice-channel satellite.

The number of credit transactions is almost double the number of checks, and the payment delay with these is much longer. In an electronic-fund-transfer society, today's credit cards would be replaced with similar-looking machine-readable cards. When a customer makes a payment in a store or restaurant a transaction would travel to a bank and a response would be received from the bank computer. If *all* credit card transactions in the United States were handled this way, 200 billion such messages would be needed, or about 10^{14} bits per year. Even if such funds are not transferred electronically a similar bit rate is needed for credit verification.

FASTER-THAN-MAIL TRAFFIC

Today, message delivery which is faster than mail costs substantially more than mail. As the cost of long-distance telephone calls has dropped, the number of telegrams sent has steadily declined. In many cases today it is cheaper to telephone than to send a telegram.

With satellites, however, the relative costs of transmitting telegrams and telephone calls change. A typical telephone call lasts 4 minutes and requires about 8 million bits (or four times this if simple PCM is used with no digital speech interpolation). A typical telegram requires 2000 bits. Furthermore, the telegram can wait to be fitted in gaps on the real-time traffic that would otherwise be unused.

Even with such a dramatic cost difference most corporate communications users would probably still use the telephone because of its convenience, its friendliness, and the cheapness of the instrument. If computerized PBXs place controls on corporate user's communications expenditures, then digital transmission may bring new life to telegraphy.

TELEPHONE MESSAGES

The total number of telephone calls in North America is far higher than the total number of written messages. AT&T alone plans a capital expenditure over the next 10 years that is more than 20 times higher than the likely capital expenditure in the U.S. Post Office. Telephone callers are often greeted with busy signals or no answer, and many of these callers would leave a brief message if they could. 32 long-distance calls out of 100 are uncompleted today [2], mostly because of busy signals or no answers. Of the business calls which are completed, on only 35 per cent does the caller reach the called party. It is estimated that this wastes 200,000 man-years of callers' time, which at $10,000 per year is equivalent to $2 billion [2].

A one-way telephone message could be digitized and stored so that the person it is sent to can retrieve it at his convenience. It could be transmitted and stored in any of the three forms mentioned earlier:

1. True-to-life: ~ 24,000 bps.
2. Vocoder: ~ 2400 bps.
3. Words from a prerecorded vocabulary: ~ 10 bits per word.

Such a service could be designed so that subscribers could leave either a coded telephone message from a list of such messages, dialed on a conventional telephone, or they could leave a brief spoken message. The system would ring the called party periodically until it could speak the message. The called party could use his telephone dial to ask for repetition of the message, give confirmation of its receipt, or dial a response. The system may be designed so that a user can dial his stored message queue from any telephone, key in a security code, and have the messages spoken to him.

If 10% of all business callers who failed to contact the individual they telephoned also left such a message, that would be more than a billion messages per year.

Telephone messages could be sent for which simple responses are required. The receiver would dial the response on his telephone after a local computer speaks to him, and the computer would receive the response and deliver it. The voice answerback unit would inform the called party what form of response was expected.

Organizations could send bulk messages in this way, in which one telephone message is sent to many individuals. Unsolicited messages could be composed by computers for verbal delivery, possibly expecting a response. One can imagine such a system being programmed to carry out opinion polls or gather statistics from individuals.

Whatever the applications or the form of the messages, it seems clear that a satellite system for corporate or government use should be designed to intermix all types of traffic, real-time and non-real-time, on digital channels.

REFERENCE

1. Roger W. Hough, "Information Transfer in 1990," a paper based on a NASA study, in *Communication Satellites for the 1970's: Systems,* Feldman and Kelley, eds., M.I.T. Press, Cambridge, Mass., 1971.

2. Harry Newton, *Communications Lines,* Business Communications Review, Sept., 1976.

14 MULTIPLE ACCESS AND DEMAND ASSIGNMENT

Telephone users make their calls at random. So do computer terminal users. A problem in telecommunications is that of assigning channels in a sufficiently flexible way that users can have a channel whenever they have the whim to communicate.

If satellite bandwidth is regarded as connecting two points, then the problem is the traditional one of allocating subchannels between those points. However, to use a satellite efficiently it should be able to interconnect *many* points. The trend toward users having their own receive/transmit antennas means that a satellite will interconnect large numbers of earth stations scattered over thousands of miles.

A unique satellite problem is: How do you allocate subchannels to users when the users are scattered across the earth and their demands vary constantly? An efficient solution to this *multiple-access* problem is necessary if users are to have their own antennas or if satellites are to serve remote locations with little regular traffic.

The geographical dispersion is one of several unique properties of satellite links, summarized in Box 14.1. These properties require that appropriate and new architectures be devised for communications systems using satellites.

VARIABLE ASSIGNMENT

When a fixed number of facilities are available for use they can be assigned to users in a fixed or a variable manner.

Suppose that a sales office has 100 employees each of whom needs a desk when he works at the office. Most of the employees, however, work at the office only occasionally. The rest of the time they are out traveling or with customers. The office therefore has only 20 desks, and when an employee comes into the office he takes any free desk. This is a *variable* allocation of desks, and this clearly needs fewer desks than fixed allocation.

In today's telephone systems the local loops are normally assigned on a fixed basis, and the trunks are assigned on a variable basis like the above office desks. The trunks are therefore used much more efficiently than the local loops.

CONCENTRATORS

Suppose that there are 100 telephone subscribers in a locality who use their telephones no higher proportion of the time than the above salesmen use their desks. There is, in theory, no need for 100 channels to connect them to their local switching office; 20 channels could be used with some means of allocating a channel to a subscriber when he needs it. This technique is called *concentration.* There are various ways in which it can be done, and hence there is a variety of devices called concentrators.

Note that there is a fundamental difference between *concentration* and *multiplexing.* With multiplexing all subscribers can have a channel simultaneously if they want one. With concentration they cannot. In the above example if all 20 channels are in use and a 21st subscriber requests a channel, he will be unlucky. He either receives a *busy* signal or else must be made to wait until a channel becomes free. Such is the nature of concentration. It takes advantage of the fact that not all the users are active all the time. Probability or queuing calculations are needed in the design of concentrator facilities.

A concentrator mechanism could be used in an apartment building or in a town street, if it were economical, to connect a large number of subscribers to a smaller number of channels to the local central office. It is sometimes used in rural areas where subscribers are a long distance from their central office to avoid having many lengthy wire-pair connections to the central office.

The design of a concentrator depends on the type of signal it is to concentrate. Concentrators are frequently used in data networks where their design is adjusted to the type of data traffic. They read data into a storage from lines with low utilization and then retransmit them over one or more lines with high utilization. A concentrator for telephone lines may be an electromechanical device which scans a bundle of lines searching for a free one. It may be a solid-state circuit which concentrates PCM traffic to travel over a digital trunk.

DEMAND ASSIGNMENT

Suppose that the sales office described above has four separate departments staffed, respectively, by sales staff, maintenance persons, design engineers, and administrative staff. Each department could have its own desks which it may

BOX 14.1 Unique properties of satellite links

A satellite channel is often used simply as a substitute for a point-to-point terrestrial channel. However, it has certain properties which are quite different from conventional telecommunications. It should not be regarded as merely a cable in the sky. New types of communications architecture are needed to take advantage of satellite properties and avoid the potential disadvantages. This is especially so in the design of interactive computer systems.

A satellite channel is unique in the following respects:

1. There is a 270-millisecond propagation delay.
2. Transmission cost is independent of distance. A link from Washington to Baltimore costs the same as a link from Washington to Vancouver. A computer center can be placed anywhere within range of a satellite without affecting transmission costs. It is becoming economical to centralize many computing operations. In an international organization worldwide links can be similar in cost to national links if the regulatory authorities so permit.
3. Very high bandwidths or bit rates are available to the users if they can have an antenna at their premises or radio link to an antenna, thereby avoiding local loops.
4. A signal sent to a satellite is transmitted to all receivers within range of the satellite antenna. Unlike a terrestrial link, the satellite *broadcasts* information.
5. Because of the broadcast property, dynamic assignment of channels is necessary between geographically dispersed users. This can give economies, especially with data transmission, on a scale not possible with terrestrial links but needs new forms of transmission control.
6. Because of the broadcast property, security procedures must be taken seriously.
7. Most transmissions are better sent in digital form. Digital techniques can therefore be used to manipulate and interleave the signals in a variety of ways. The high bit rates make possible new uses of telecommunications not economical on terrestrial links.
8. A transmitting station can receive its own transmission and hence monitor whether the satellite has transmitted it correctly. This fact can be utilized in certain forms of transmission control.

or may not assign dynamically. The demand for desks, however, will vary from time to time; sometimes the sales staff will want many desks and sometimes not. A high demand for salesperson desks may occur when the demand for design-engineer desks is low. If the desks could be quickly reassigned from one department to another on a basis of the demand at that instant, then the organization could use fewer desks than if they were permanently assigned to the four departments. This is referred to as demand assignment.

The toll telephone network uses demand assignment. Its trunk-switching offices assign different groups to different destinations on a demand basis.

A particularly important form of demand assignment occurs on satellite channels. It has the difference that although there is one satellite and hence a single cluster of channels to assign, the users of this cluster are widely scattered geographically. To give efficient utilization of the facility the channels should be constantly reassigned between one earth station and another. It is rather as though the sales organization we discussed had many offices and capability existed to magically switch desks from one office to another as the demand for desks varied.

DEMAND-ASSIGNED MULTIPLE ACCESS

The capability to switch channels between multiple-access points on a demand basis is referred to as *demand-assigned multiple access.* When traffic fluctuates widely, fixed assignment of satellite channels to separate geographical locations will lead to inefficient utilization of the satellite capacity. The satellite is sufficiently costly that it is economic to use elaborate control equipment to achieve demand-assigned multiple access.

MULTIPLE SPOT BEAMS

As with simple multiplexing, demand assignment can be achieved by frequency division, time division, or space division. Space division, in this context, implies multiple spot beams from the satellite, as in Fig. 14.1, and some capability to switch between the beams. The phrase "switch in the sky" has been used to refer to a satellite which can switch spot beams on and off and switch channels between the beams.

Space-division demand assignment is of limited value today. It takes an exceptionally large antenna to produce spot beams like those in Fig. 14.1. Even then a beam covers several hundred miles of earth, so the technique relates to common carrier earth stations or television distribution covering a large geographical area. It is of value for allocating a few hours of television to a remote country. It is not of value for enabling many users with small earth stations to share a domestic satellite. Switching equipment on board the satellite adds to the risk of an unrepairable failure in space. Furthermore, many satellites do not have narrow spot beams. Domestic satellites such as ANIK and WESTAR

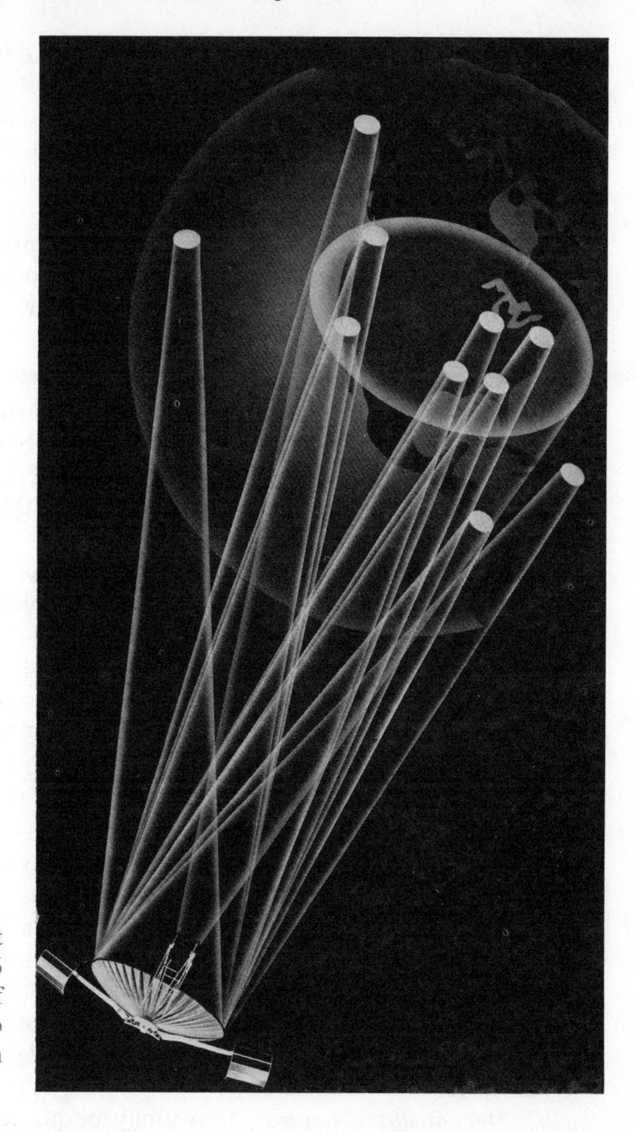

Figure 14.1 Multiple spot beams from NASA's ATS-6 satellite. To make full use of such beams it is desirable to have switching capability on the satellite.

have only one major beam. Hence some other form of demand assignment is needed—frequency division or time division.

MULTIPLE TRANSPONDERS

The simplest way to subdivide satellite capacity by frequency is to give different users different transponders. A television organization uses one transponder, Muzak uses another, a common carrier leases another, a large corporation uses a fourth, and so on. The trouble with this approach is that the transponders are of fixed capacity, whereas many users want variable channel assignment. Furthermore, the transponder capacity is much too big for many users even though transponders are smaller now than on INTELSAT II or III It may be a worthwhile trade-off for future satellites to carry at least some still smaller transponders. Small transponders, however, mean more weight on the satellite for a given channel capacity.

Even when a corporation leases a whole transponder, it still has a demand assignment problem in using that transponder. Some means is needed for geographically dispersed users to share a transponder.

FDMA AND TDMA

When each of many earth stations has access to the same transponder, the bandwidth of that transponder may be shared by *frequency-division multiple access* or *time-division multiple access*. These are referred to as FDMA and TDMA, respectively. Either can be employed with any existing satellite.

An FDMA system makes available a pool of frequencies and assigns these, on demand, to users. A TDMA system makes available a stream of time slots and assigns these, on demand, to users.

One channel, derived either by frequency division or time division, may be reserved to function as a control channel. The signals on this channel convey each station's requests for capacity and inform each station about the channel assignments.

THREE TYPES OF CONTROL

A mechanism is necessary for maintaining control over channel allocations. There are three ways of controlling the transmission:

1. Central control. A computer at a central control point can accept requests for channels, allocate channels to the requestors, and inform the interested parties which channels are allocated to whom. The requests and allocations will be transmitted on a common control signaling channel which all stations listen to.

2. Decentralized control. There may be no central control location; each station may have its own form of control. Any station requiring channel space makes a request for it on the common control signaling channel. Every other station hears the request, and some form of joint protocol determines how the channels are allocated.

3. Contention. A high-capacity channel may be shared in a free-for-all fashion. Stations are permitted to send only a short burst of information at a time. They do so at random. Sometimes the bursts from different stations collide and damage each other. Each station can detect when this happens and so can retransmit its bursts. This seemingly reckless form of operation has advantages in certain circumstances, which we shall describe in Chapter 20.

CENTRALIZED CONTROL

Many multiple-access systems will use centralized control.

In theory, the satellite itself could be the controlling location. The reader might imagine a demand-assignment system operating with a little old lady sitting in the satellite talking to the users like a PBX operator and pushing plugs into a panel to connect them. In reality, if the satellite *were* the control point, an on-board computer would be used. It is desirable, however, not to complicate the satellite too much because it must have extremely high reliability. Furthermore, demand assignment is needed with *today's* satellites. Therefore, if a controlling location is used, it must be on earth, and a transmission to and from it must be via the satellite. In Fig. 14.2 the centralized control is performed by a computer at earth station 2.

If earth station 1 wants to transmit to earth station 10, it sends a request on the common control channel. The request is received by earth station 2, where a controlling computer examines a list of free channels and allocates one to satisfy earth station 1's request. Earth station 2 sends a message on the common control channel to earth stations 1 and 10 saying that a certain channel has been allocated for transmission between them. When the transmission ends, either earth station 1 or 10 signals earth station 2, indicating that the transmission is over. The controlling computer updates its list of free channels so that the channel which was used can now be allocated to another user.

Such a scheme may be designed for allocating channels of equal capacity, for example telephone channels. On the other hand, it may be designed so that earth stations can request different capacity channels. An earth station may request a low-speed data channel at one time, a high-speed data channel at another, a telephone channel, a video channel, or channels of other capacity.

When a station requests a channel, it cannot have it immediately. It usually has to wait at least 540 milliseconds for its request to reach the controlling location and the response to be returned. For much transmission this delay does not matter. It is trivial compared with the time it takes a telephone user to obtain a dialed connection.

If a transmission is very short, for example a few characters traveling between a computer and a terminal, then the overhead of reserving a channel for it and terminating the reservation may be excessively high. As we shall discuss later, different techniques have been used for computer terminal users employing satellite channels.

DECENTRALIZED CONTROL

With decentralized control of channel reservation, the list of available channels must be maintained at every station rather than at a controlling location. A station wishing to transmit to another station selects a free channel and sends a control message to that station requesting permission to transmit on that channel. The control message is again sent on a common control signaling channel.

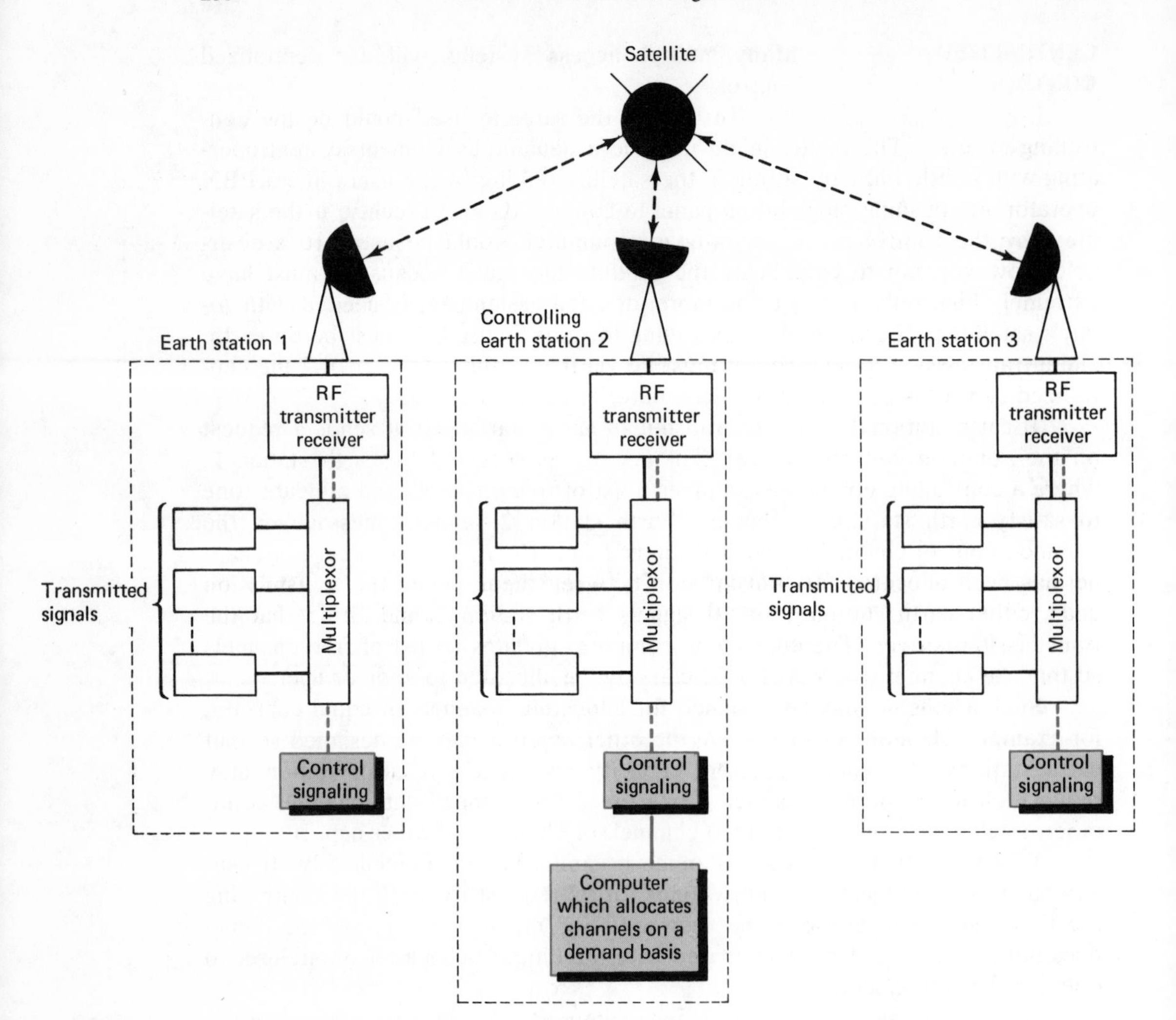

Figure 14.2 Centralized control of demand-assignment performed by a computer at earth station 2.

The recipient station responds if the selected channel is still free, saying whether it is ready to receive. If the originating station receives this go-ahead correctly, it transmits. When the stations are finished with the channel a control message will be sent to that effect so that all stations can update their list of available channels.

Two separate stations may, by chance, select the same channel and request permission to transmit on it. In that case some preprogrammed protocol must determine which station received permission to go ahead. Possibly both requests are negated and the stations in question must make new requests.

One advantage of decentralized control is that the system is not vulnerable to the failure of one control station. A disadvantage is that it may be more expensive, especially if there are many earth stations. Centralized control may be more appropriate when elaborate allocation schemes are used such as the allocation of many different channel capacities. With either form of control a centralized location may have the function of billing users for the channel time they use.

Both centralized and decentralized control can operate with either frequency-division or time-division multiple access.

A multiple-access system provides a complete set of interconnections between the nodes which use it, as shown in Fig. 14.3. If it is flexibly designed, these connections can be for a widely varying number of channels and may provide channels of widely varying capacity.

IMPROVING TRANSPONDER THROUGHPUT

There are three ways to improve the throughput of a transponder already in orbit. First, analog signals such as speech can be encoded into the minimum number of bits. Second, the modulation and multiple-access techniques can be chosen so as to maximize the number of channels the transponder can handle. Third, traffic can be dynamically allocated to the available channels from all earth stations so that the channels are utilized as fully as possible.

All three techniques are performed by the earth-station electronics.

Figure 14.3 A multiple-access satellite system provides a complete set of interconnections between the earth station locations. This would require $1/2\ N(N-1)$ paths if terrestrial links were used without switching.

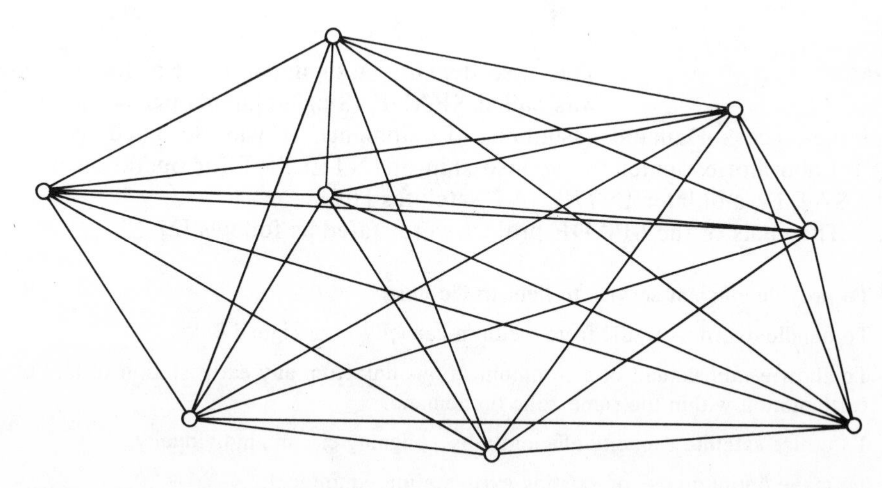

15 FREQUENCY-DIVISION MULTIPLE ACCESS

With frequency-division multiple access (FDMA) the transponder bandwidth is divided into smaller bandwidths. An earth station transmits on one or more of these subdivisions. The control mechanism makes sure that no two earth stations transmit on the same subdivision at the same time. The subdivisions can, however, be reallocated from one earth station to another as the demand for channels varies.

In a typical FDMA system the bandwidth subdivisions are capable of carrying one voice channel. The spacing between channels is 45 kHz. A carrier at the center of this band is modulated with the voice channel. Every earth station knows, at a given instant, which carriers it will use for transmission and on which carriers it will receive and demodulate. The result, in effect, is a multipoint interconnection as in Fig. 14.3 with each of the links on the diagram varying from one to several hundred voice channels according to the demand.

SPADE

The first demand assignment system for satellites was called SPADE (Single-channel-per-carrier PCM multiple-Access Demand assignment Equipment). It was designed at COMSAT Laboratories under the sponsorship of INTELSAT for operation on INTELSAT IV and later INTELSAT satellites [1,2].

The goals of the SPADE project were stated as follows [2]:

1. To provide efficient service to light traffic links.
2. To handle overflow traffic from medium-capacity preassigned links.
3. To allow establishment of a communications link from any earth station to any other earth station within the same zone on demand.
4. To utilize satellite capacity efficiently by assigning circuits individually.
5. To make optimum use of existing earth-station equipment.

SPADE increased the traffic over existing INTELSAT IV links and made the establishment of new links more economical. By permitting variability in channel assignment and routing, SPADE enabled INTELSAT transponders to be used more efficiently. The INTELSAT system is characterized by large earth stations separated by large distances. For a domestic system characterized by large numbers of small earth stations, a technique to enable these stations to share the satellite flexibly is a key to economic viability.

MULTIPLE CARRIERS Figure 15.1 shows the frequency allocations of SPADE. At the center of the transponder bandwidth is a pilot, used in receiving the RF signal and converting it to IF. On either side of the pilot are 400 channel carriers spaced 45 kHz apart. At the extreme left is a carrier with a wider bandwidth allocation than the others, which is used to carry the 128,000-bps common signaling channel. This is the channel used to control the continually changing allocation of voice carriers to earth stations.

There are thus 800 carriers, of which 794 are used to provide 397 two-way telephone calls. The carriers are allocated in pairs for two-way transmission.

In general, more information can be transmitted via a transponder *if only*

Figure 15.1 The frequency assignments of SPADE, giving 397 pairs of usable voice channels in one 36 MHz transponder.

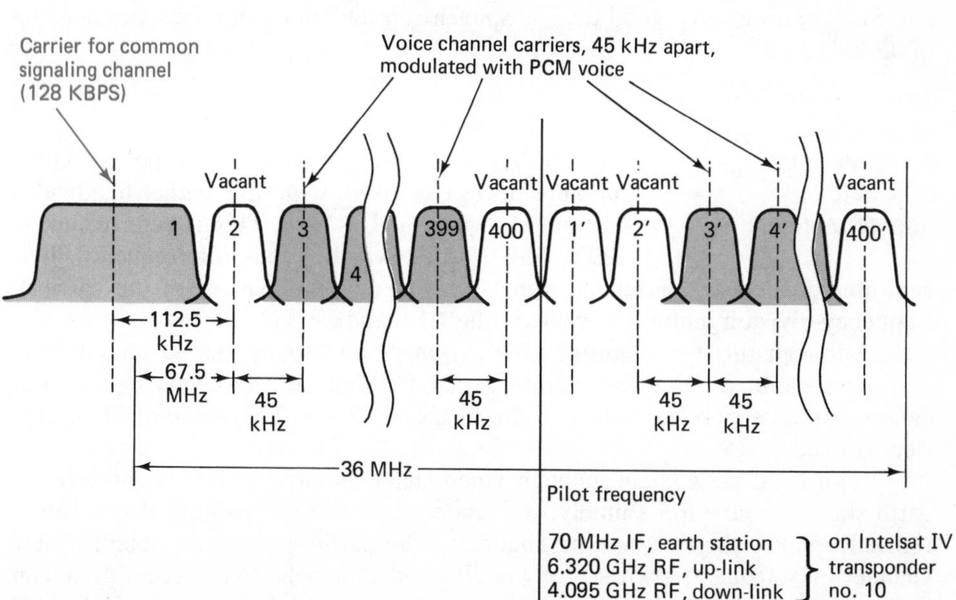

one carrier is used. The more carriers share a transponder, the lower the overall capacity. Table 15.1 illustrates this, showing the four modes of accessing an INTELSAT IV satellite prior to the use of SPADE. Much INTELSAT equipment still uses these four modes. The figures are for the global antenna beams.

There are two reasons throughput drops as the number of carriers increases. First, a guard band is needed between the carrier bands; the more carriers, the more guard bands. Second, the carriers tend to modulate one another because of the nonlinear characteristics of the traveling wave-tube amplifiers. To avoid the intermodulation products causing interference the transmission is operated below the full power possible so that no carrier saturates the amplifier.

SPADE uses 800 carriers. Why, then, does not SPADE have less throughput than the worst situation in Table 15.1? There are two reasons. First, when there is one voice channel per carrier the carrier can be switched off when nobody is speaking. TASI or DSI is not used on SPADE, so even when filled with telephone conversations the SPADE carriers (one-way) can be switched off at least half the time. If there are less than the full complement of users, more power is saved. Carriers cannot be switched on and off in this way when they are modulated by mastergroups or other blocks of speech channels. Power, rather than bandwidth, is the factor limiting throughput of INTELSAT IV transponders, so the power saving means that more channels can be transmitted.

The second reason is that SPADE uses *digital* voice transmission. The carriers in Table 15.1 are modulated using FM. The SPADE carriers are modulated using quaternary phase shift keying by a 64,000-bps PCM voice channel. Such carriers give good-quality speech reproduction with a carrier spacing of 45 kHz.

FIXED VERSUS DYNAMIC ASSIGNMENT

On INTELSAT transponders not using SPADE, multiple access is used with the higher-bandwidth carriers shown in Table 15.1. This is referred to as FDM/FM/FDMA, which stands for frequency-division multiplex (of the baseband signals), frequency modulation (of the carrier), frequency-division multiple access (in the RF band).

Although this technique permits a transponder to be shared among several earth stations, it does not permit demand assignment, i.e., varying the allocation of the carriers according to the traffic variation. The assignment of carriers is fixed.

With fixed assignment, enough voice channels have to be allocated to an earth station to give it a suitably low grade of service (probability that a call is refused because there is no free channel). The earth station can obtain a new channel only from its own set of preallocated channels. With SPADE it can have any of the 397 pairs of channels in the transponder it uses. A COMSAT study of the Atlantic region traffic and earth stations showed that to give a

Table 15.1

Number of Carriers Per Transponder	Carrier Bandwidth (MHz)	Number of Channels Per Carrier	Number of Channels Per Transponder
1	36	900	900
4	3 of 10	132	456
	1 of 5	60	
7	5	60	420
14	2.5	24	336

grade of service $P = 0.01$ about four times as many channels would be needed with fixed assignment (FDM/FM/FDMA) as with dynamic assignment (SPADE). The 800 telephone channels of one transponder using SPADE are equivalent to about 3200 channels in transponders without demand assignment.

DYNAMIC ASSIGNMENT

Figure 15.2 illustrates dynamic assignment of channels on the SPADE system. The 36-MHz transponder at the top of the diagram is, in effect, a pool of 397 usable two-way voice channels. A free channel can be seized by any earth station when it needs it. Earth station C in the diagram is using channels f_1 and f_3. It ignores all other channel frequencies. Channel f_1 is connected at this instant with earth station B, and channel f_3 is connected to earth station D.

If earth station C finishes with channel f_3, it will send a signal that that channel is free again. Every earth station receives the signal and knows that channel f_3 is again in the pool of available channels. Earth station A may then request that it use channel f_3 to communicate with earth station B. Earth station B acknowledges the request, and A and B switch a voice channel to modems operating with the carrier frequencies of channel f_3.

The earth station may be connected to a telephone exchange with many thousands of lines coming into it. The SPADE equipment thus performs a concentration and a demand assignment function.

CONTROL SIGNALING

The SPADE system uses decentralized control. No specific stations are in charge of allocating channels. If any of the earth stations are inoperative, the system can keep operating on the remaining stations.

The system is controlled by means of a common signaling channel which every station receives. This signaling channel transmits a bit stream of 128,000 bps which modulates a carrier at the bottom of the transponder bandwidth (Fig. 15.1). Two-phase PSK modulation is used, designed to give a low error rate of 1 bit error in 10^7.

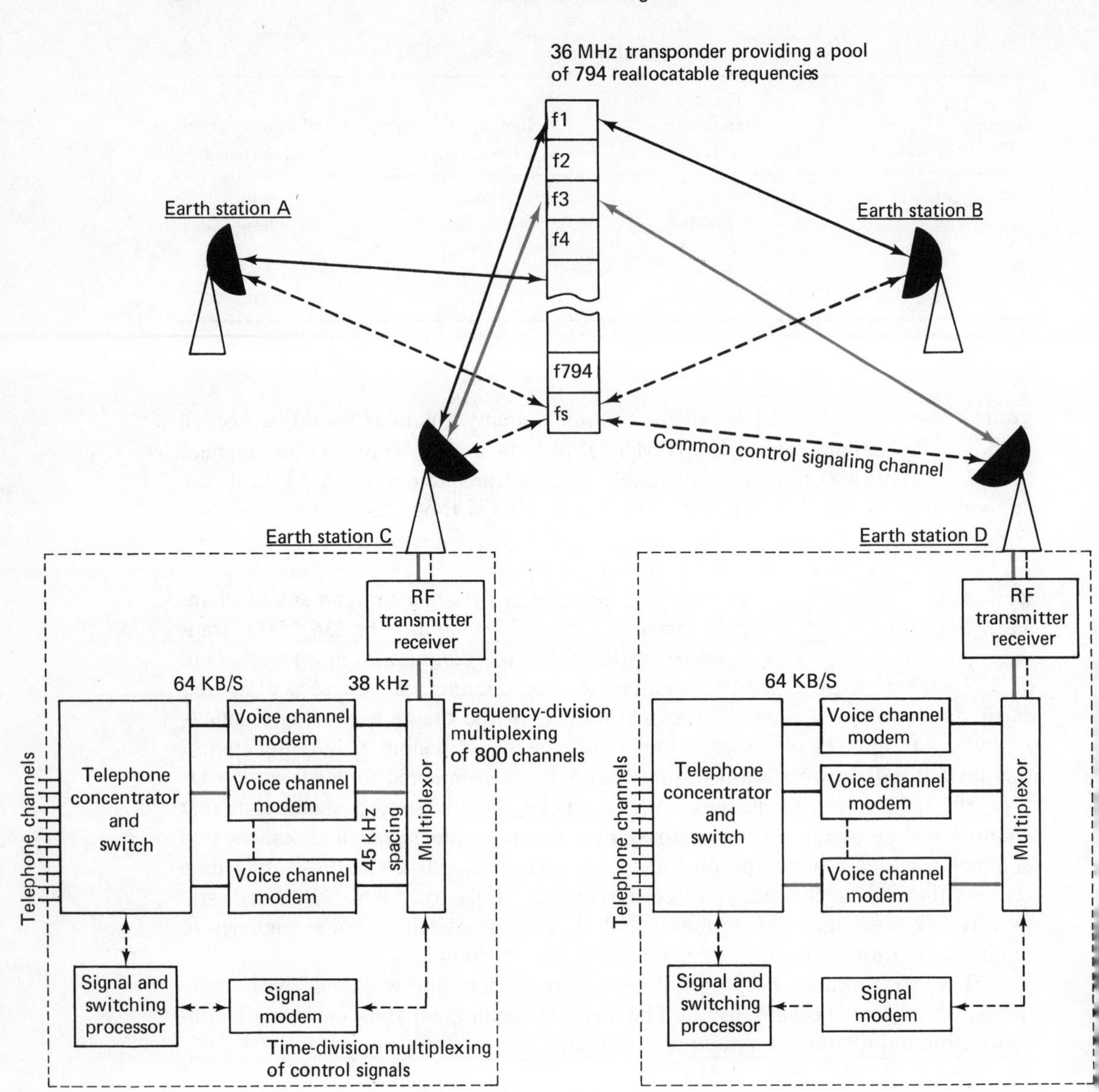

Figure 15.2 The operation of the SPADE frequency-division demand-assignment system.

This signaling channel must itself be shared by all earth stations and so presents a multiple-access problem in miniature. Time-division multiple access is used for sharing the signaling channel. Each earth station is allocated a time slot of 1 millisecond, permitting it to transmit a burst of 128 bits, some of which are synchronization and error-detection bits. Each station has one such time slot every 50 milliseconds, as shown in Fig. 15.3. This permits up to 49 earth stations to be linked.

Referring to the situation in Fig. 15.2, let us suppose that earth station A wants to set up a connection with earth station B. Earth station A examines its table of available frequencies and selects an unused frequency at random. It selects channel f_2 and so encodes a request to use that channel to earth station B. It transmits this request in its time slot; 270 milliseconds or so later, station B receives it. If channel f_2 is still available, station B sends a message to station A, in the B time slot, telling A to go ahead. Stations A and B then assign to modems the two frequencies of channel f_2 and proceed to test the connection. If the test is successful, the appropriate telephone links are switched to the connection.

Occasionally earth station A will request a certain channel and before its request reaches its destination that channel will be allocated to some other station. Station A will detect this new allocation coming shortly after its own request, will realize that its own request cannot be met, and will make a new request for a different channel. To lower the probability of such duplicate requests each station selects the channels it requests at random from the pool of available channels.

The signaling processor at each station has a small memory in which the status of all 397 channels is recorded. This processor is used to select free

Figure 15.3 Use of the SPADE signaling channel of 128,000 bits per second.

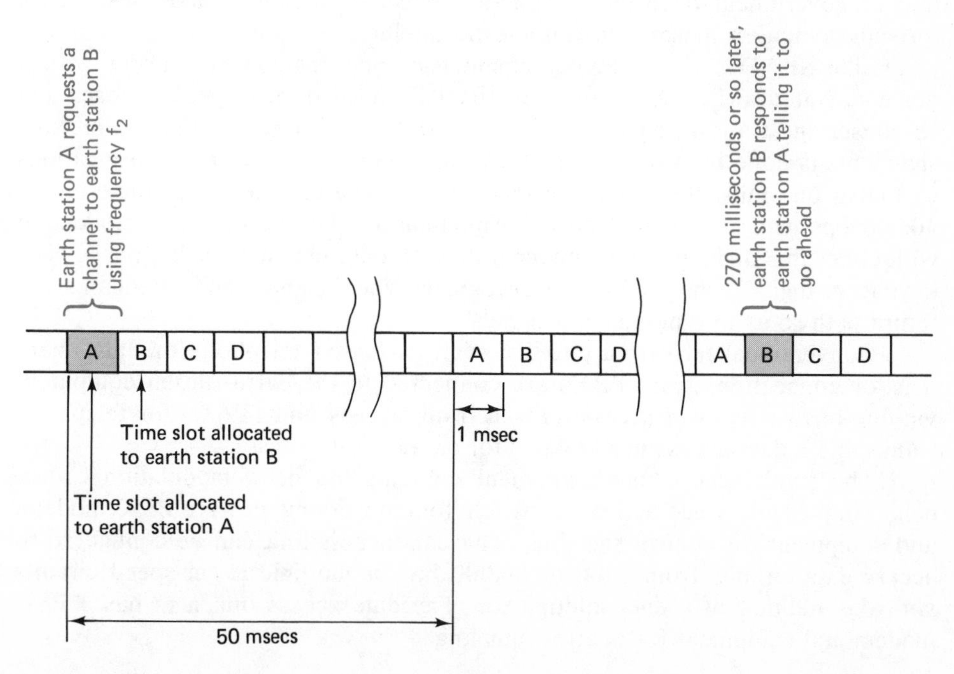

channels and to respond to other stations requesting channels. As soon as a channel allocation is made, the processor at every earth station deletes that from its list of available channels. As soon as a telephone call ends a signal is transmitted saying that that channel is free again.

CHANNEL CHARACTERISTICS

Table 15.2 shows details of channels used by SPADE. The communications channels transmit PCM voice using 64,000 bps. The common signaling channel is of higher quality and is designed to have a lower error rate.

EQUIPMENT FOR CORPORATE NETWORKS

The SPADE system was designed for the large earth stations of INTELSAT, which typically have a 30-meter antenna. Systems using similar principles have been designed for small domestic earth stations with small antennas and uncooled electronics. One such is the General Electric System ES-144 [3].

General Electric provides a range of complete low-cost earth stations. It includes a 5- to 10-meter antenna, which can be pointed at the satellite with a hand crank, and electronics in a small transportable shelter. The electronics includes the RF transmit and receive equipment, codecs, modems, and FDMA demand assignment equipment. Such a system would permit a larger corporation or government department to interconnect its locations and local PBXs (private branch exchanges) via a domestic satellite.

Like SPADE, the G.E. equipment uses one channel per carrier with a voice-activated switch for cutting off the RF signal during speech pauses and so conserving transmitted power. Unlike SPADE, control of the channel assignments is done by a computer at a central location—or rather two locations so that if one fails, the other can take over. Voice channels are converted to 40,000 bps using variable-slope delta modulation. The carriers are modulated with either this digitized voice channel or with data channels, using phase shift keying. A digital echo suppressor recognizes voice signals and attenuates the return path so as to remove echo signals.

Conventional terrestrial circuits, such as leased telephone or data channels, or connections from PBXs are connected to the earth-station equipment via one of two types of access unit—a trunk access unit (TAU) for telephone trunks and a data access unit (DAU) for data circuits.

The trunk access unit equipment contains the delta modulation codec, echo suppressor, voice-activated switch for conserving power, PSK modem, and equipment for control signaling. The data access unit can be configured to access data circuits from 2400 to 50,000 bps or multiple lower-speed circuits with the addition of a data multiplexor. The data access unit also has a PSK modem and equipment for control signaling.

Table 15.2 SPADE channel characteristics

a. Communications Channel Characteristics	
Channel encoding	PCM
Modulation	4-Phase PSK (coherent)
Bit rate	64 kbps
Bandwidth per channel	38 kHz
Channel spacing	45 kHz
Stability requirement	±2 kHz
Bit error rate at threshold	10^{-4}
b. Common Signaling Channel Characteristics	
Access type	TDMA
Bit rate	128 kbps
Modulation	2-Phase PSK
Frame length	50 ms
Burst length	1 ms
Number of accesses	50†
Bit error rate at threshold	10^{-7}

† 49 stations plus 1 reference.

Figure 15.4 shows the equipment in an earth station, and Figs. 15.5 and 15.6 show details of the trunk and data access units.

The supervisory and control information produced by the trunk and data access units modulates a carrier in the digital transmission unit, and this carrier is received continuously by the *system routing center*. The most essential function of the system routing center is to tell the trunk and data access units what frequencies they must transmit and receive on. The system routing center answers their requests for channels. However, in addition to centralized network control the system routing center monitors the network, collects the data necessary for centralized message accounting and billing, gathers traffic statistics, and has a console for maintenance functions and a console for giving operator assistance to callers. The maintenance console is used to conduct manual or automated checks of system parameters, carry out loopback tests of circuits or modems, and bring backup functions on-line when necessary.

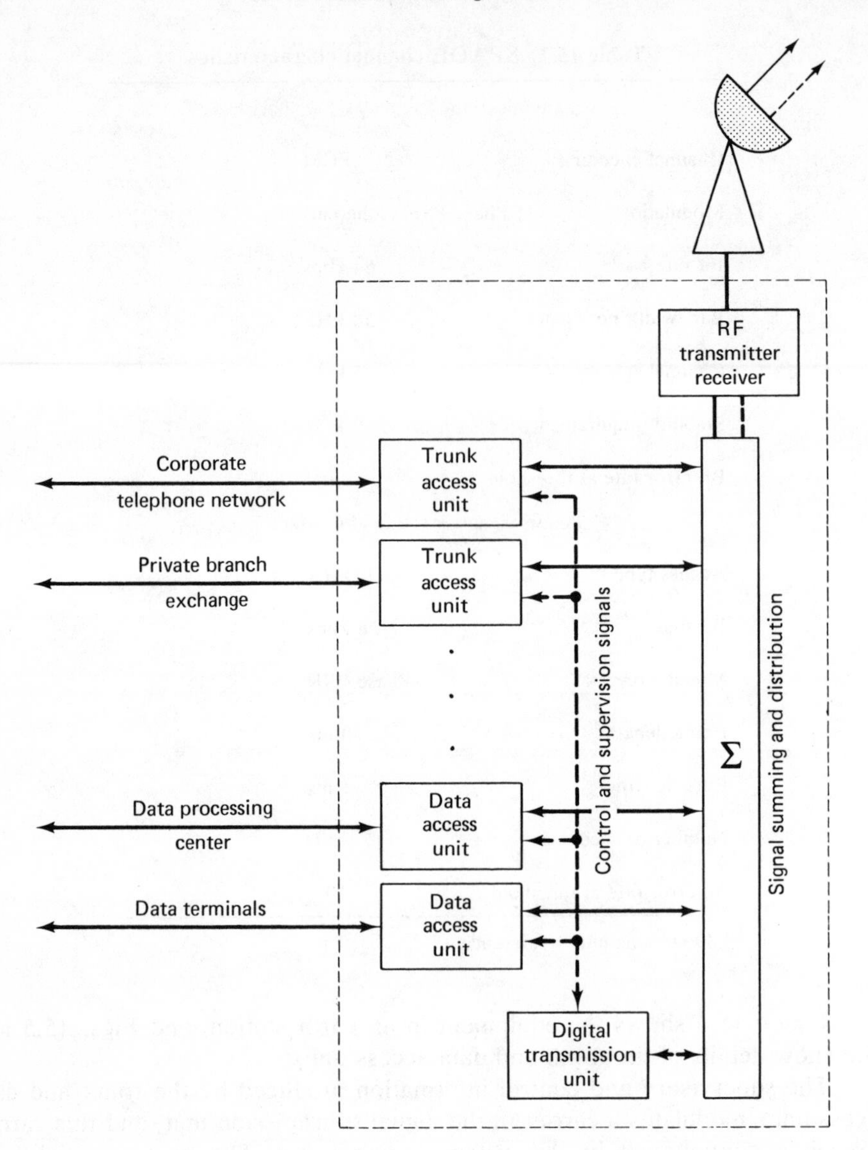

Figure 15.4 FDMA in a low cost corporate earth station—the General Electric ES-144.

The system routing center is located at a primary earth station, sharing its equipment. There would normally be two such centers at separate geographical locations in case of failure.

Small-scale demand-assignment equipment makes the prospect of corporate networks sharing transponders very attractive. The communications manager in large- and medium-sized corporations has a complex job ahead of him.

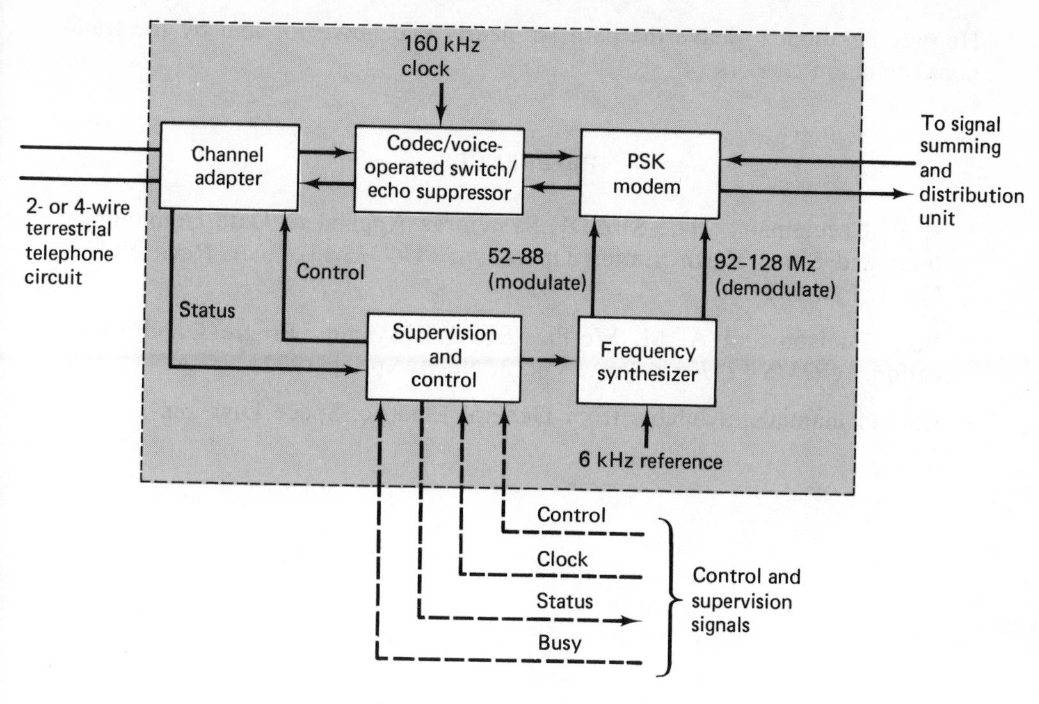

Figure 15.5 The trunk access unit in Fig. 15.4.

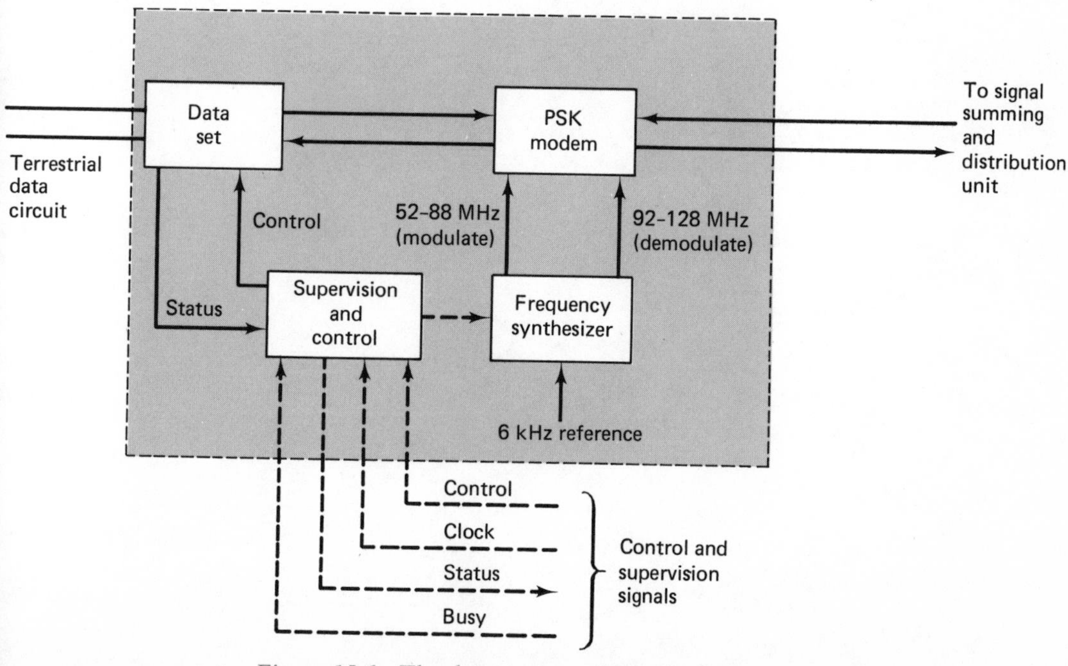

Figure 15.6 The data access unit in Fig. 15.4.

He may be about to leave the path so clearly laid down for him by the traditional common carriers.

REFERENCES

1. E. R. Cacciamani, "The SPADE System as Applied to Data Communications and Small Earth Station Operation," *COMSAT Tech. Rev.,* Vol. 1, No. 1, Fall 1971.
2. B. I. Edelson and A. M. Werth, "SPADE System, Progress and Application," *COMSAT Tech. Rev.,* Vol. 2, No. 1, Spring 1972.
3. ES-144 manuals, available from General Electric, Space Division.

16 TIME-DIVISION MULTIPLE ACCESS

Interesting though equipment such as that in Chapter 15 is, there may be better ways to tackle demand assignment.

With time-division multiple access (TDMA) each earth station is allowed to transmit a high-speed burst of bits for a brief period of time. The times of the bursts are carefully controlled so that no two bursts overlap. For the period of its burst the earth station has the entire transponder bandwidth available to it.

In the simplest form of TDMA, each station in turn is allocated an equal-length burst (in a round-robin fashion), as on the control-signaling channel shown in Fig. 15.3. To be efficient, however, the stations must be able to vary their transmission rate, and so either the bursts will be of variable length or else the scheme must permit some stations to transmit more often than others.

The burst can carry voice, video, data, or anything which is digitally encodable. An earth station may be transmitting many voice channels or possibly a video channel, in which case it is allocated frequent bursts or large bursts. On the other hand, it may be allocated infrequent bursts or small bursts because it has relatively few bits to send.

Most such systems will have *demand assignment,* meaning that the earth stations can continually vary their demand for channels. A control channel is used to convey the requests for channel assignment and inform all stations what assignments have been made. The control channel is sometimes called an *order wire*—a term dating from the early days of manual telephone switching.

SINGLE CARRIER

One of the objectives of time-division multiple access is to employ a *single carrier* for the transmission via one transponder. This, in a sense, is the opposite philosophy to that of the SPADE system, which has 800 separate carriers, one for each voice channel. Every earth station in a TDMA system receives the entire bit stream and extracts those bits which are addressed to it. As we commented in Chapter 15, more data can be sent via a transponder if a single carrier is used (see Table 15.1). When two or more carriers are sent via a transponder they tend to intermodulate one another. Figure 16.1 illustrates the effect of intermodulation [1]. Two tones, 3.95 MHz apart, are transmitted via the ANIK satellite. The signal strength is adjusted so that there is just sufficient up-link power to saturate the transponder, and the higher-frequency tone is then reduced in strength by 3 decibels. The bottom illustration in Fig. 16.1 shows the signal received. It contains multiple tones which result from the two tones intermodulating one another.

Such intermodulation products can cause interference when multiple carriers are transmitted as in Fig. 15.1. The problem can be made less serious by running the transponder at less than its full power, and this is done on systems such as SPADE. However, today's transponders are power limited rather than bandwidth limited, and as we saw in Chapter 15 the power limitation restricts the channel throughput.

At 12/14 GHz single-carrier operation causes much less problem with interference with other satellites. It has been estimated that 2° orbital spacing from a television broadcast satellite would be needed, rather than 6° spacing, if single-channel-per-carrier operation were used as in FDMA systems [2].

TDMA avoids intermodulation problems by using a single carrier and is highly efficient in using satellite power. The up-link also can be run at full transmitter power.

SYNCHRONIZATION

TDMA has other types of problems, associated with the synchronization and control of the high-speed digital bit stream.

When a high-speed digital path is used via satellite, careful synchronization is needed. Channels to different earth stations are of different path length and hence propagation time. The propagation time changes slightly as the satellite drifts from its station. How much the propagation delay varies from its nominal value depends on how frequently the satellite position is adjusted. The design of some digital equipment specifies that the deviation from the nominal delay should not exceed 250 microseconds.

In addition to long-term drift the satellite experiences a daily oscillation in its position due to the pull of the sun and moon, rather like the tides on earth. Figure 16.2 shows this tidal oscillation.

The path length to and from an accurately positioned satellite can change by up to 50,000 feet in half a day, and hence the propagation time can change by 50 microseconds in half a day. The bottom chart of Fig. 16.2 shows the rate of change. The tidal movement of the satellite can cause it to be moving to or from the earth at speeds up to about 2 feet per second. This means that in 1

TWO TONES ARE TRANSMITTED TO THE SATELLITE AS SHOWN:

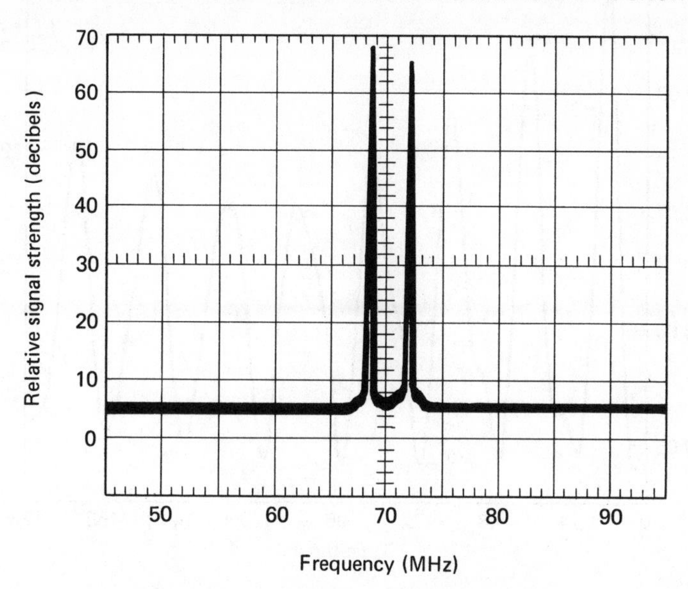

THE FOLLOWING SIGNAL IS RECEIVED AT EARTH STATIONS, SHOWING HOW INTERMODULATION HAS OCCURRED BETWEEN THE TWO TONES:

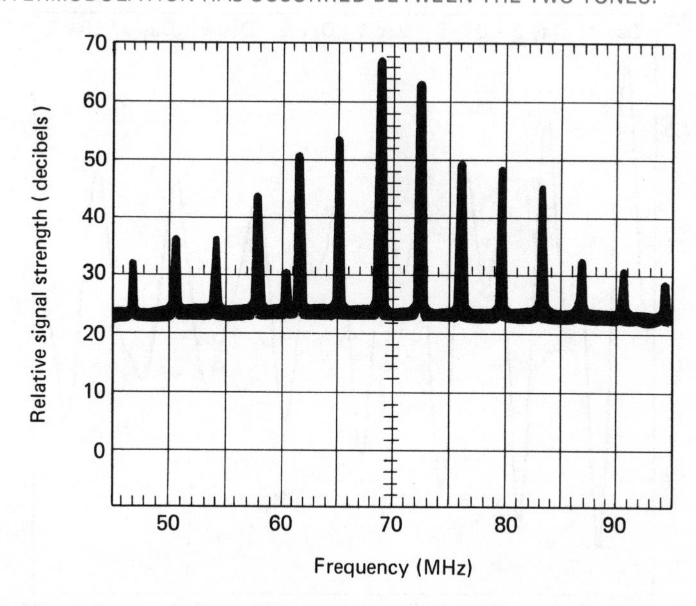

Figure 16.1 An example of the intermodulation that occurs when multiple carriers are transmitted via the same transponder. (*Drawn from cathode ray tube photographs in reference 1.*)

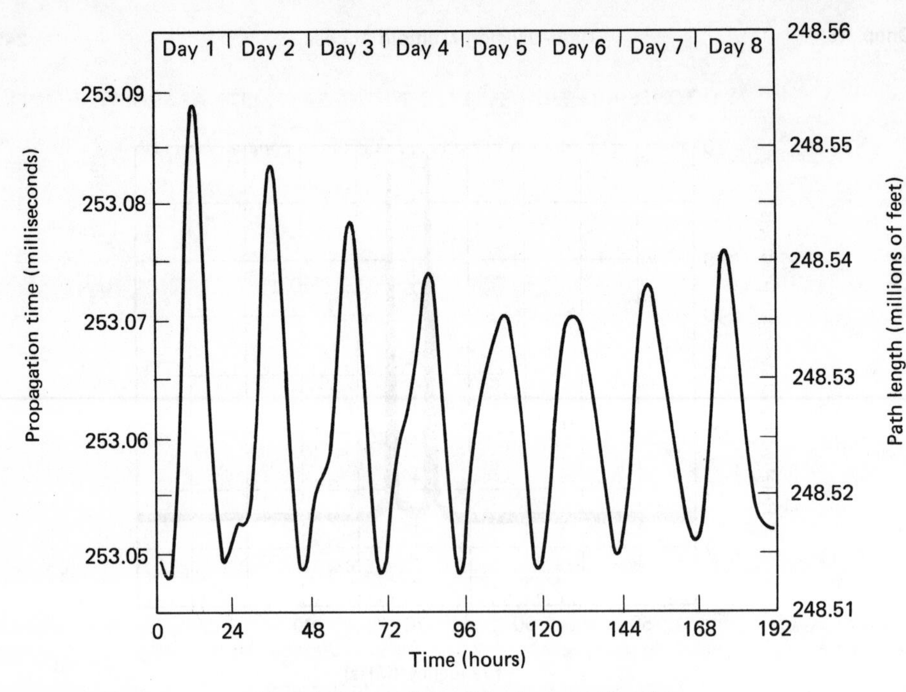

Variations in path length and propagation time via a satellite, measured over an 8-day period

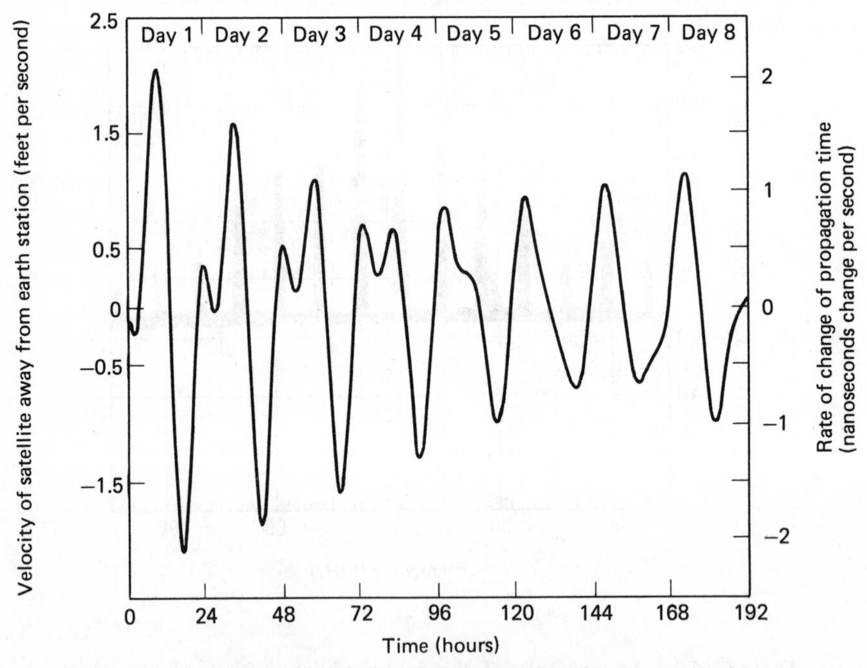

Variations in velocity of the satellite relative to an earth station, for the same 8 days

Figure 16.2 "Tidal" variations in satellite position. A high-speed digital satellite link needs careful synchronization.

second the propagation time can change by up to 2 nanoseconds (2×10^{-9} seconds). The change is slightly different for different earth stations.

At data rates of 60 million bps the time between bits is 16.67 nanoseconds, so frequent resynchronization is needed. Much higher data rates have been proposed for future satellites, including a Bell Laboratories proposal for 630 million bps over a transponder of wide bandwidth.

The tidal movement of the satellites also slightly changes the frequencies of signals reaching it and transmitted to earth. This is the Doppler effect—the effect that causes a person on a railroad station to hear a train whistle drop in pitch as it rushes past him. Transmission at 6 GHz is changed by up to ±13 Hz by the satellite's meanderings. This has to be taken into consideration in the design of multiplexing and modulation equipment.

BURST MODEMS

TDMA systems use modems designed especially to transmit high-speed bursts. Each burst carries its own means of synchronization, so it can be transmitted and received in isolation. Each burst starts with a synchronization pattern which permits the receiving modem to recover the carrier and demodulate it correctly, to recover the clock timing, and know which is the first of the data bits. The burst contains a powerful error-detecting code which permits the receiving device to know whether the data have been received correctly. Such modems are operating successfully, transmitting 60 million bps via today's 36-MHz satellite transponders, using quaternary phase shift keying (QPSK).

FRAMES AND MASTER FRAMES

A small time gap is left between bursts to make sure that they do not interfere (the guard time of Fig. 10.1). Each burst contains synchronization bits and a preamble containing control information. These all constitute an overhead which lessens the total information-carrying capacity. The overhead will be a small proportion of the total if the system is designed to have large burst sizes. On the other hand, large burst sizes increase the cost of the earth-station equipment because large buffers are needed to hold the bursts.

The set of bursts, one from many of the earth stations, is referred to as a *frame* and is illustrated in Fig. 16.3. The first burst in the frame contains no traffic but serves to synchronize and identify the frame. It is transmitted by a controlling earth station. On a typical system a frame is on the order of a millisecond in duration and hence on the order of 60,000 bits in length.

Many frames of similar structure follow one another and form a *master frame,* as shown in Fig. 16.4. For the duration of a master frame an earth station transmits the same burst length and starts its burst at the same time after the frame burst. Periodically, say every second or so, the burst durations that are allocated will be changed to meet the changing channel demands.

Each of the columns in Fig. 16.4 represents a continuous stream of bits,

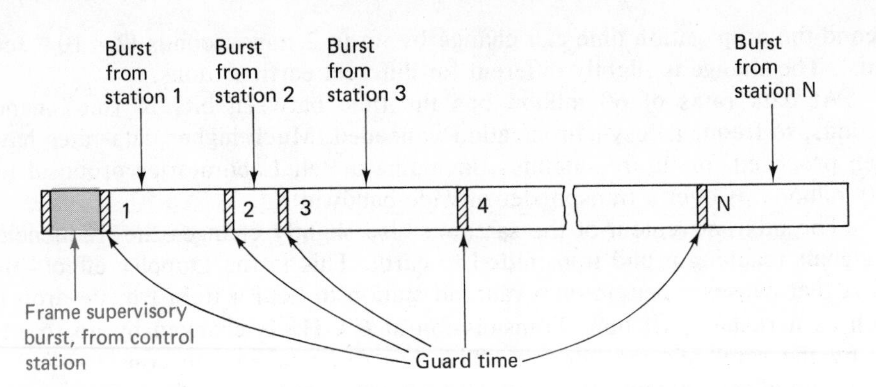

Figure 16.3 A frame in a TDMA system. A typical frame length is on the order of 1 millisecond.

after demultiplexing, from one earth station. This stream of bits may itself be split up using time-division multiplexing into many separate channels going to different earth stations, or alternatively separate bursts addressed to different destinations. If speech channels are needed, the bit stream from one earth station will be time-multiplexed into continuous subchannels. If interactive computer or message delivery channels are needed, the bit stream may be subdivided into individually addressed bursts.

BURST STRUCTURE The information in each TDMA burst in Fig. 16.3 preceded a group of bits referred to as a *preamble* or *header,* which synchronizes the burst, identifies the station sending it, and assists in controlling the network. The bit structures have differed substantially from one TDMA design to another but nevertheless serve a similar set of functions. Figure 16.5 shows the bit structures used on a Comsat system called MATE [3].

Each burst begins with 30 modem symbols which have the function of establishing the carrier reference for demodulation and regenerating the bit timing. A quaternary phase shift modem is used, so 30 symbols are equivalent to 60 bits. The gap between bursts can be variable, but each burst is enclosed in a guard time cushion of at least 12 symbols (24 bits), with at least 6 symbols before and after the burst. This represents a worst-case burst timing tolerance of ±200 nanoseconds when the transmission speed is 60 million bps.

The 60-bit synchronization pattern is followed by a 20-bit unique word which establishes the frame starting point. Every tenth frame the complement of the unique word is sent to establish the channel timing (the position in the vertical column in Fig. 16.4).

Next follow 6 bits which contain the address of the station which transmitted the burst and 2 bits giving the status of that station (indicating whether the station is the reference station or a standby reference station).

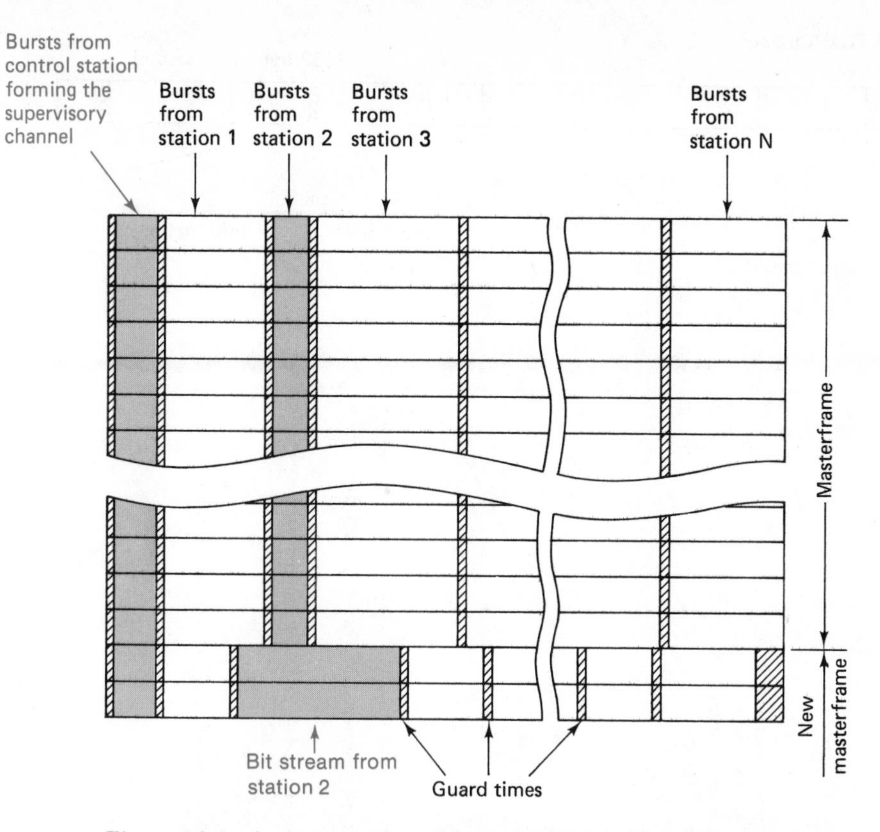

Figure 16.4 A given number of frames follow one another, forming a masterframe. The vertical columns form a continuous bit stream from each earth station containing channels to different earth stations. The burst size allocations will be different in each masterframe if the channel demands vary. Only the supervisory channel will remain the same.

The first burst in each frame identifies the start of the frame and contains only the above bits. The other bursts have more bits in the preamble, including 4 control bits and several *order wires.*

Order wire is a term used in manual telephone switching, meaning a circuit on which operators or maintenance men can talk to one another. Operators use the order wire for setting up calls. The term is used today for data channels on which instructions are passed as well as for voice channels. Each traffic burst on the Comsat system carries a 4-bit fragment of a group of teletype channels and two 24-bit fragments of voice channels, these channels being used as order wires. The frame repeats every 750 microseconds, so the bit rate of the teletype order wire channel is 5333 bps and of the speech order wires 32,000 bps. Two speech channels are derived by data modulation, and multiple teletype channels are multiplexed in the 5333-bps.

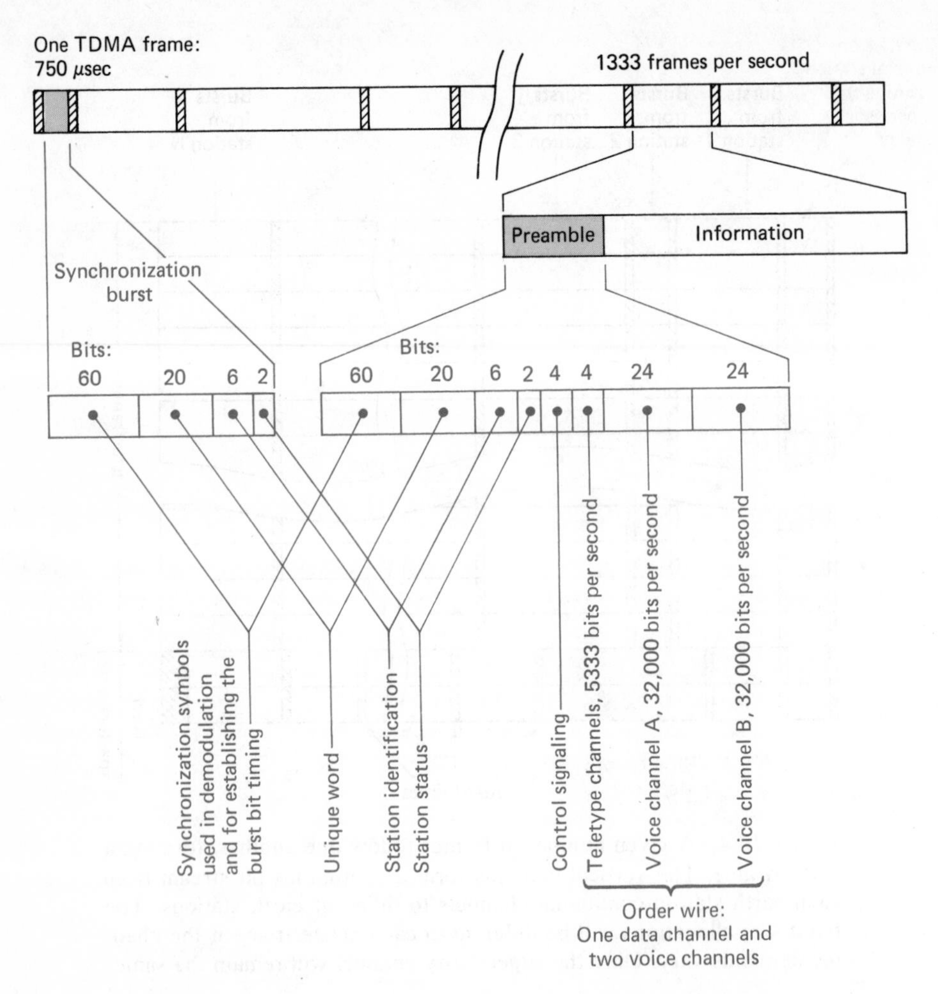

Figure 16.5 Bits used for synchronization and control of TDMA frames on the Comsat MATE system [2].

LOWER OVERHEAD A TDMA system can be devised with fewer overhead bits than in Fig. 16.5. In Fig. 16.5, each burst carries several order wires, conforming to international CCITT recommendations for signaling between telephone exchanges. A system interconnecting private branch exchanges or providing corporate facilities can be designed more tightly. The order wire and control channels can reside in the first burst of each frame (the shaded bursts of Fig. 16.4) rather than form part of every information burst. This first burst can be transmitted by each earth station in turn. If there are 50 earth stations, station A may transmit the frame control burst of the 1st, 51st . . . and so on, frames.

In the control bursts the stations make requests for the channel space they require. The allocation of channels can be done either in a centralized or decentralized fashion, as described in Chapter 14. If burst times are allocated by a central facility, then one or two "hot standby" facilities should be waiting to take over the job if the central facility fails.

ESSENTIAL FUNCTIONS OF A TDMA SYSTEM

Figure 16.6 shows the essential functions of a TDMA system which handles voice and data traffic.

The control units, which encode the signal into the digital form in which it is transmitted and decode it after transmission, could be remote from the earth station. They relay a bit stream to the multiple-access units, possibly over a terrestrial T1 or other PCM carrier link or possibly over a private microwave or millimeterwave link or other wideband facility.

The unit which processes speech ready for transmission is separate from that which prepares data. Data require an error-correcting code, whereas speech may not. Digital speech interpolation is shown being used to condense the telephone traffic. Data, therefore, cannot be sent over the voice channels because of the clipping of speech that occurs. Digital echo control is of major importance in the speech processing.

High-speed bit streams from both the speech and data access units are sent to the multiple-access facility, which buffers this traffic and transmits it in appropriately timed bursts. Video traffic may also be sent.

The speech and data access units are designed to interconnect to existing terrestrial facilities such as private branch exchanges, corporate tie-line networks, data network concentrators, and so on.

Various network management facilities are needed, discussed in Chapter 22 (see Box 22.1). The signal control unit which digitizes the signals could be remote from the DAMA unit, as shown in Fig. 16.7. In addition to the functions shown in Fig. 16.6, earth-station equipment may also include a PBX and network management facilities.

THE SBS SYSTEM

The SBS system is designed around the TDMA concept. SBS will operate 12/14-GHz satellites, each with ten transponders of 49 MHz, as shown in Fig. 16.8, plus several standby transponders for reliability.

The earth stations will send bursts at 43 mbps via the transponders. These bursts can carry speech, data, images, or any other digitized information. There will be many hundreds of such earth stations. They will be unmanned, easy to install at a customer location, and designed along with the overall system for high availability. It would be possible to transmit at a much higher data rate through the 49 MHz transponders, but to do so would increase the earth station cost, and a major objective of the design is to obtain low-cost earth sta-

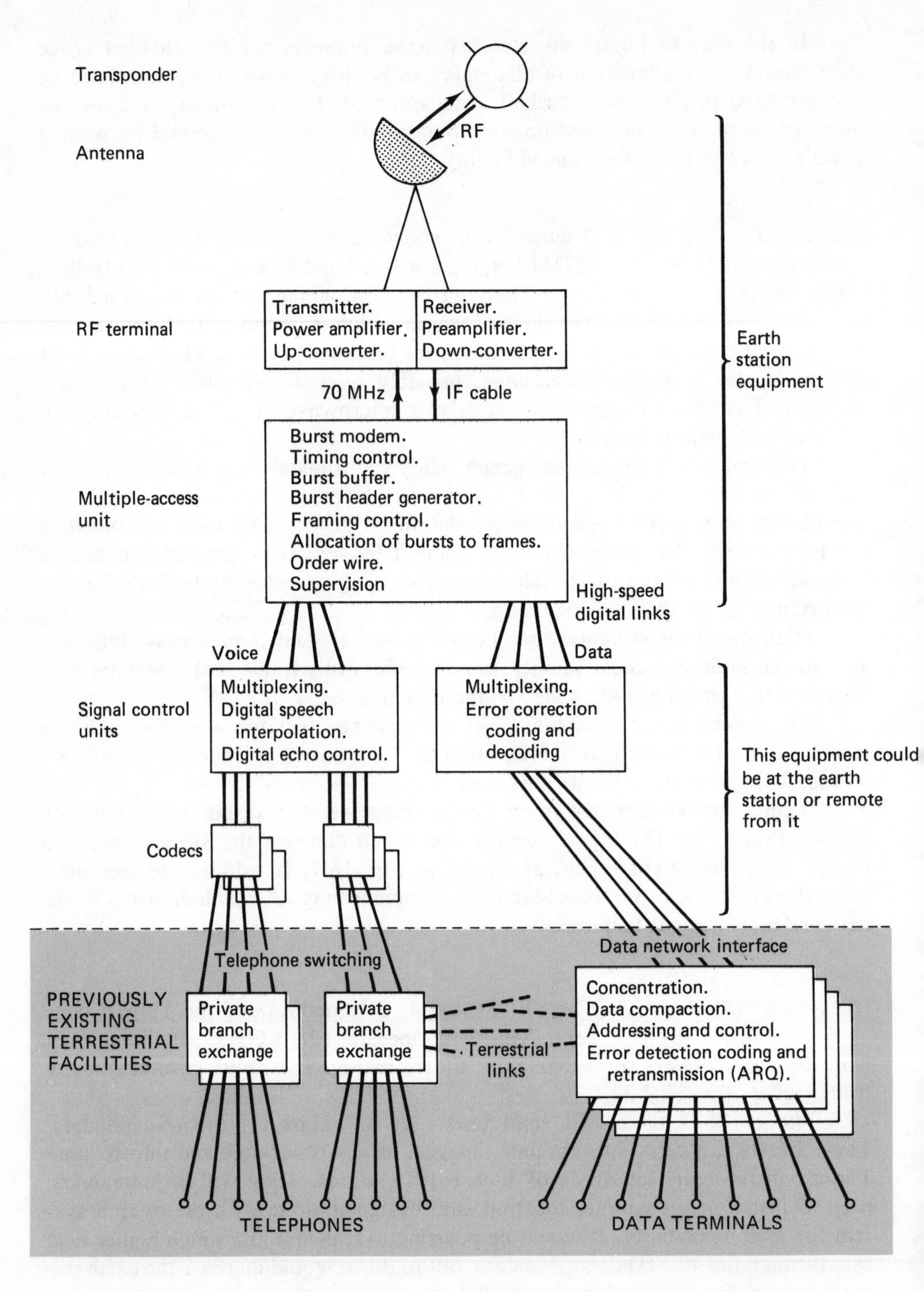

Figure 16.6 Basic functions of a TDMA system used for telephone and data traffic. The SBS system operates as shown in this diagram.

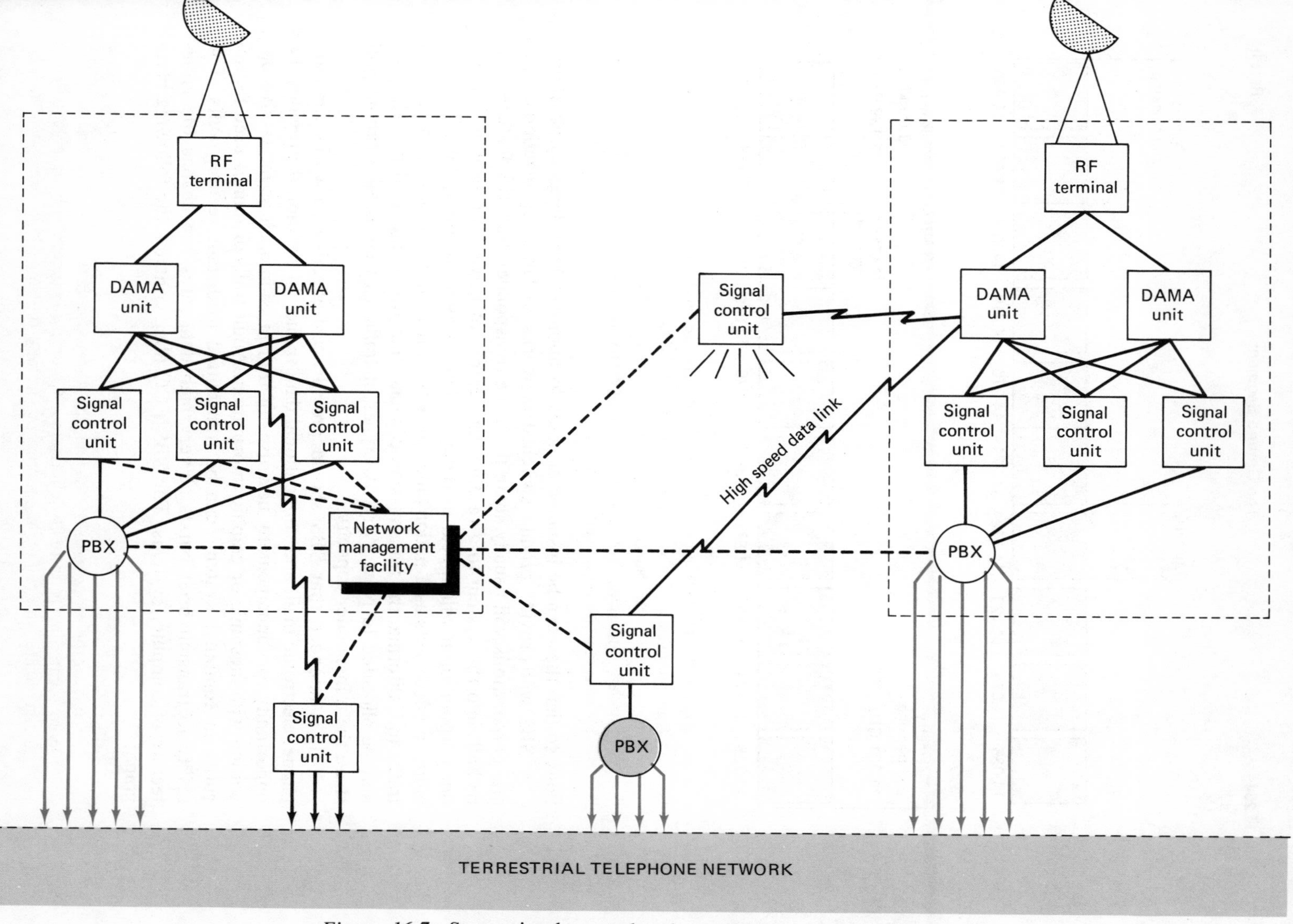

Figure 16.7 Some signal control units could be remote from the earth station.

Uplink frequencies (GHz):

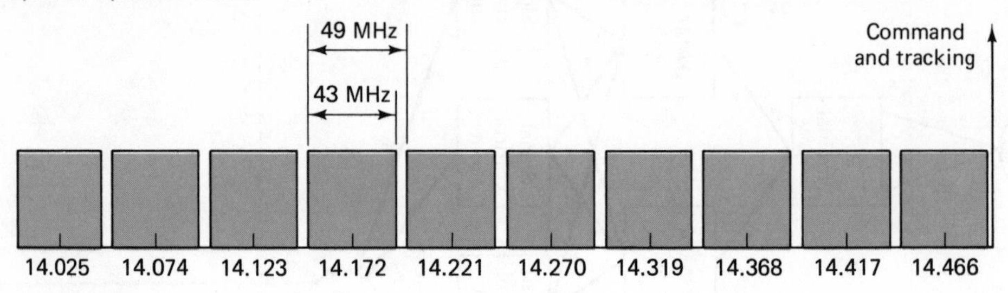

Downlink frequencies (GHz):

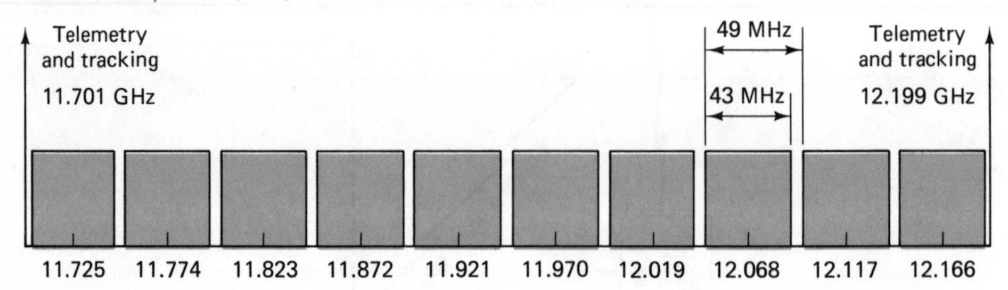

Figure 16.8 SBS plans to use 12/14 GHz satellites with ten active transponders and six spares each, of 49 MHz bandwidth. The earth stations will use burst modems which transmit 43 mb/s via a transponder, and relay voice, data, or images by means of TDMA. This bit rate may be increased later.

tions so that they can be installed at many locations within a large corporation.

SBS will provide private switched networks to large organizations. Several corporations will usually share the same transponder. Security features will be built into the system to give different degrees of privacy protection. To any one corporation it will appear as though it has its own private system. The satellite, in effect, acts as a nationwide concentrator. If at one moment there is no traffic in California, the transmission capacity that was being used in California may be allocated to New York. All types of traffic as well as all locations will share the transponder capacity.

It is intended that SBS should provide cost-effective networks for telephone, electronic mail, and conventional data traffic. In addition it provides the opportunity for fundamentally new uses of transmission such as interactive operation with facsimile or graphic information. Audit trails or dumps used for security in computer systems can be transmitted to distant secure vaults. The ability to transmit brief bursts at very high data rates can change the architecture of computer systems. The SBS FCC filing contains the following paragraph:

> With the proposed SBS system, the distinction between central and remote computing is virtually eliminated. Central computers will be able to communicate with remote computers at virtually the same high data rates at

> which they process data internally. A company's data base can, in effect, be moved out to remote processors at all traffic concentration points, and thus be much closer to the company's most remote operations. This new accessibility will reduce the customer's terrestrial communications expense and improve the data processing service that his remote locations receive.

The access time on a mass storage unit is substantially longer than the access time on an SBS channel. Mass storage units or library systems could therefore be far away from the computers or data base management systems which use them. One mass storage unit could be shared by many distant computers.

THE SBS CAPACITY POOL

A corporate network derived from SBS transponders is designed to have a certain capacity which its various earth stations share. At certain times, however, a corporation may need, momentarily, a much higher capacity. It may need to read a section of a data base into a local computer. It may want to hold a video conference. To accommodate such needs the SBS control mechanisms retain a *pool* of unallocated capacity for each transponder which can be allocated *on demand* to any authorized user whose earth stations employ that transponder. A user can request instantaneous capacity of *not less than 448 kbps* from the pool.

ADVANTAGES OF TDMA

Demand assignment systems in commercial use with satellites in the first half of the 1970s were all FDMA, not TDMA. The reason was that TDMA needed very high-speed burst modems, high-speed manipulation of bit streams, stringent system timing, and sizable data buffers at each earth station. As the 1970s progressed the cost of high-speed digital equipment dropped and its reliability improved. Given appropriate cost and reliability, TDMA offers significant advantages over FDMA, as follows:

1. TDMA is highly flexible. Channels of widely differing capacities can be used and intermixed.
2. The maximum throughput with a TDMA system is greater than with an FDMA system.
3. There are no interference problems caused by the intermodulation of carriers. Consequently, the transponder can be used at its full power.
4. On an FDMA system the maximum number of users or the total earth terminal power output must be controlled to avoid driving the transponder into saturation. There is no such concern on a TDMA system.
5. There is less problem with interference between satellites when TDMA is used. The orbital spacing from a broadcast satellite can be much less.

6. The digital equipment used with TDMA can be designed to employ DSI (digital speech interpolation) to more than double the capacity for telephone users.

The simplest FDMA earth stations are likely to remain lower in cost than TDMA stations. This is especially so with fixed-assignment stations, which, in effect, permanently lease one or more channels.

REFERENCES

1. "Report to the Federal Communications Commission on a Satellite Link Test Conducted by IBM, 1974"—available from the Federal Communications Commission, Washington, D.C.

2. *FCC Filing of Satellite Business Systems,* Part II, Federal Communications Commission, Washington, D.C., 1976, pp. 4–23.

3. W. G. Schmidt, "The Application of TDMA to the INTELSAT IV Satellite Series," *COMSAT Tech. Rev.,* Vol. 3, No. 2, Fall 1973.

17 ERRORS IN TRANSMISSION

Some of the line control procedures used for data transmission on terrestrial links are inappropriate for satellite links. This is an important observation because today many data transmission systems are being installed with the simple substitution of a satellite channel for a terrestrial channel. A leased long-distance voice-grade circuit is often cheaper from a satellite common carrier such as Western Union or RCA, but the line control equipment for the voice channel may have been designed before the advent of domestic satellites.

The effect of inserting a satellite voice channel into existing systems is often to increase their response time or decrease their throughput somewhat. Both of these effects are acceptable on many systems, and the cost saving is worthwhile. Nevertheless, a minor modification of the line control mechanisms would enable the satellite channel to be used without a decrease in efficiency.

The space link, as we have discussed, has a poor signal-to-noise ratio, especially if inexpensive earth stations are used. Protection of the end user from the noise begins with the earth-station equipment which derives a channel for the user. How efficiently a data user employs that channel depends on the line control mechanisms of the machines he connects to it.

DELAY

We have already commented on the satellite transmission delay. It causes little concern in telephone conversations between persons who are used to it, and tends to make them slightly more polite in their speech. With data transmission it can cause delays and throughput degration when line control procedures designed for terrestrial links are used indiscriminately over satellite links. Polling a multipoint line with multiple terminal locations, for example, can cause an unacceptable increase in response time. On the other hand, network control procedures designed specifically for satellite channels can give both efficient channel utilization and a response time fast enough for psychologically effective dialogues with computers.

ERRORS

Data channels provided to users by satellite common carriers usually have error rates as good as or better than equivalent terrestrial links. To achieve the good error rate in spite of a signal-to-noise ratio much worse than on terrestrial links requires appropriate modem design and possibly the use of error-correcting codes.

When digital telephone channels are used the speech can sound acceptably good even when the channel has an error rate of between 1 error bit in 100 and 1 bit in 1000. Earth stations producing this high error rate are much cheaper than earth stations with an error rate comparable to terrestrial links, say 1 bit in 10^5, so it can be economically viable to operate at such high error rates.

Whether on a noisy channel or not, computer data need to be protected with efficient *error-detecting* codes. These codes are now well understood and can protect data with a high level of safety [1]. When data are found to be in error, the sending location is notified, and it retransmits the data in question.

Error detection and retransmission can give efficient operation when the error rate is not too great—say 1 bit error in 10^5 or better. With error rates of 1 bit error in 10^4 or worse, the error detection alone is generally a poor technique for most purposes and needs to be supplemented with the use of *error-correcting* codes. For transmission of high data rates 1 error in 10^5 is generally inadequate as shown in the following chapter. When forward error correction is employed it is normally transparent to the users of the circuit. The users merely perceive the resulting circuit speed and error rate.

Many satellite circuits are engineered to have an error rate of about 1 error bit in 10^7.

CHANNEL EFFICIENCY

The efficiency with which a link operates is related by the probability of a block of data having to be retransmitted.

Suppose that a block which is transmitted contains N_d bits of data.

The block also has to contain a number of overhead bits such as address information, synchronization bits, control bits, and the bits required for error detection.

Let the number of overhead bits be N_h.

The length of the block is then $N_d + N_h$.

Let the probability of a bit in error be P_B.

The probability that a bit is correct is $1 - P_B$, so the probability that all the bits in a block are correct if bit errors occur independently is $(1 - P_B)^{(Nd + Nh)}$.

That is, the probability of the block having an error in it is

$$P_E = 1 - (1 - P_B)^{(Nd + Nh)} \tag{17.1}$$

On terrestrial links the bit errors are *not* all independent. Errors tend to arrive in bursts, often caused by the effects of man-made noise, and hence P_E is best assessed by empirical measurements. On satellite links the noise caused by the poor quality of the space link is Gaussian ("white") noise, and hence Eq. (17.1) gives a good approximation of P_E.

The probability of a block having to be retransmitted once because of error is P_E. The probability of it having to be retransmitted a second time is P_E^2. If it is unlucky, it may have to be retransmitted many times; hence the mean number of bits required for transmission is

$$N_t = (N_d + N_h)(1 + P_E + P_E^2 + \cdots)$$
$$= \frac{N_d + N_h}{1 - P_E} \qquad (17.2)$$

Substituting from Eq. (17.1),

$$N_t = \frac{N_d + N_h}{(1 - P_B)^{(Nd + Nh)}} \qquad (17.3)$$

The efficiency of transmission can be described as the number of *information* bits sent, N_d, divided by the number of bits used in transmitting them, N_t:

$$\text{Efficiency} = \frac{N_d(1 - P_B)^{(Nd + Nh)}}{N_d + N_h} \qquad (17.4)$$

Figure 17.1 plots this efficiency figure for different bit error rates, assuming 100 overhead bits per block ($N_h = 100$). It will be observed that for any given error rate there is an optimum size of transmission block. For error rates better than 1 bit in 10^5, i.e., $P_B < 0.00001$, the transmission efficiency can be better than 90%. For error rates of 1 bit in 10^3 or worse, transmission efficiency is below 50%, and often well below it. If earth-station equipment is designed for these low error rates, which are appropriate for speech transmission, then forward error correction should be used for data transmission.

RESPONSE TIME

Retransmission of error blocks has a more serious effect on response times and control mechanisms than it does with terrestrial links because of the long propagation time. A sending station has to wait at least 540 milliseconds before it receives an acknowledgment from the receiving station saying that the block was received correctly. It must hold the block until that acknowledgment is received. If it has many blocks to transmit, it will be transmitting new blocks before the acknowledgment is received. It must therefore have a buffer large enough to hold many blocks, all awaiting acknowledgment.

Let t_F be the time taken to transmit a block if it is received correctly the first time. t_F will be approximately 270 milliseconds for a short block.

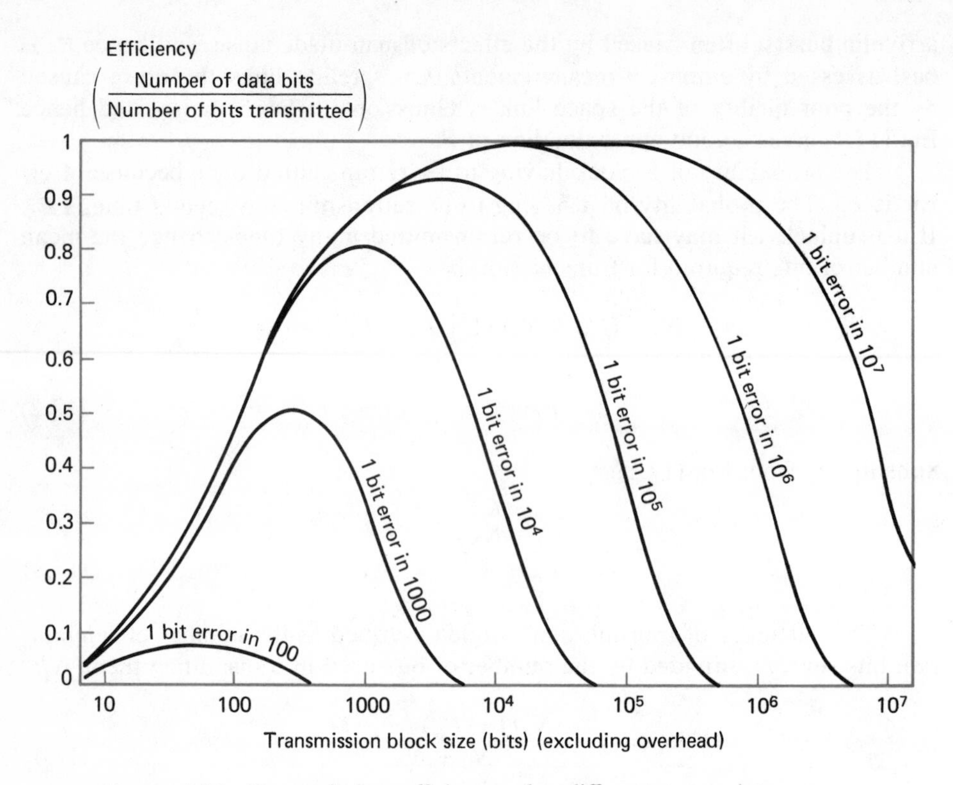

Figure 17.1 Transmission efficiences for different error rates, assumed 100 overhead bits per block.

Let t_R be the time taken between the time a block is received incorrectly and the time it is received again after retransmission. The sending station might typically wait 560 milliseconds after sending a block for an acknowledgment of successful transmission and then try again; t_R would be approximately 560 milliseconds for a short block.

Let P_E again be the probability of a transmitted block having an error.

$$t_t = t_F + t_R(P_E + P_E^2 + P_E^3 + \cdots)$$
$$= t_F + \frac{t_R P_E}{1 - P_E} \tag{17.5}$$

Substituting P_E from Eq. (17.1),

$$t_t = t_F + t_R\left[\frac{1-(1-P_B)^{(Nd+Nh)}}{(1-P_B)^{(Nd+Nh)}}\right]$$
$$= t_F + t_R\left[\frac{1}{(1-P_B)^{(Nd+Nh)}} - 1\right] \tag{17.6}$$

If the satellite protocol is designed so that the user is given an extremely high-speed channel for a brief moment, then t_F and t_R will be primarily related to the propagation delay and the multiple-access technique, for assigning time slots. Often today, however, the user is given a channel derived by multiplexing which is comparable in speed to terrestrial channels. The time to send the bits then becomes significant.

Suppose that the channel permits the user to transmit at S bps.

The time to send the block will then be $(N_d + N_h)/S$ seconds.

Let N_d be the propagation delay in seconds. Then

$$t_F = t_D + \frac{N_d + N_h}{S} \tag{17.7}$$

and

$$t_R = t_D + t_A + t_F \tag{17.8}$$

where t_A is the time the receiving location takes to formulate and send an acknowledgment.

Substituting these two equations into Eq. (17.6), we find that the mean time taken to transmit a block is

$$t_t = t_D + \frac{N_d + N_h}{S} + \left(2t_D + t_A + \frac{N_d + N_h}{S}\right)\left[\left(\frac{1}{(1 - P_B)^{(N_d + N_h)}}\right)^{-1}\right] \tag{17.9}$$

Figure 17.2 plots t_t for some typical values:

Channel speed, $S = 4800$ bps.

Overhead per block, $N_h = 100$ bits.

Propagation delay, $t_D = 270$ milliseconds.

Acknowledgment time, $t_A = 20$ milliseconds.

RETRANSMISSION PROBABILITIES

If the satellite channel operated at very high speed, the significant component of the total transmission time would be the sum of the propagation delays, i.e., 270 milliseconds if no error occurred, 270 + 540 milliseconds if one retransmission is needed, and $270 + 540N$ if N retransmissions occurred.

The probability of having N or more retransmissions is P_E^N. Consequently the probability of having exactly $N - 1$ retransmissions is $P_E^{N-1} - P_E^N$. Table 17.1 lists these probabilities.

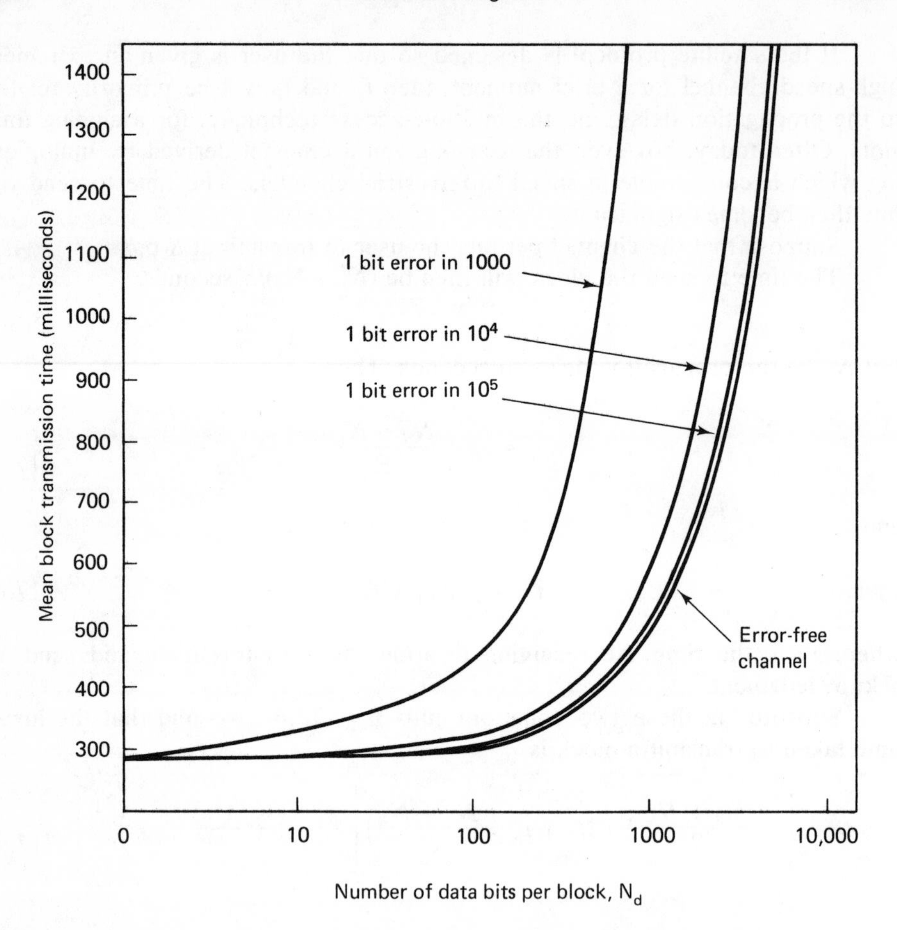

Figure 17.2 Mean transmission time on a 4800 bit per second channel after error detection and retransmission.

Figures 17.3 and 17.4 plot the retransmission probabilities for different bit error rates.

ARQ

A system which detects an error in data and has those data automatically retransmitted is called an ARQ (automatic repeat request) system. Most terrestrial data transmission links use ARQ.

ARQ systems are of two types: *stop-and-wait ARQ* and *continuous ARQ*.

Stop-and-wait ARQ is the most widely used. After sending a block the transmitting terminal waits for a positive or negative acknowledgment from the receiving terminal. If the acknowledgment is positive, it sends the next block. If it is negative, it resends the previous block. Data transmission codes such as

Table 17.1

Number of Retrans- missions	Sum of Propagation Delays (msec)	Probability	Probability for Different Values of P_E				
			0.1	*0.03*	*0.01*	*0.003*	*0.001*
0	270	$1 - P_E$	0.9	0.97	0.99	0.997	0.999
1	810	$(1 - P_E) P_E$	0.09	0.029	0.0099	0.00299	0.001
2	1350	$(1 - P_E) P_E^2$	9×10^{-3}	8.7×10^{-4}	9.9×10^{-3}	9.0×10^{-6}	10^{-6}
3	1890	$(1 - P_E) P_E^3$	9×10^{-4}	2.6×10^{-5}	9.9×10^{-5}	2.7×10^{-8}	10^{-9}
4	2430	$(1 - P_E) P_E^4$	9×10^{-5}	7.9×10^{-7}	9.9×10^{-7}	8.1×10^{-11}	10^{-12}

the ASCII code contain special characters for positive acknowledgment (an *ACK character*) and negative acknowledgment (a *NAK character*).

With *continuous* ARQ the transmitting terminal does not wait for an acknowledgment after sending a block; it immediately sends the next block. While the blocks are being transmitted the stream of acknowledgments is examined by the transmitting terminal. When the transmitting terminal receives a negative acknowledgment, or fails to receive a positive acknowledgment, it must determine which block was incorrect. The blocks are therefore numbered, often on terrestrial links with a 3-bit binary number (modulo 8). The acknowl-

Figure 17.3 Probabilities of retransmission on a channel with an error rate of 1 bit in 1000.

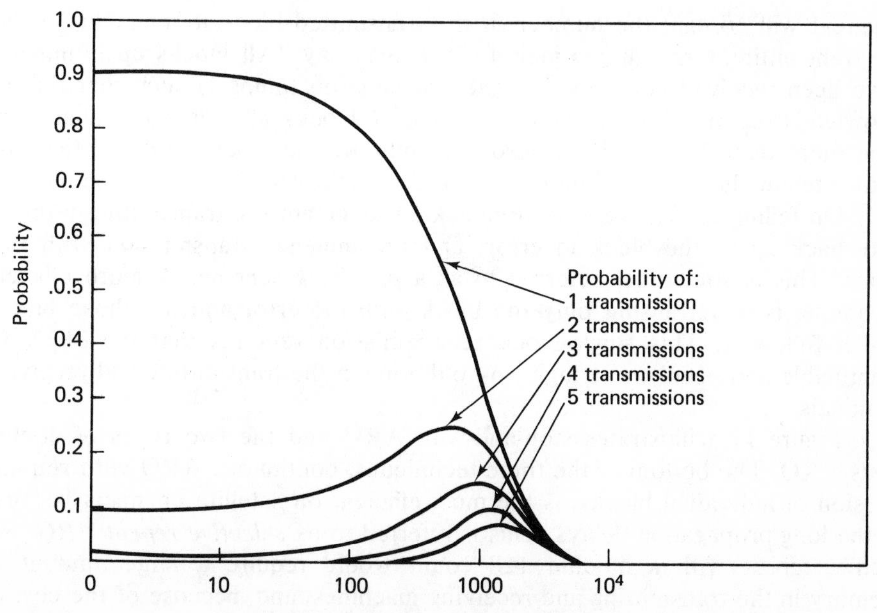

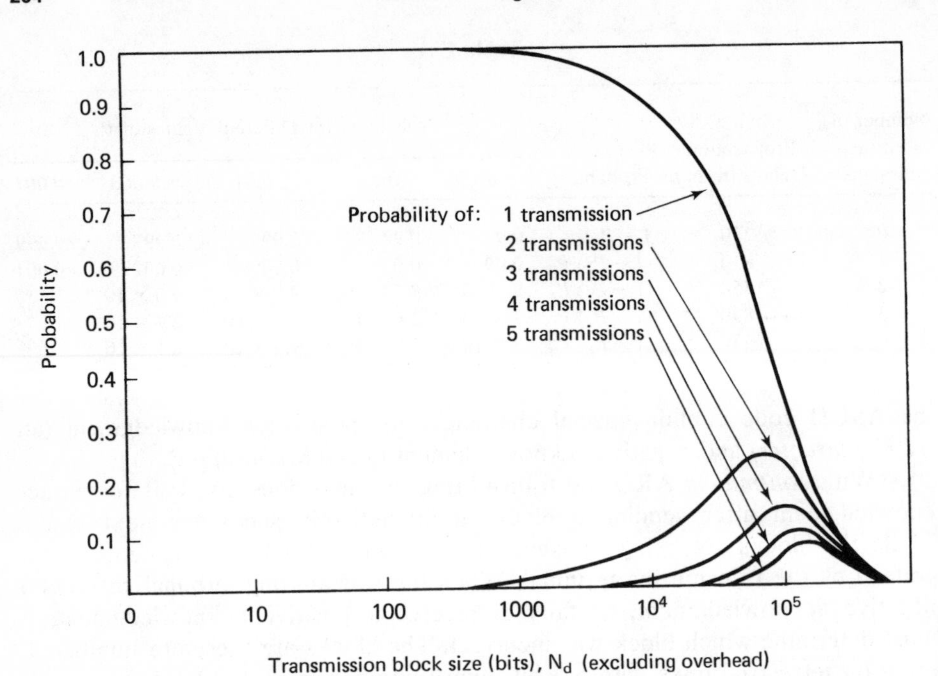

Figure 17.4 Probabilities of retransmission on a channel with a bit error rate of 1 in 10^5.

edgment will contain the number of the transmitted block it refers to so that the transmitting terminal can identify it. It may say "All blocks up Number *n* have been received correctly." If the transmission timing is such that the acknowledgment may be received more than 7 blocks after it was transmitted, then more than 3 bits will be needed to number the blocks. Such is the case with satellite channels, and here a count of modulo 128 (7 bits) may be used.

On failing to receive a positive acknowledgment the transmitting terminal may back up to the block in error, and recommence transmission with that block. This is sometimes referred to as a *pull-back* scheme. A more efficient technique is to retransmit only the block with the error and not those blocks which follow it. This single-block retransmission requires that the block be identifiable and needs more logic and buffering in the transmitting and receiving terminals.

Figure 17.5 illustrates stop-and-wait ARQ and the two types of continuous ARQ. The bottom of the three techniques, continuous ARQ with retransmission of individual blocks, is the most efficient on satellite channels because of the long propagation delays. This is referred to as *selective repeat ARQ*. Selective repeat with a modulo 128 count would require a large amount of memory in the transmitting and receiving machines, and, because of the cost of

1 Stop-and-wait ARQ (half-duplex line)

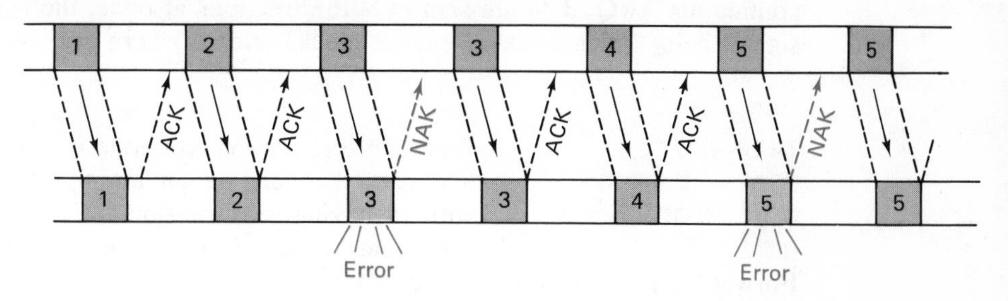

2 Continuous ARQ, with pull-back (full-duplex line)

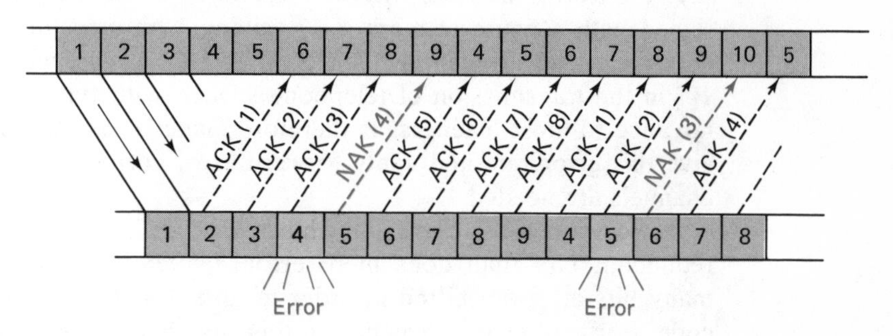

3 Continuous ARQ, with retransmission of individual block (full-duplex line) (sometimes called selective repeat ARQ)

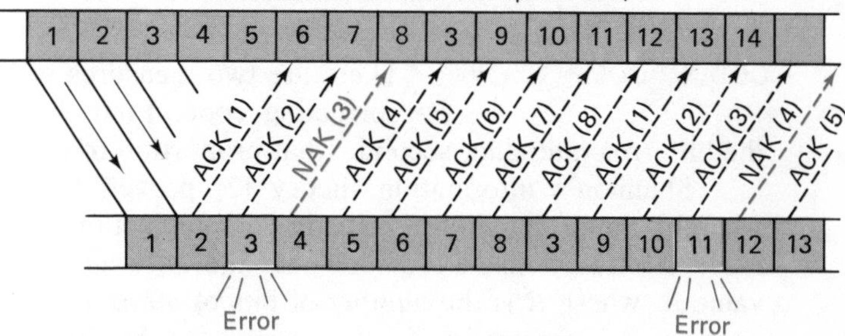

Figure 17.5 Three types of ARQ (automatic repeat request) used on today's teleprocessing equipment. The continuous ARQ here uses a modulo 8 (3-bit) count. The bottom technique is more expensive, and usually not necessary, as illustrated in the following chapter.

this, pull-back ARQ is usually preferred. On some full duplex circuits with continuous ARQ, data are sent in both directions at once, the acknowledgment signals being "piggybacked" on the blocks of data where possible.

THE NEED FOR FORWARD ERROR CORRECTION

When an error-*correcting* code is employed the system uses the code in an attempt to correct errors without having to retransmit the data. It is therefore sometimes called *forward error correction* (FEC). Forward error correction is not so safe as the use of error-detecting codes. Some errors slip through. It therefore needs to be backed up by the use of error detection and ARQ.

Forward error correction can, however, improve the bit error rate from, say, 1 bit error in 1000 bits to 1 bit error in 10 million. An earth station designed with a figure of merit, G/T, giving 1 bit error in 10^3 is significantly less expensive than one giving 1 bit error in 10^7. As we commented, 1 bit error in 10^3 in the transmission of telephone speech with typical codecs causes no annoyance to most listeners. It therefore makes sense to design earth stations to give this error rate and apply forward error correction when data transmission channels are needed.

Forward error correction, however, needs a much greater proportion of redundant bits than does mere error *detection*. In some schemes, twice as many bits are transmitted in order to give a suitably powerful error-correcting code. Forward error correction in this case halves the channel throughput. Satellites have a prolific data capacity—usually more than can be used—so the reduction in channel throughput is an acceptable trade-off when it permits voice channels to be obtained less expensively.

CODING COMPLEXITY

There are two measures of effectiveness of an error-correcting code. First, how much does it improve the bit error rate, and, second, what proportion of redundant bits is needed?

Shannon's information theory [2] proved that a coding method exists whereby the probability of error in communication over a noisy channel can be made arbitrarily small provided that the *information rate* is less than a given value R, where R is the number of bits of *information* received per bit of transmission. R is referred to as the *rate* of a code. In a 3/4-rate code, three-quarters of the bits transmitted are information and the rest redundant.

A channel with an equal probability of an information bit being a 0 or 1 is referred to as a *binary symmetric* channel. Shannon showed that in a binary symmetric channel with an equal probability P_B of any bit being in error, the probability of error in communication can be made smaller than any assignable value provided that the information rate is less than

$$R = 1 + P_B \log_2 P_B + (1 - P_B) \log_2 (1 - P_B) \qquad (17.10)$$

This equation gives values of R which are close to 1 for most channels; thus,

$$P_B = 10^{-3}; \quad R = 0.9886$$

$$P_B = 10^{-4}; \quad R = 0.9985$$

$$P_B = 10^{-5}; \quad R = 0.9998$$

In reality, however, to reduce a bit error probability of 10^{-3} to, say, 10^{-6} without retransmission and with a value of R close to the Shannon limit would need an impossibly complex coding scheme. A third factor is therefore critical in the choice of error-correction method: How complex can the encoding and decoding process be? Powerful error-correcting codes can be devised but need a special-purpose high-speed computer at each end of the link to make them operable.

A compromise is needed between degree of protection, information rate R, and complexity of encoding.

ERROR PATTERNS

The bit errors of terrestrial links are not all independent of one another. They tend to come in bursts, sometimes with several adjacent bits being damaged, the bursts usually being caused by terrestrial disturbances such as noise impulses from switching equipment, electrical machines, lightening flashes, etc. A code which is efficient for correcting bursts with error-free intervals between the bursts is different from a code which is efficient when the transmission is damaged by independent random errors.

The noise on space links is mainly Gaussian "white" noise which tends to damage the transmitted symbols at random. The damage to one symbol is largely independent of the damage to others. By *symbol* we refer to the information which is transmitted at one instant in time. With a binary modulation process this will be 1 bit, and so the bit errors will be independent. With a quaternary, modulation process such as QPSK (quaternary phase shift keying) 2 bits will be encoded at one instant. 8-phase modulation encodes 3 bits.

With quaternary modulation the pairs of bits will be damaged largely independently of one another. Some types of damage will be more probable than others. For example, a phase change of 45° will be more probable than a phase change of 135° (see Fig. 11.9). With 8-phase modulation a change of 22.5° will be the most likely. To be most effective on a space link, the error-correction process should take these probabilities into consideration. *The error-correction process therefore needs to be linked to the modulation process.*

Some of the classical error-correcting codes assume an equal likelihood of any bit being damaged. These are sometimes referred to as *hard-decision* codes. A code which takes into consideration that some errors are more probable than others is called a *soft-decision* code.

TWO TYPES OF CODE

There are two types of error-correcting codes, referred to as *block codes* and *tree codes.*

In a block code the input stream of bits is broken into fixed-length blocks. Every block contains a fixed number of information bits and redundant bits. The block, referred to as a *code word,* is composed, transmitted, and decoded, independently of all other blocks.

When a tree code is used, the input stream of bits is not broken up into independent blocks but is processed continuously. The encoder takes the bits, or small groups of bits, and treats them as though they represent branches of a tree. For each branch in the input stream, a coded group of bits is transmitted. The coded group will depend not on *one* branch but on the previous branches also, possibly a very large number of previous branches. After transmission the decoder has the task of determining what was the most probable set of tree branches and hence what was the input stream. The decoding of tree codes can be far more complex than block codes, but a higher degree of safety can be provided for a given amount of redundancy.

BLOCK CODES

A block code breaks the bit stream into blocks of k bits, each block containing n bits of information, and hence $n-k$ bits of data. The *rate* of the code is thus n/k.

Sometimes the letters n and k are used to refer to *symbols* transmitted, rather than bits. The letter q is used to denote the number of distinct symbols transmitted. Thus for quaternary phase shift keying, $q=4$.

Figure 17.6 shows a simple block code for binary symbols ($q=2$) with 5 bits per block; $n=5$ and $k=2$. Thus the rate $R=2/5$.

Because $k=2$, there are $2^2=4$ valid blocks which can be transmitted. $5^2=32$ possible blocks might be received after noisy transmission. The possible blocks are listed in four columns of eight words in Fig. 17.6. At the top of each column is the valid bit combination, the first 2 bits of which are the information. Underneath this are the five words which are *1 bit different* from a valid word. The remaining words in the column are 2 bits different from the valid word.

When the decoding mechanism receives a word it assumes that the word transmitted was that at the top of the column. If there is only one error, that error will be corrected. If there are two errors in a word, these *might* be corrected but often they are not.

The error-correction scheme shown in Fig. 17.6 is not very efficient because the block size is small. Consider larger blocks of n bits, containing k in-

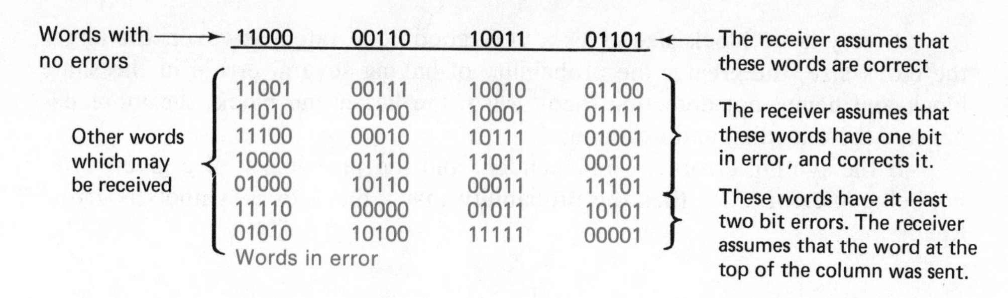

Figure 17.6 An illustration of a simple error-correcting block code, with a rate of 2/5. The first two bits of each word are the *data;* the other three are the redundancy bits. Hence the rate is 2/5. (*Redrawn from reference* [2].)

formation bits. There would be 2^k valid words and hence 2^k columns if they were listed as in Fig. 17.6. Each column would contain n words which have 1 bit error. If all 1-bit errors are to be corrected, there must then be $2^k(n+1)$ words in the diagram. 2^n combinations of bits are possible. A code can thus correct all words with a single error of $2^k(n+1) \leqslant 2^n$, i.e.:

$$k \leqslant \log_2\left(\frac{2^n}{n+1}\right) = n - \log_2(n+1)$$

The rate of the code is

$$R = \frac{n}{k} \leqslant 1 - \frac{1}{n}\log_2(n+1) \tag{17.11}$$

For a given value of k the smallest value of n that will give complete single-error correction is that which gives $n+1$ entries in each of the 2^k columns, i.e., $2^k(n+1)$ possible words, so that

$$2^n = 2^k(n+1)$$

$n+1$ is then a power of 2, i.e., $n = 2^{M-1}$, where M is an integer.

The larger the block, the better the rate of the code. Thus,

Block Size (bits)	R
3	1/3
7	4/7
15	11/15
31	26/31
•	
•	
•	
1023	1013/1023

A very large block size gives a very good code rate. However, the larger the block size, the greater the probability of having several errors in the same block and hence not detecting them. Also, the larger the block, the more expensive the encoding and decoding.

If the symbol errors are independent and the probability of a given symbol being in error is P_B, then the probability that a block of n symbols is error-free is

$$(1 - P_B)^n$$

The probability that a block will contain one error is $nP_B\,(1 - P_B)^{n-1}$.

If the code corrects only blocks with one error, the probability of an error remaining after the decoding is

$$1 - (1 - P_B)^n - n^P{}_B\,(1 - P_B)^{n-1}$$

This is plotted in Fig. 17.7 for bit error rates, P_B, of 0.001 and 0.0001. If P_B

Figure 17.7 Capability of a block code which corrects single bit errors, when the errors are randomly distributed.

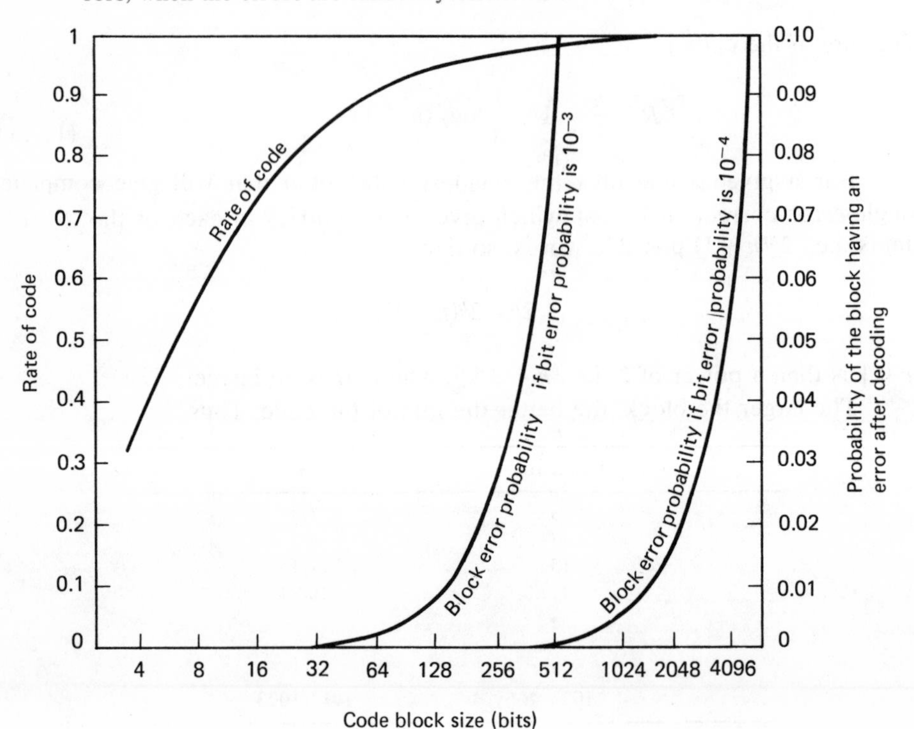

is 0.001, the block size selected might be between 16 and 64, and the error-correcting code would be backed up with powerful error-*detection* capability.

If the bit errors occur in bursts rather than at random, then the code described is of little value. Different types of error-correcting codes are designed for situations where the errors tend to be clustered, as on terrestrial telephone lines.

A table somewhat like that in Fig. 17.6 can be devised which corrects two errors per block in all cases as well as single errors. The columns in such a table will be longer than those in Fig. 17.6. If there were 5 bits per block as in Fig. 17.6, the columns would have to have 16 entries. The code would thus have a rate of 1/5. As before, when the block size is larger, the rate improves greatly. Some block codes correct up to 3 bit errors per block.

TREE CODES

Figure 17.8 shows how a simple tree code is formed [3]. The tree illustrated is a balanced binary tree, and each branch of the tree represents bits which form part of the encoded bit stream. The input bit stream directs the encoding process down one path through the tree, each input causing a branch as shown.

The decoding process attempts to decide from the received bit stream which is the most probable path that was taken through the tree. This can be a complex operation. Tree codes are generally more complex to decode than block codes, and much of the ingenuity in the design of tree codes goes into devising suitably inexpensive encoding and decoding mechanisms. The tree used could be almost infinite in theory, but in practice the coding is restricted to manageable trees of fixed length. The larger the tree, the greater the probability of correcting errors, but the greater the cost.

The tree in Fig. 17.8 operates with *bits* in the input stream. Trees used in practice often employ the *symbols* which are the output of a modem (for example, 4-phase or 8-phase conditions); symbols cause the branching. The error-correction design is therefore linked to the modem design. The rate of the code illustrated in Fig. 17.8 is 1/2, because 1 input bit is converted into 2 output bits. Other rates can be used; for example, the 3-bit symbols may cause branches which are encoded with 4 bits, giving a 3/4 rate code.

When symbols are used for determining the branches, the decoding process can take into consideration which types of symbol damage are the more probable—a soft-decision code.

In the general case a symbol with q possible values reaches each node of the tree. For the binary bit stream in Fig. 17.8, $q = 2$. For symbols from an 8-phase PSK modem, $q = 8$. With a simple tree code each node handles one symbol, and so there are q branches from each node. In a more complex code each node handles k symbols at a time, and hence there are q^k branches from the node. The number of output symbols from each branch is n. The rate of the code is therefore k/n. In Fig. 17.8, $q = 2$, $k = 1$, and $n = 2$.

Figure 17.9 shows the general form of encoder for a tree code. k symbol

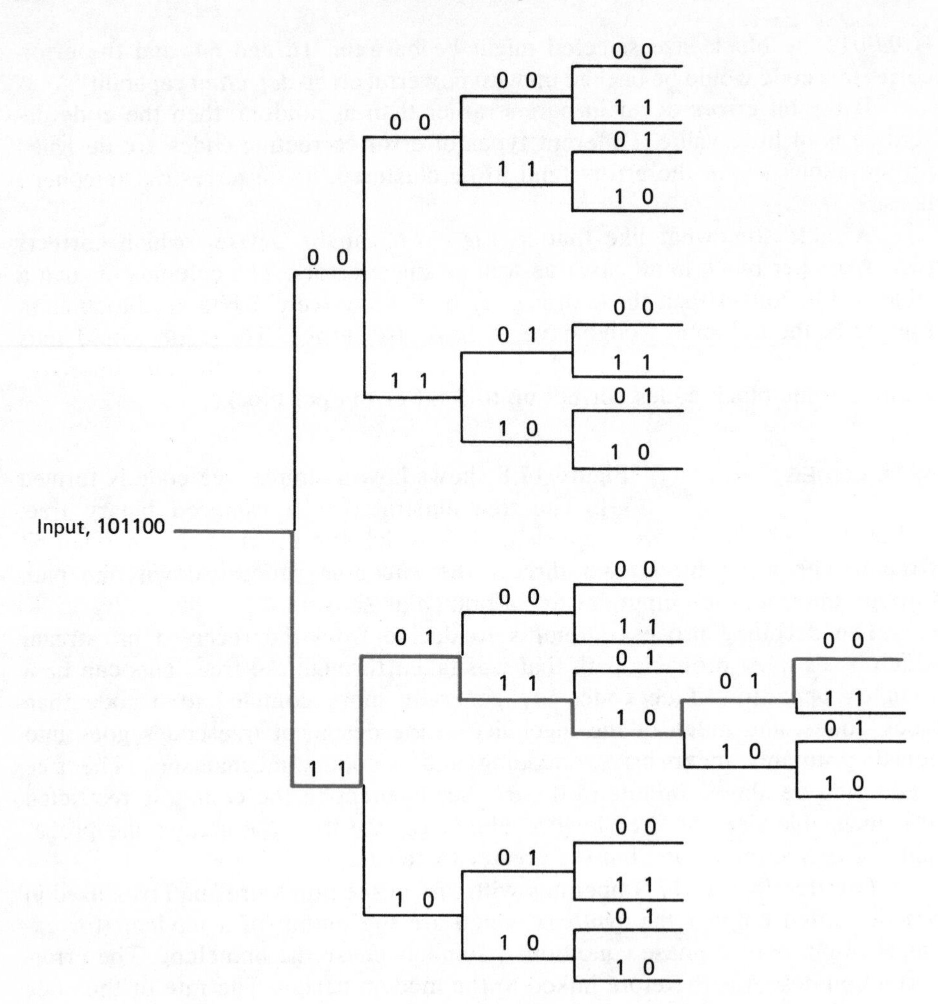

Figure 17.8 A simple code tree, with a rate, $R = 1/2$. Each bit in the input stream causes a branch in the tree. 1s cause a downward branch; 0s cause an upward branch. Thus the input stream 101100 is encoded as 110111100100. (*Redrawn from reference* [2].)

lines form the input to the circuit and n symbol lines form the output. The circuit can be of many forms, containing shift registers and logic gates to do the encoding.

The decoder in some cases is much more complex than the encoder.

Many decoders progress in a forward direction through the input stream, attempting to guess the path that was taken through the tree. A more expensive type of decoder has the ability to backtrack when it needs to and change the decisions it made earlier, if necessary. How often it backtracks depends on the error probability. If the encoding tree is large, the backward search could be

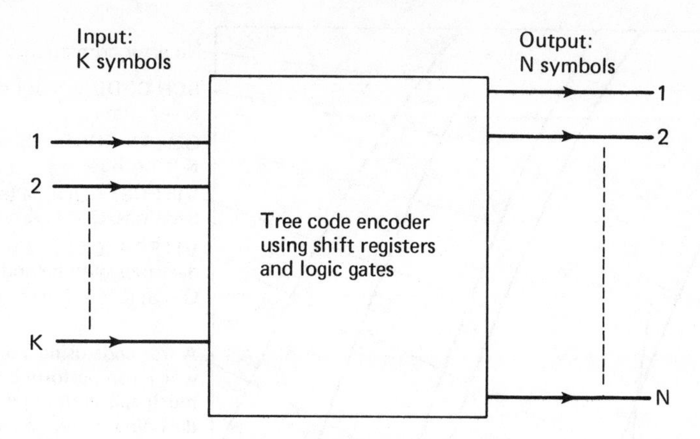

Figure 17.9 The general form of the tree code encoder, with a rate $= K/N$.

deep. If the background search is large enough, the probability of error becomes negligible. The limiting factor is the number of computations needed in searching back and forth through the tree.

The cost of computation is falling, and with satellites it is economic to use complex error-correction techniques, although they would not be used on terrestrial links of the same capacity. It is economic to allow a satellite channel to be noisier than the channels on earth, and correct the errors when necessary.

EFFECT OF ERROR CORRECTION

Figure 17.10 shows the effect on a typical satellite channel of some error-correcting codes of moderate complexity. The channel employs a PSK modem. It will be seen that the tree codes are more effective than the block codes. The soft-decision tree code is better than the hard-decision one. The decoder with the backward search capability gives a near-vertical line on the chart, indicating that it could be used to give highly error-free channels—at a cost. Details of the codes shown are given in Ref. 1.

The Viterbi soft-decision tree code gives a line with a signal-to-noise ratio about 4.5 decibels less than with no error correction. It permits earth stations to be used with a figure of merit (G/T) 4.5 decibels less (see Fig. 7.7). Alternatively, a higher data rate can be used. 4.5 decibels is equivalent to about 2.8 times the data rate, but the data rate is halved in using the code of rate $= 1/2$.

The 4- to 5-decibel saving has been achieved in practice. Error correction is primarily of value when a transponder is used below saturation.

Figure 17.11 illustrates this trade-off, showing how the channel capacity of an INTELSAT IV transponder using spot beams with SPADE multiple access varies with the figure of merit, G/T, of the earth-station antenna [5]. The curve for uncoded transmission shows that for large antennas the capacity of

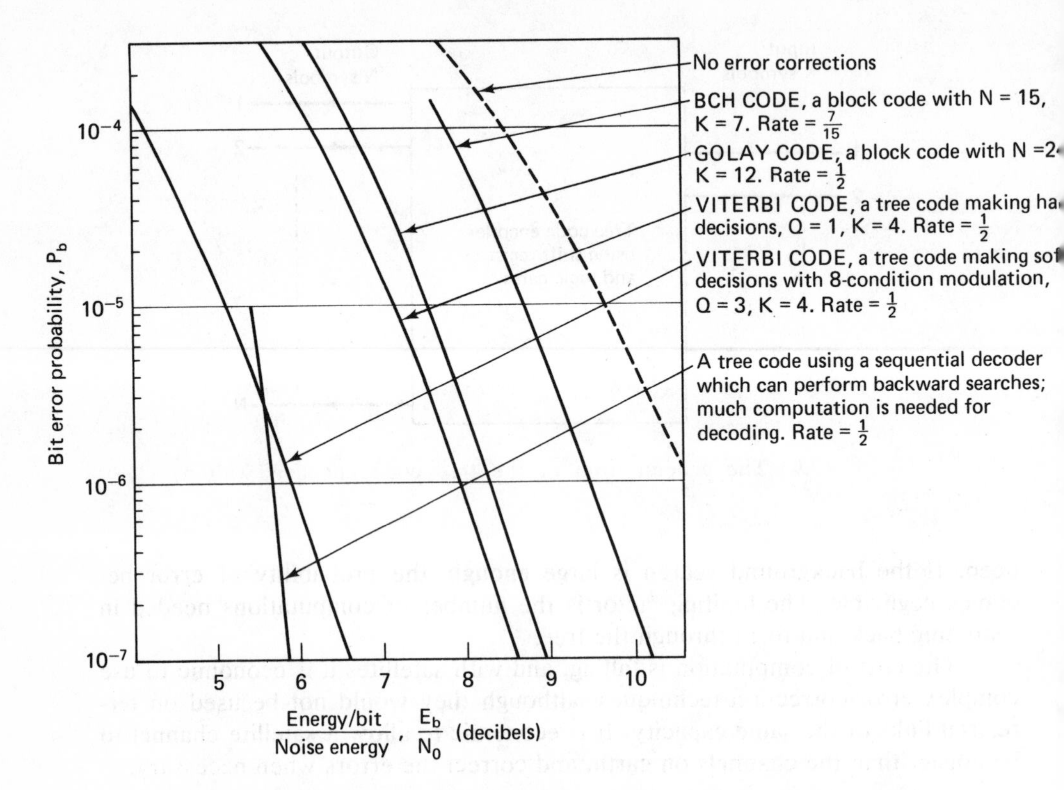

Figure 17.10 The effect of different error correcting codes on a typical satellite channel. (*Data from reference* [4].)

the satellite is 800 voice channels. However, when the earth station *G/T* falls below 28, the capacity of the transponder falls rapidly. (A figure of merit of *G/T* = 28 is approximately that of a 30-foot antenna with an uncooled amplifier.) The transponder must apply more power to each channel to preserve an acceptable signal-to-noise figure with the smaller earth antenna. The transponder is power limited, not bandwidth limited, and the power restriction means that fewer channels can be used.

When an error-correcting code is used, a greater level of noise can be tolerated. Figure 17.11 shows the curve for a *three-quarter-rate* code and a *half-rate* code. A half-rate code halves the number of channels usable with the large earth stations. However, with small earth stations with a *G/T* below 19 (25-foot antenna) the half-rate code permits more channels to be sent than on a system without an error-correcting code.

REFERENCES

1. James Martin, *Security, Accuracy, and Privacy in Computer Systems,* Prentice-Hall, Englewood Cliffs, N.J., 1974, Chap. 8.

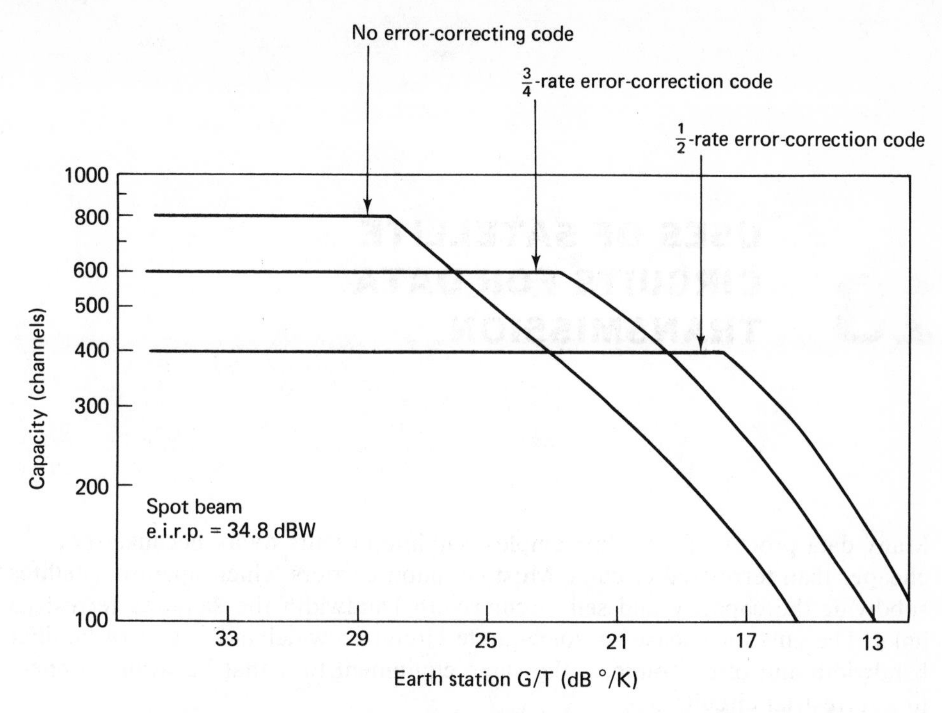

Figure 17.11 The tradeoff between channel capacity and the use of error-correcting codes on INTELSAT IV. (*Reproduced from reference* [5].)

2. C. E. Shannon, "A Mathematical Theory of Communication," *Bell System Tech. J.*, Vol. 27, July and Oct. 1948.

3. W. W. Peterson and E. J. Weldon, Jr., *Error-Correcting Codes,* 2nd ed., M.I.T. Press, Cambridge, Mass., 1972. An excellent treatment of error-correcting codes.

4. Defense Communications Agency, "Satellite Communications Reference Data Handbook," *NTIS Document No. AD-746-165,* U.S. Department of Commerce, Springfield, Va., 1972.

5. I. Edelson, and A. M. Werth, "SPADE System Progress and Application," *COMSAT Tech. Rev.*, Vol. 2, No. 1, Spring 1972.

18 USES OF SATELLITE CIRCUITS FOR DATA TRANSMISSION

Many data processing systems employ satellite circuits today because they are cheaper than terrestrial circuits. Most common carriers which operate satellites subdivide the capacity and sell circuits with bandwidth the same as terrestrial links. The customer leases a voice-grade circuit or wideband circuit of familiar bandwidth and often connects the same equipment to it that he would connect to a terrestrial circuit.

The circuit is used in this case as a point-to-point link. It does not take advantage of the broadcasting nature of satellites which permits one circuit to interconnect many points.

Substituting a satellite circuit for a terrestrial circuit without changing the equipment or software used can have three disadvantages, all caused by the increase in propagation delay:

1. Throughput can be degraded, sometimes to a half or less of that with an equivalent terrestrial circuit.
2. Mechanisms which regulate the flow or pacing of data can be interfered with, in some cases causing a device such as a remote printer to cease functioning correctly.
3. Response time can be lengthened, sometimes by about half a second, but sometimes by several seconds if a technique like polling is used.

It is important to note that the propagation delay need have *no* harmful effects on data transmission if the protocols and system design have taken it into consideration. The throughput of a satellite channel *ought* to be approximately as high as that of a conventional channel of the same bandwidth. The pacing of distant machines *ought* to work well. Response time *ought* to be fast enough for efficient interactive computing. However, to achieve this, the software, and possibly hardware, of the machines which communicate may need to be slightly different than that designed for conventional circuits on earth.

A problem may arise on public data circuits if the common carrier can occasionally switch to a satellite link instead of a ground circuit. If this possibility exists it needs to be considered when a data transmission system is designed.

LINK CONTROL PROTOCOLS

Since the mid-1970's a type of line control protocol has come into use which employs continuous ARQ with pullback (see Fig. 17.5). The International Standards Organization (ISO) refers to this protocol as HDLC (high-level data link control). Other standards organizations have a similar protocol (different in subtle details) and various manufacturers have their own version of such a protocol such as IBM's SDLC, Burroughs' BDLC, Univac's UDLC, etc. DEC (the Digital Equipment Corporation) has its own line protocol, which is similar in its ARQ behavior but entirely different in detail, called DDCMP (digital data communications management protocol).

With many older protocols for the control of a data transmission link, the sending machine transmits a frame of data and must wait for an acknowledgment from the receiving machine before it sends the next frame. On a satellite link it will have to wait about 540 milliseconds. If the transmission speed is reasonably high and the frame size is not very large, a sending machine could spend most of its time waiting. The utilization efficiency of the link would be very low. This corresponds to the *stop-and-wait ARQ* diagram of Fig. 17.5.

In HDLC and any protocols with continuous ARQ, the sending machine can send up to a given number of frames before receiving a response. In most implementations it can send up to 7 frames. *Modulo 8 counts,* requiring 3 bits, are used to number each frame sent and to refer to frames that have been received. Most of these protocols can be extended with one 8-bit byte so that two *modulo 128 counts* can be used, meaning that 127 frames can be sent before a response is received.

Continuous ARQ is highly desirable for satellite channels. Users of binary synchronous line control (in which each frame needs a response) have found that the substitution of a satellite circuit for their existing terrestrial circuit lowered the throughput dramatically. An HDLC link, on the other hand, gave a throughput that was similar on satellite and terrestrial circuits. How much degradation occurs on a satellite channel depends upon the parameters used, such as the transmission rate, mean frame size, and the size of the count used for frame acknowledgment. If the parameters are selected appropriately the throughput of a point-to-point satellite circuit is the same as that of an equivalent terrestrial circuit.

THE EFFECT OF THE FRAME COUNT

Let us suppose that the sending machine can send M frames before receiving a response. In other words, it uses modulo $M + 1$ counts. In *binary synchronous* line control, $M = 1$. In HDLC, $M = 7$ or 127. Let F be the frame size (in bits), and S be the transmission speed (in bits per second).

When a frame is sent it is desirable that a response should be received within the time it takes to send $M - 1$ more frames, i.e. within $(M - 1)\ F/S$ seconds. If the overall response delay exceeds $(M - 1)\ F/S$, the link will be used inefficiently.

The satellite propagation delay is about 0.54 seconds. The total response delay will be somewhat longer than that, due to buffering requirements, to the responsing machine and often because a small delay occurs on the terrestrial tail circuits to the satellite earth station. A typical figure for the overall delay is 0.7 seconds.

It is desirable, then, that $(M - 1)F/S > 0.7$. This condition cannot be satisfied with $M = 1$.

With $M = 7$: the frame size should exceed $0.12S$ bits
With $M = 127$: the frame size should exceed $0.0056S$ bits

If a voice-grade circuit is used to transmit at 9600 bps, the frame size should exceed 1120 bits if $M = 7$, and 76 bits if $M = 127$. When a 56,000 bps circuit is used, the frame size should exceed 6533 bits if $M = 7$ and 311 bits if $M = 127$.

Such figures indicate that for voice-grade circuits via satellite HDLC and similar protocols are suitably efficient if implemented correctly with a sufficiently large frame size. For high transmission speeds the frame size needs to be higher. Binary synchronous and start-stop protocols are generally inadequate for satellite circuits.

Figure 18.1 plots the minimum frame size that should be used for different circuit speeds with $M = 7$ (3-bit count fields) and $M = 127$ (7-bit count fields).

INEFFICIENT TRANSMISSION

Let t_D be the overall response delay. For efficient transmission,

$$(M - 1)F/S > t_D \tag{18.1}$$

If this condition is not met, the sending machine will transmit M frames and then pause.

The time to transmit one frame is F/S seconds. The sending machine transmits one frame and receives a response t_D seconds later. In the time $F/S + t_D$, it sends M frames.

In time t, where $t \gg (F/S + t_D)$, it sends

$$\frac{Mt}{F/S + t_D} \text{ frames} \tag{18.2}$$

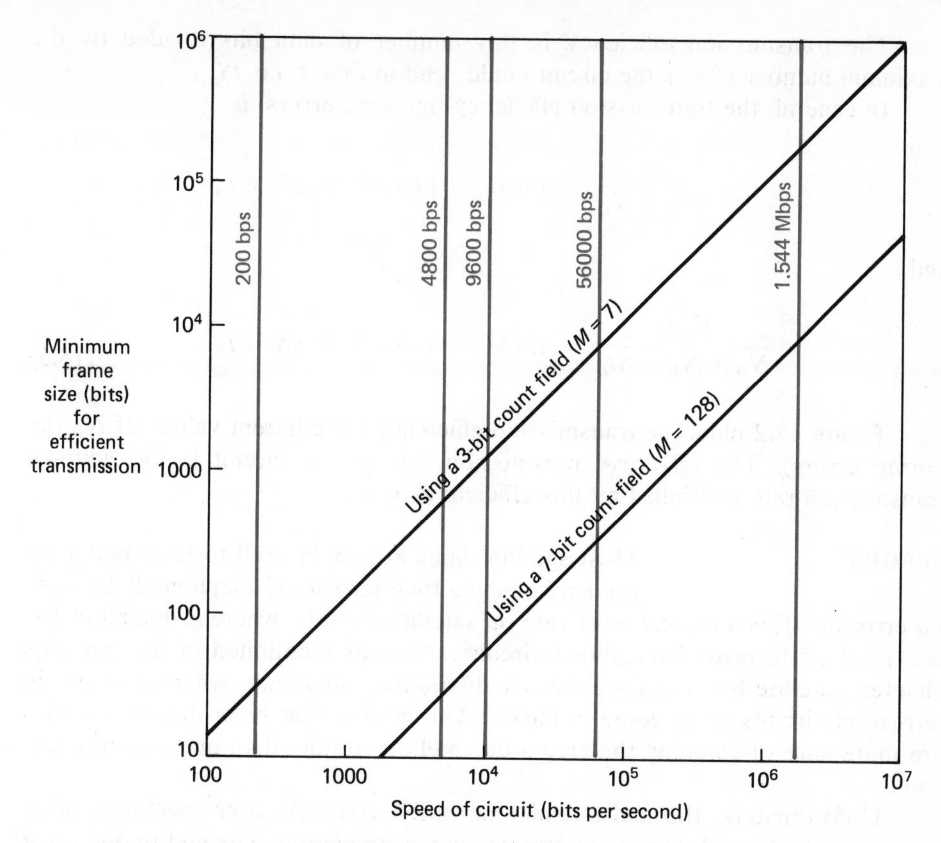

Figure 18.1 The minimum frame size for efficient transmission via satellite at different speeds using HDLC, SDLC, and similar line control protocols.

Let N_D be the number of data bits in a frame; let N_H be the number of overhead bits in a frame, then

$$F = N_D + N_H$$

If $(M - 1)F/S > t_D$ the sending machine can transmit without pauses and the transmission efficiency with continuous ARQ will be

$$\frac{N_D}{N_D + N_H}$$

If $(M - 1)F/S < t_D$ there will be pauses, and, substituting into Eq. (18.2), in time t seconds the sending machine transmits

$$N_D\left(\frac{Mt}{((N_D + N_H)/S) + t_D}\right) = \frac{N_D MtS}{N_D + N_H + t_D S} \text{ data bits}$$

The transmission efficiency is this number of data bits divided by the maximum number of bits the circuit could send in time t, i.e. tS.

In general, the transmission efficiency (ignoring errors) is

$$\frac{N_D}{N_D + N_H} \quad \text{if } (M - 1)\,(N_D + N_H)/S > t_D$$

and

$$\frac{N_D M}{N_D + N_H + St_D} \quad \text{if } (M - 1)\,(N_D + N_H)/S < t_D \tag{18.3}$$

Figure 18.2 plots the transmission efficiency for different values of N_D (ignoring errors). The *effective* transmission rate of the circuit is the nominal transmission rate multiplied by this efficiency figure.

ERRORS

Most satellite links are engineered to have better error performance than terrestrial telephone links. One bit error in 10^7 is a typical error rate on satellite circuits, whereas one bit in 10^5 is typical on terrestrial telephone circuits. Also, as mentioned in the previous chapter, satellite bit errors are usually distributed randomly, whereas errors in terrestrial circuits often come in bursts. This means that error-detecting codes are more sure of catching the errors on satellite circuits than on terrestrial circuits.

Unfortunately, the end-to-end link which connects user machines often travels via terrestrial circuits to the satellite earth station. The end-to-end error control therefore has to deal with terrestrial circuit errors as well as satellite link errors.

Because of the long propagation delay, it is desirable to minimize the end-to-end error rate. Tail circuits with a relatively high error rate should be avoided if there is any choice. The Bell System DDS circuits (Dataphone Digital System) have an error rate of about one error in 10^7. It would be better to use DDS circuits than telephone circuits to connect to the earth stations. Often, however, unless the earth stations are on the users' premises, an error rate of one error in 10^5 has to be assumed.

If there were no propagation or response delay and the sender was notified *immediately* that an error was detected, the error would cause the loss of one frame. However, there is a lengthy response delay. With conventional use of link control procedures like HDLC the frames transmitted during the response delay period are lost. They would not all be lost with selective repeat ARQ, but continuous ARQ with pull-back (Fig. 17.5) is normally used.

Let t_E be the mean time between errors; let t_D be the overall response delay. To assess the transmission efficiency of the link we are concerned with

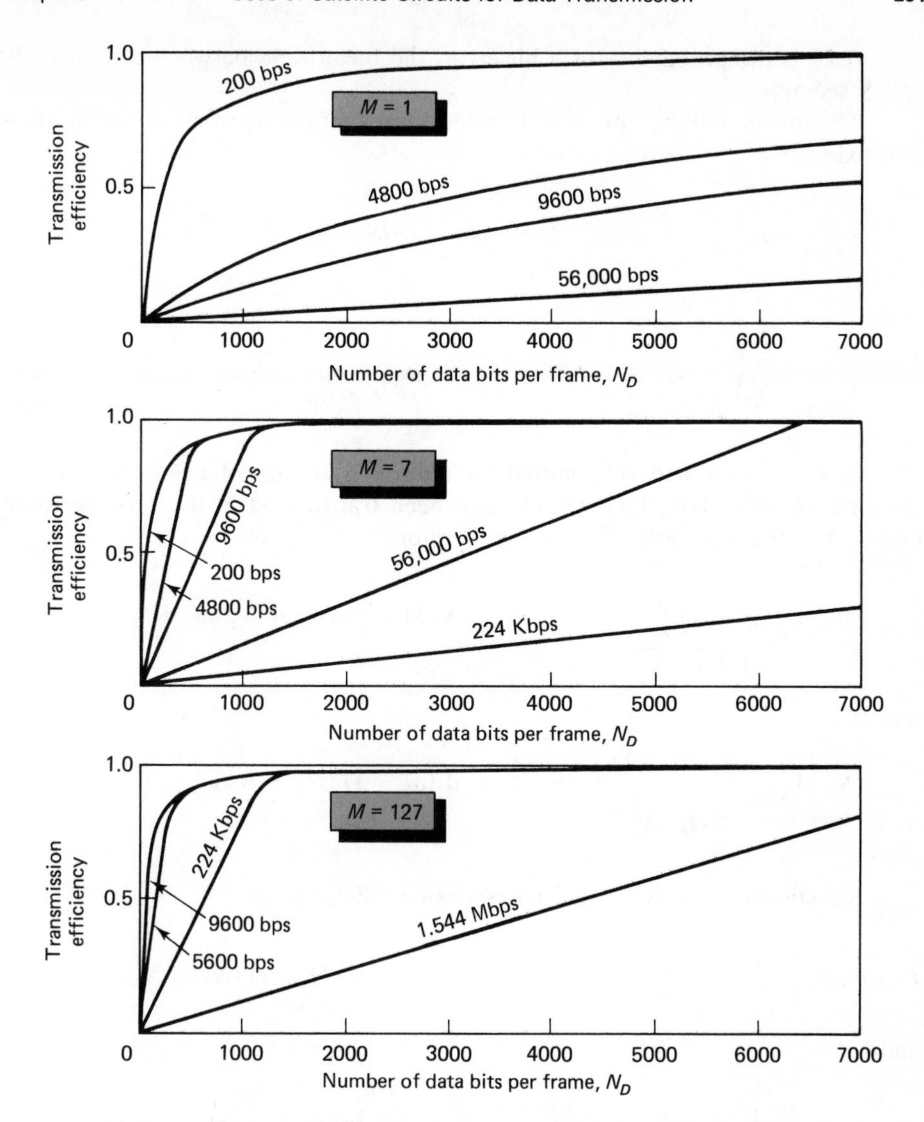

Figure 18.2 Transmission efficiency of satellite circuits (ignoring errors) with common data control protocols. Protocols with $M = 1$ are generally inadequate. HDLC, SDLC, etc. with $M = 7$ are efficient if a large enough frame size is used. For high data rates $M = 127$ is desirable.

how many frames can be transmitted in the time $t_E - t_D$. One of these frames will be lost (the one with the error).

If $(M - 1)F/S > t_D$, then the line will transmit $(t_E - t_D)S/F$ frames in time $t_E - t_D$. The number of useful frames will be $(t_E - t_D)S/F - 1$.

If P_B is the probability of a bit error, the mean time between errors, t_E, is $1/P_BS$ seconds.

The mean number of useful frames transmitted between errors is then approximately:

$$\left(\frac{1}{P_BS} - t_D\right)\frac{S}{F} - 1 \qquad \text{if } (M-1)F/S > t_D$$

and

$$\left(\frac{1}{P_BS} - t_D\right)\frac{M}{F/S + t_D} - 1 \qquad \text{if } (M-1)F/S < t_D \tag{18.4}$$

If the circuit had transmitted an unbroken stream of useful bits at the nominal bit rate, $1/P_B$ bits would have been transmitted in the time between errors. The transmission efficiency is therefore

$$\frac{N_D\left[\left(\frac{1}{P_BS} - t_D\right)\frac{S}{F} - 1\right]}{1/P_B} \qquad \text{if } (M-1)\,(N_D + N_H)/S > t_D$$

and

$$\frac{N_D\left[\left(\frac{1}{P_BS} - t_D\right)\frac{M}{F/S + t_D} - 1\right]}{1/P_B} \qquad \text{if } (M-1)\,(N_D + N_H)/S < t_D$$

Substituting $F = N_D + N_H$, transmission efficiency =

$$E = N_DP_B\left[\left(\frac{1}{P_B} - St_D\right)\frac{1}{N_D + N_H} - 1\right] \qquad \text{if } (M-1)\,(N_D + N_H)/S > t_D$$

and

$$E = N_DP_B\left[\left(\frac{1}{P_B} - St_D\right)\frac{M}{N_D + N_H + St_D} - 1\right] \qquad \text{if } (M-1)\,(N_D + N_H)/S < t_D \tag{18.5}$$

(These equations are valid only for positive values of transmission efficiency.)

The effective throughput of a satellite circuit is the above efficiency multiplied by the nominal bit rate of the circuit.

In attempting to recover from an error the number of bits that are sent is $St_D + F$. If *another* error is likely to occur in this number of bits then the transmission efficiency will be very low. It is therefore desirable that $St_D + F$ is much smaller than $1/P_B$. This implies that when satellite links are used to transmit a data rate of 56,000 bps or higher an error rate much better than one error bit in 10^5 is needed. One error bit in 10^7 can give efficient transmission of data rates of 1.544 Mbps or higher.

CURVES SHOWING THE EFFECTS OF ERRORS

Figure 18.3 shows some of the curves in Fig. 18.2 redrawn but now showing the effect of errors. The top of each red band is for error-free transmission. The bottom is for an error rate of 1 bit in 10^5. An error rate of 1 in 10^7 is close to the top of the red band.

No red band is shown in Fig. 18.3 for the transmission rate of 1.544 Mbps. The transmission efficiency is approximately zero at this speed with an error rate of 1 bit in 10^5. In the time taken to respond more than a million bits are transmitted and there is an error every 100,000 bits, on the average.

High bit rates *can* be transmitted over satellite circuits with a 10^{-5} error rate, but not using pull-back ARQ. Selective repeat ARQ is needed, or else transmissions on multiple parallel channels of low speed can be combined. To transmit high data rates with conventional protocols a lower error rate than 10^{-5} is needed, which means the analog tail circuits to the earth stations should be avoided.

Figure 18.4 shows the effects of an error rate of 1 bit in 10^7, full-duplex continuous ARQ with pull-back. High transmission efficiencies can be achieved, approximately as high as those on terrestrial circuits. An appropriate frame size and value of M needs to be selected.

Figure 18.5 shows the general shape of the transmission efficiency curve. The region to the left of point A needs to be avoided because the sending machine is kept idle, waiting for responses. The region on the right-hand side needs to be avoided because the frame size is too large and a high proportion of frames contain errors.

MAXIMUM THROUGHPUT

Point B in Fig. 18.5 is that of maximum efficiency, i.e. maximum data throughput for a given circuit. From the equation

$$E = N_D P_B \left[\left(\frac{1}{P_B} - St_D \right) \frac{1}{N_D + N_H} - 1 \right] \tag{18.6}$$

we can calculate the maximum transmission efficiency by setting $dE/dN_D = 0$. This gives

$$N_D = \sqrt{\left(\frac{1}{P_B} - St_D \right) N_H} - N_H \tag{18.7}$$

This is valid only when $(M - 1)(N_D + N_H)/S > t_D$, i.e. when $N_D > [St_D/(M-1)] - N_H$, when point B in Fig. 18.5 is to the right of point A. If this condition does not hold, then point A will be that of maximum efficiency.

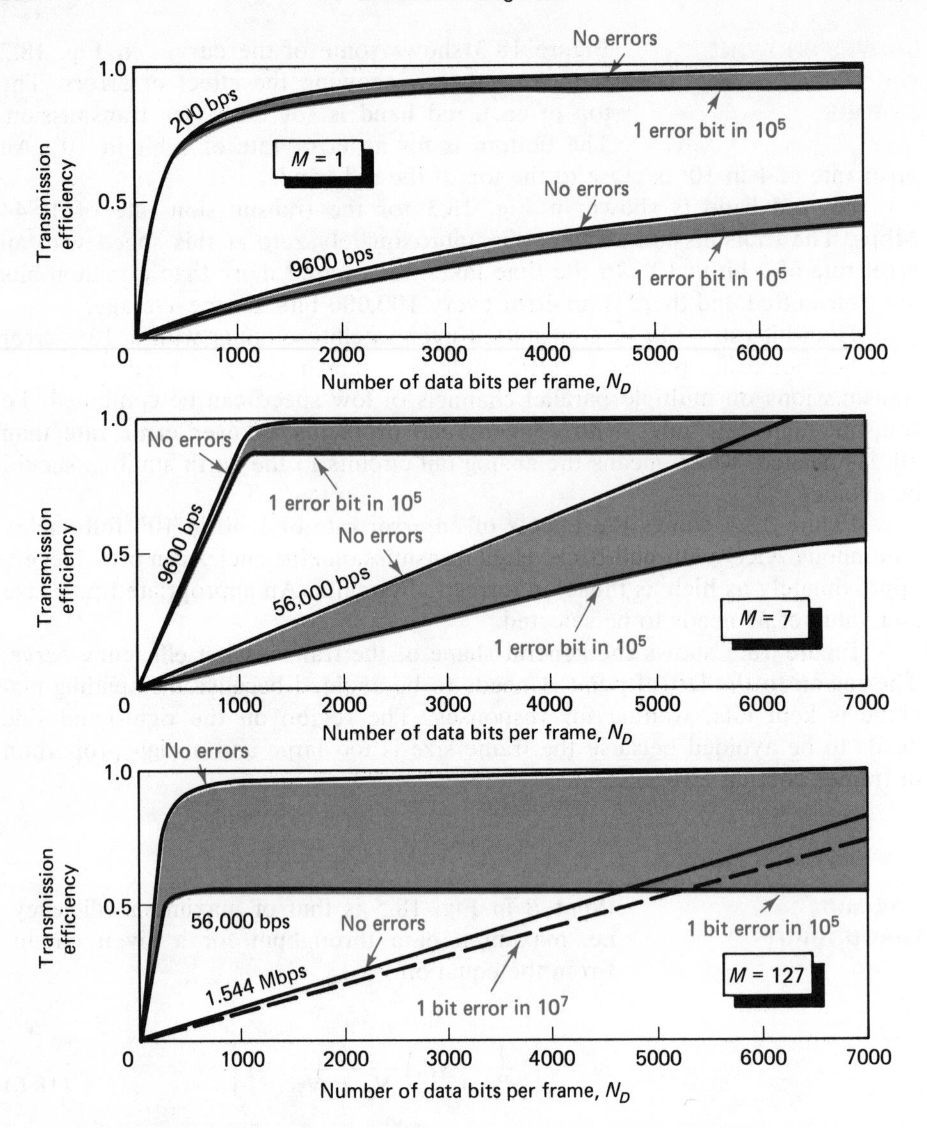

Figure 18.3 Some of the curves of Fig. 18.2 are redrawn here to show the effect of errors. The top of each red band is for error-free transmission; the bottom is for an error rate of 1 bit in 10^5. An error rate of 1 bit in 10^7 would be close to the top of the red band.

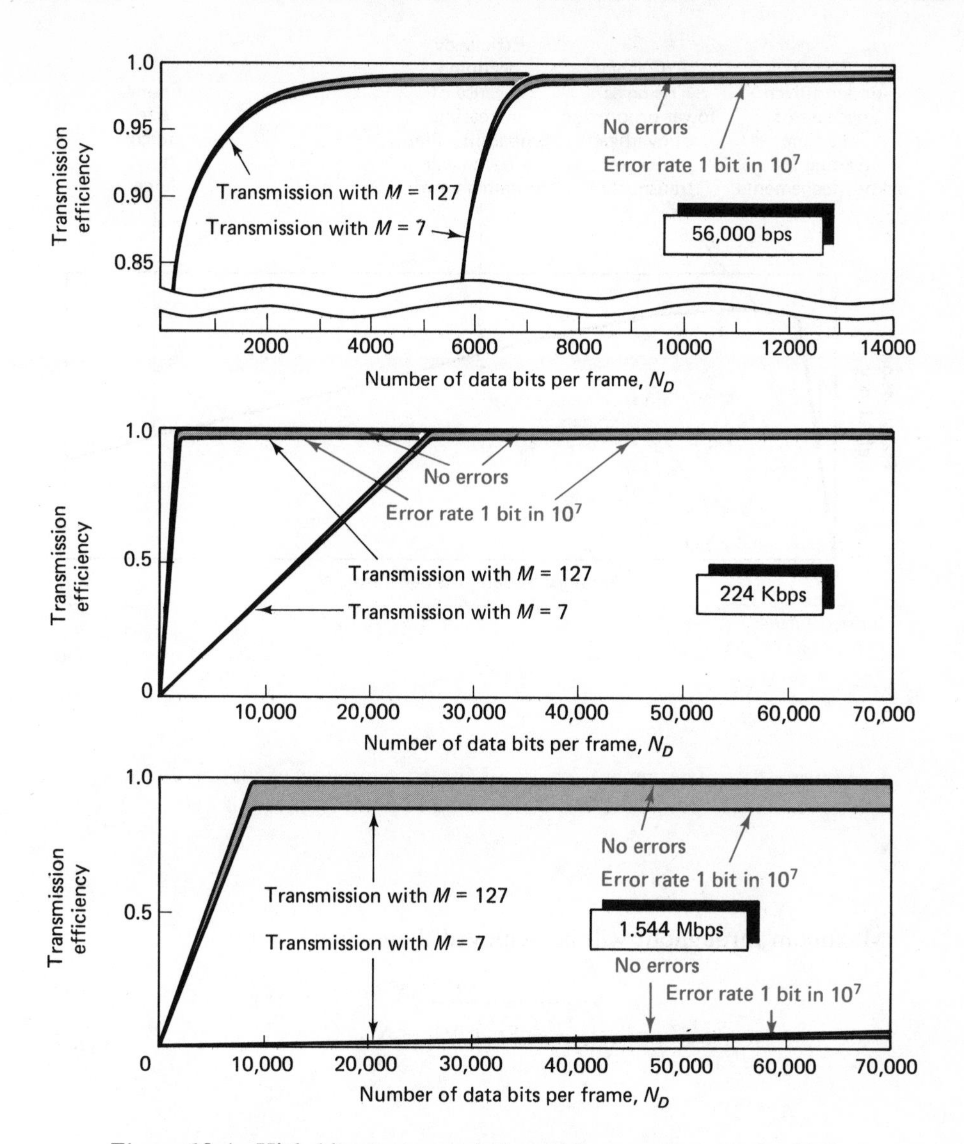

Figure 18.4 High bit-rate transmission with an error rate of 1 bit in 10^7, using full duplex continuous ARQ with pull-back.

High bit-rate transmission over links with 1 error bit in 10^5 needs selective repeat ARQ. For transmission at 1.544 Mbps, $M = 7$ is inadequate; $M = 127$ is needed.

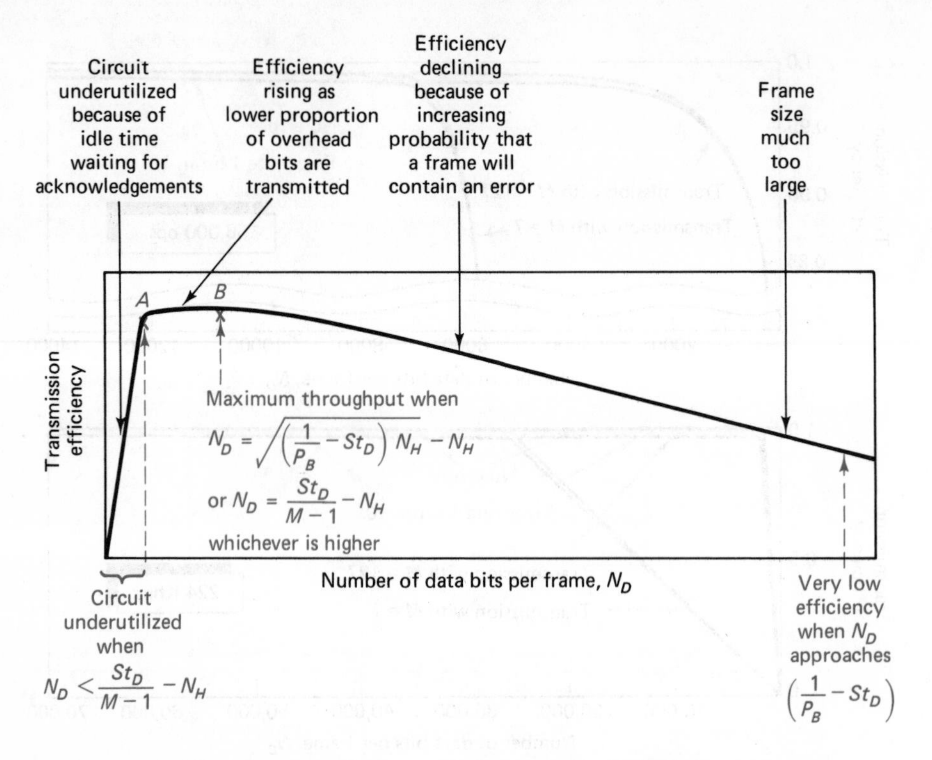

Figure 18.5 The general shape of the transmission efficiency curve. In some cases point A is the maximum and point B does not exist.

Maximum throughput will be achieved then when

$$N_D = \sqrt{\left(\frac{1}{P_B} - St_D\right) N_H} - N_H \qquad (18.8)$$

or

$$N_D = \frac{St_D}{M-1} - N_H$$

whichever is greater.

Figure 18.6 shows the optimum frame size for different circuits and different values of M.

The transmission efficiency curves have very flat peaks as can be seen from Figs. 18.3 and 18.4. There are, therefore, wide ranges of value of N_D which give throughput close to the maximum.

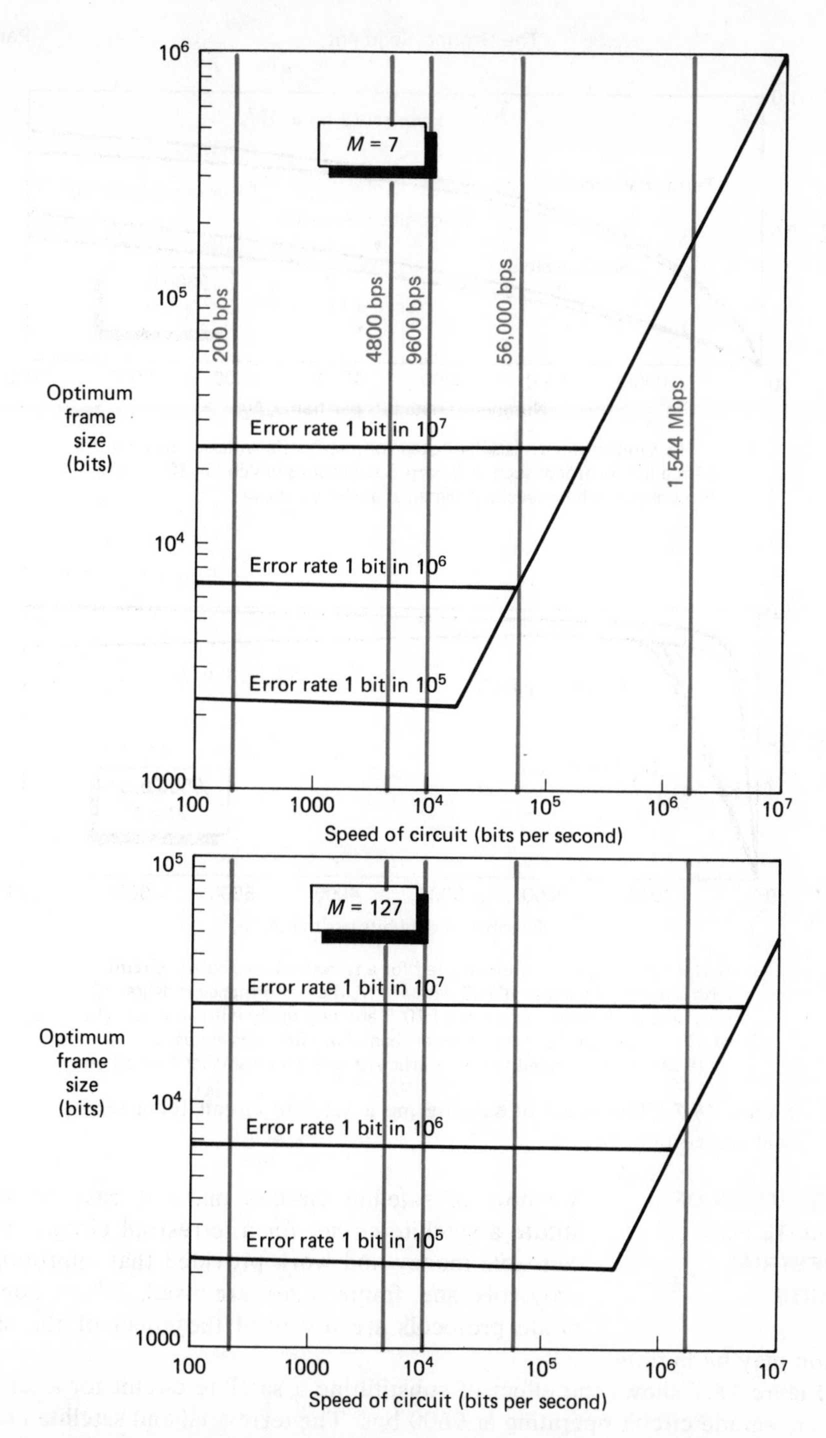

Figure 18.6 The frame sizes which give the highest throughput using full duplex HDLC and similar link protocols.

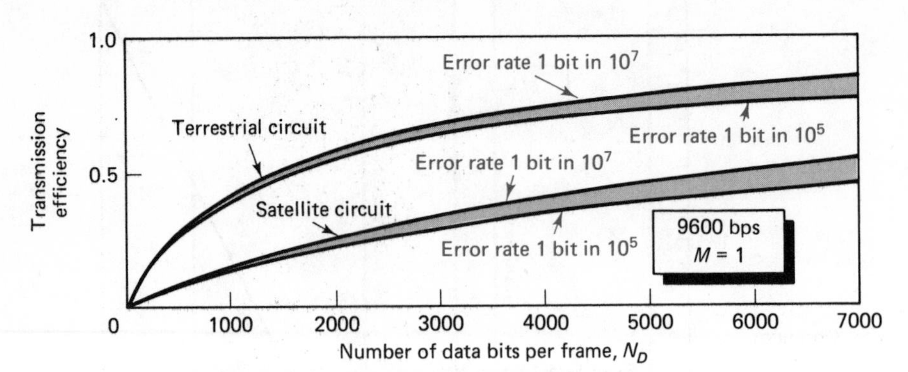

If a satellite circuit is substituted for a terrestrial voice-grade circuit with a link protocol such as *binary synchronous* which has $M = 1$, the throughput will be severely degraded as shown above.

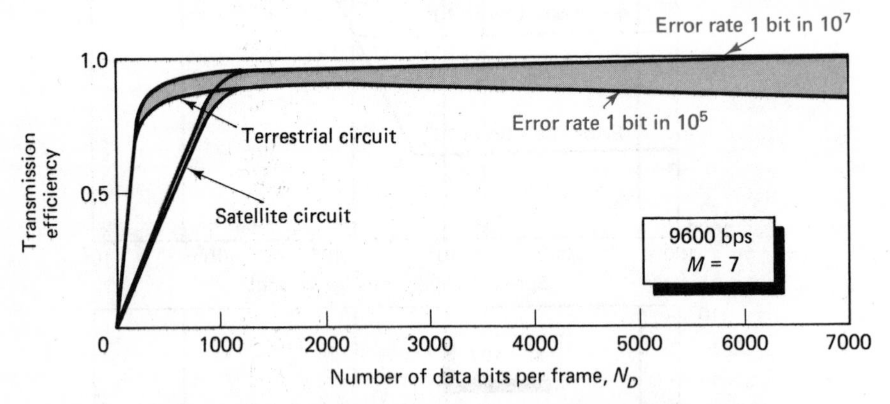

If a satellite circuit is substituted for a terrestrial voice-grade circuit with a protocol such as HDLC with $M = 7$, the performance is degraded only slightly with an error rate of 10^{-5} and negligibly with an error rate of 10^{-7} provided that frames larger than about 960 bits are used.

If $M = 127$ the satellite link is efficient with much smaller frame sizes.

Figure 18.7 The effect of substituting a satellite circuit for a terrestrial voice-grade circuit.

SUBSTITUTION OF SATELLITE FOR TERRESTRIAL CIRCUITS

Vendors of satellite circuits make it easy to substitute a satellite circuit for a terrestrial circuit. This can save money and work provided that appropriate protocols and frame sizes are used. When appropriate protocols are not used the effect of the substitution may be harmful.

Figure 18.7 shows the effect of substituting a satellite circuit for a terrestrial voice-grade circuit operating at 9600 bps. The terrestrial and satellite usage is the same except that $t_D = 150$ milliseconds on the terrestrial link, and 650 milliseconds on the satellite link. 110 milliseconds of each of these is a delay other than propagation delay. The top diagram is for protocols such as binary synchronous with $M = 1$. The throughput is greatly reduced. If frames of 1000 bits are sent, it is reduced to about one-third.

The bottom diagram relates to HDLC and similar protocols. These give more efficient utilization *of the terrestrial circuit.* Also, when a satellite circuit is substituted, there is a negligible drop in throughput provided that a frame size greater than about 140 bytes (1120 bits) is used.

Some communications managers and computer system designers avoid using satellite circuits because of experience of throughput degradation. A better attitude would be to adapt the protocols used so that satellite circuits can be employed efficiently.

FLOW CONTROL AND PACING

Some software, especially that for computer networks such as the Digital Equipment Corporation's DECNET, IBM's SNA, and Univac's DCA, contain flow control or pacing mechanisms. These permit a sending machine to send a certain number of blocks of data before it receives a response. The sending machine is regulated so that it keeps slightly ahead of the receiving machine, but not too much.

When a two-way delay of 540 milliseconds is inserted into the link, havoc can result with the regulation. Or, to state that more constructively, the parameters which control the regulation need to be adjusted to accommodate the delay.

Consider a computer which is using a remote printer. The computer sends a line at a time to the printer and the printer has, say, buffers for four lines of print. If the sending machine sends too slowly print cycles will be missed. If the sending machine sends too fast it will overload the buffers and data will be lost and will have to be retransmitted. In either case the printing operation is slowed down. The printer does not always print at a constant rate. It sometimes skips lines, or skips to the next page, responding to editing instructions. The rate of flow is therefore regulated by the printer controller transmitting control messages to the sending machines, telling it to send, say, up to two more lines of print.

The printer might take 100 milliseconds per line of print. When an extra 540 milliseconds delay is introduced into the two-way path the printer may print two lines and then miss five or so print cycles, print two more lines, and so on.

To keep the printer going at full speed the sending machine needs to take the propagation delay into consideration and keep ahead of the printer. When this is done the printer may need a larger buffer because it cannot tell the sending machine when it has finished line skips and page skips with sufficient speed. The sending machine may be programmed to anticipate the time for line and page skips.

The faster the remote printer the more serious the effects of the dalay. A satellite channel can be fast enough to operate a high-speed electrostatic printer which prints thousands of lines per minute.

A similar problem applies to any process which requires pacing. This includes most processes involving electromechanical components, for example, the use of a file on a remote disk storage unit. This has variable access times and end-to-end control.

Software which is intended to work on *either* satellite or terrestrial links should be designed to *measure* the propagation delay, and adjust the pacing parameters accordingly. This is particularly desirable for use on dial lines when a satellite link may or may not be present in the circuit.

CONTROL PROCEDURES FOR DATA TRANSMISSION

We may now summarize some of the desirable features of the link control procedures used with satellite circuits.

1. Stop-and-wait ARQ should be avoided. Ptorocols such as binary synchronous line control are inefficient.
2. Continuous ARQ with pull-back, as used in HDLC, SDLC, DEC's DDCMP, etc., is efficient provided that a suitable range of frame sizes are employed, with $M = 7$ or higher.
3. Selective repeat ARQ makes link control equipment more expensive and is not necessary except when a high bit rate is sent over a channel with a high error rate.
4. When a very high bit rate is sent with an HDLC-like protocol, a high value of M (e.g. $M = 127$) is needed, along with a circuit of low error rate (e.g. 1 error bit in 10^7).
5. Care is needed with pacing and flow control mechanisms such as those used in IBM's SNA and other protocols for distributed intelligence, distributed processing, and computer networks. Correct choice of parameters for control of pacing is necessary.
6. Polling of satellite circuits should be avoided as a technique used on interactive systems. If an appropriate technique is used, as discussed in the following chapter, response times appropriate for interactive systems can be achieved.
7. Although terrestrial protocols work well over satellite links, they do not take advantage of the broadcasting, multilocation nature of satellite links. To do this, multiple-access protocols need to be designed specially for satellites.

19 USES OF SATELLITES FOR INTERACTIVE COMPUTING

A system in which a man communicates with a distant computer in a conversational fashion is called an *interactive* computer system. The fastest-growing use of telecommunications, although still small in total, is for interactive computer systems.

Satellites could provide effective low-cost channels for interactive computing because, as we have stressed, the digital throughput of a satellite transponder is high enough for a very large number of terminals to share it. The techniques by which the terminals share the bandwidth, however, may be fundamentally different from techniques in common use with terrestrial links.

DATA PROCESSING USERS

The problem of selecting appropriate techniques for computer transmission via satellite applies at two levels. First, it applies to common carriers or satellite operators, who design earth-station equipment to give users the types of services they want. Second, it applies to the user organizations who can lease a satellite subchannel and must organize their methods of using it so as to best serve their data processing needs. Related to this, it applies to the manufacturers of computers, terminals, and the equipment and software which control their data transmission.

Those in the latter categories find themselves provided with a two-way data channel via the satellite. If they wish, it can be a channel with a much higher data rate than they would normally lease on terrestrial circuits. They are confronted with the problem: How should data transmission be organized so that the satellite is taken advantage of? This chapter and the following chapters relate to that question.

DIFFERENCES IN SATELLITE LINKS

The satellite link is different from a terrestrial link, such as a telephone link, in several important respects, and these differences make teleprocessing line control procedures used on earth inefficient if applied to satellites. As we have stressed, the differences in a satellite link are

1. 270-millisecond transmission delay in each direction.
2. A very large number of terminals can share the facility.
3. The terminals can be scattered across one-third of the world without increase in transmission cost.
4. The channel is a broadcast facility, not a point-to-point channel.
5. A transceiver can receive its own transmission to observe whether data have been correctly relayed.

NEEDS OF INTERACTIVE SYSTEMS

Not only is the satellite channel different. Interactive systems make different types of demands on the transmission and switching facilities from the traditional uses of telecommunications such as telephony, telegraphy, broadcasting, and batch data transmission.

Today's telecommunications facilities are largely oriented to what the common carriers refer to as POTS – "Plain Old Telephone Service." The ingrained wisdom of decades in the telecommunications administrations relates to POTS, and most administrations consider that their bread-and-butter revenue for the foreseeable future will come from POTS. It is sometimes difficult for experienced telephone engineers to accept that with computers conversations are very different from telephone conversations. In reality the differences are major and may be taken advantage of to make satellite channels spectacularly more cost-effective in interactive use.

We shall first illustrate some examples of interactive terminal usage and shall then point out how they differ in their requirements from other transmissions. This will enable us to compare the effectiveness of different satellite transmission control procedures.

TERMINAL DIALOGUES

To bring the power of computers and the information in their data banks to the maximum number of people, careful attention must be paid to the person-machine interface. Increasingly, in the next decade, users must become the prime focus of computer systems design. The computer is there to serve him, to obtain information for him, and to help him do his job. The ease with which he communicates with it will determine the extent to which he uses it. Whether or not he uses it powerfully will depend on the terminal dialogue programmed for him and how well he is able to understand it—in other words, what the user says to the machine and what the machine says to the user.

The user-computer dialogue should be a starting point in the design of the data communications facilities in many interactive systems. A systems analyst must design the structure of the dialogue. In some cases a formal programming language is used such as FORTRAN or BASIC. The majority of terminal users, however, are nonprogrammers; they are clerks, travel agents, brokers, factory workers, managers, and persons in all walks of life. Some are intelligent and highly trained; some are not. The dialogue they employ for communicating with a terminal must be more straightforward than a programming language and must be tailored to their psychology and ability.

"Dialogue," then, is a generic word for a preplanned person-machine interaction; it encompasses formal programming languages, languages for interrogating a data base, and innumerable nonformal conversational interchanges many of which are designed for one specific application.

The dialogue differs greatly from one user to another and from one machine to another. Different applications will need fundamentally different types of dialogue structures [1]. Some are very complex and need a high level of intelligence; others are simple. In some applications today, one can observe the "man-in-the-street" who has never touched a terminal before sitting down at one and carrying on a successful, if simple, dialogue with a computer. On the other hand, one also finds terminals being thrown out a few months after installation because the intended user never learned to communicate successfully with the system.

SIX EXAMPLES OF DIALOGUES

Figures 19.1 through 19.6 give six typical examples of dialogues in which a person is using a distant computer. They are chosen to illustrate types of dialogues in common use. The last of the six is less likely to take place over terrestrial communication links than the other five, largely because the communications channels in existence are not yet organized, in most cases, in a suitable fashion for it.

TRANSMISSION SPEEDS

The reader should relate the transmissions taking place in Figs. 19.1 through 19.6 to the data rates of communications channels that might be used. Typical speeds for terminals used in the figures would be as follows:

Specimen	Application	Terminal Type	Typical Circuit Speed
1	Calculation in BASIC	Teletype	10 characters/s
2	Stock analysis	Typewriter-like device	15 characters/s

Specimen	Application	Terminal Type	Typical Circuit Speed
3	Airline reservations	Application-built display	2,400 or 4,800 bps
4	Sales order	Visual display unit	4,800 bps
5	Data entry	Data entry unit	9,600 bps
6	Circuit design	Graphics terminal	56,000 bps

Dialogue Specimen 1 (Fig. 19.1)

The first example takes place at a teletype machine at the frustratingly slow speed of 10 characters per second. The operator is using a programming language, BASIC, to do a simple calculation in order to plot a set of curves of the function

$$P(T) = e^{-[(1-R)T/S]}$$

The first action is to inform the computer that BASIC is the language to be used. The operator makes several mistakes as he keys in statements. Some he notices and corrects. Some the computer catches and tells him. He obtains his results much quicker than with a desk calculator or with a noninteractive computer to which he has to submit a job and receive the results hours later.

Dialogue Specimen 2 (Fig. 19.2)

The second example shows a financial analyst using a typewriter-like terminal [2]. He is not a programmer but is using a dialogue programmed for him which permits him to do various kinds of stock analyses. The terminal responds at 15 characters per second, which is still frustratingly slow and which places constraints on the designers to limit the lengths of responses.

Dialogue Specimen 3 (Fig. 19.3)

Figure 19.3 shows an airline agent's dialogue. He is talking to potential passengers on the telephone at the same time as conversing with the terminal to handle the passenger's requirements. He needs to be able to act promptly, obtaining quick responses from the distant computer. The computer may be thousands of miles away, perhaps at the other end of a satellite link.

In this case the terminal is a visual display unit designed for the application. Some of the keys have been specially labeled to assist in the dialogue. The terminal can display responses at a rate of several hundred characters per second, so the operator is not frustrated by slow responses as in the previous two specimens.

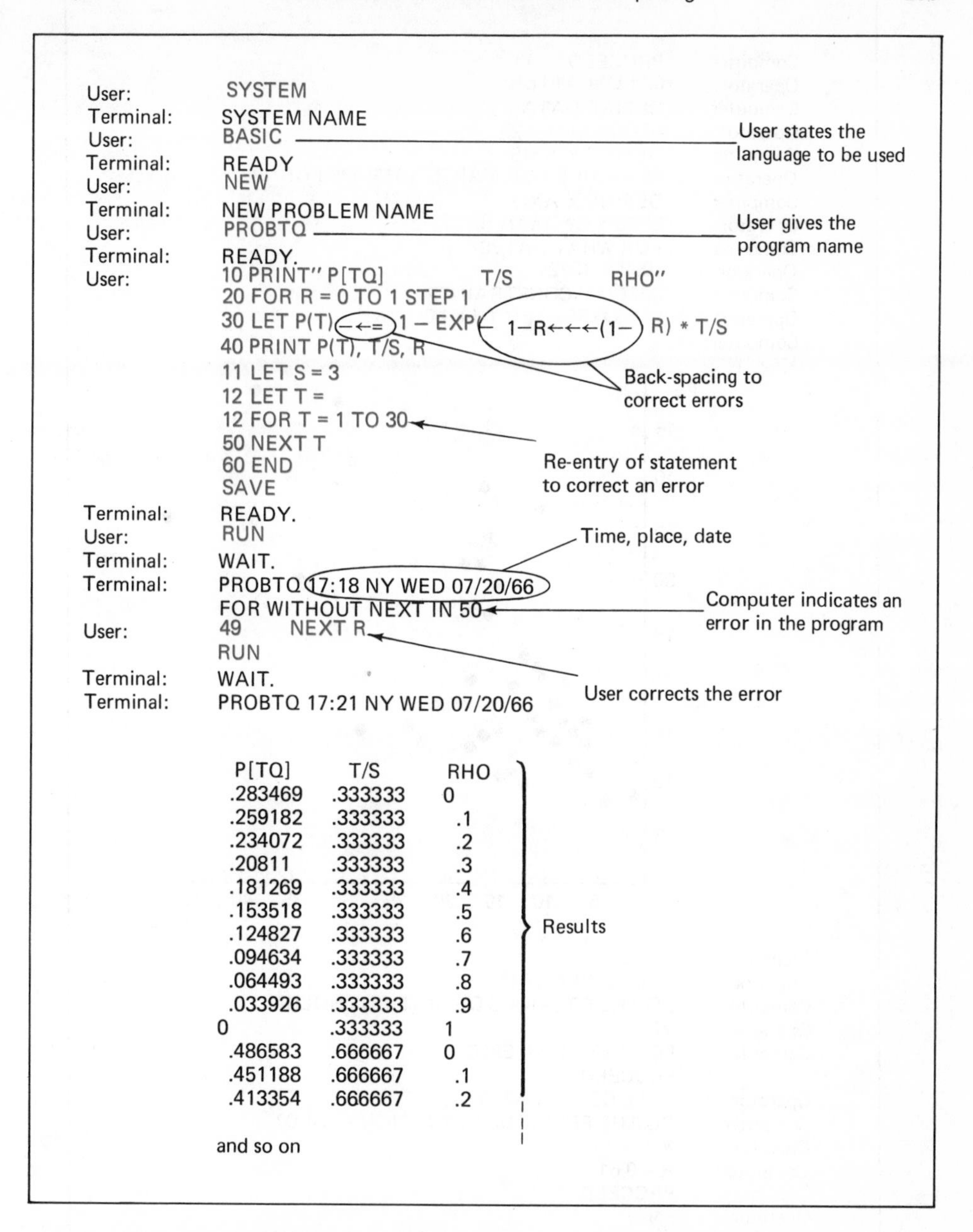

Figure 19.1 Dialogue Specimen Number 1. A user at a teletype machine doing a calculation using the programming language BASIC.

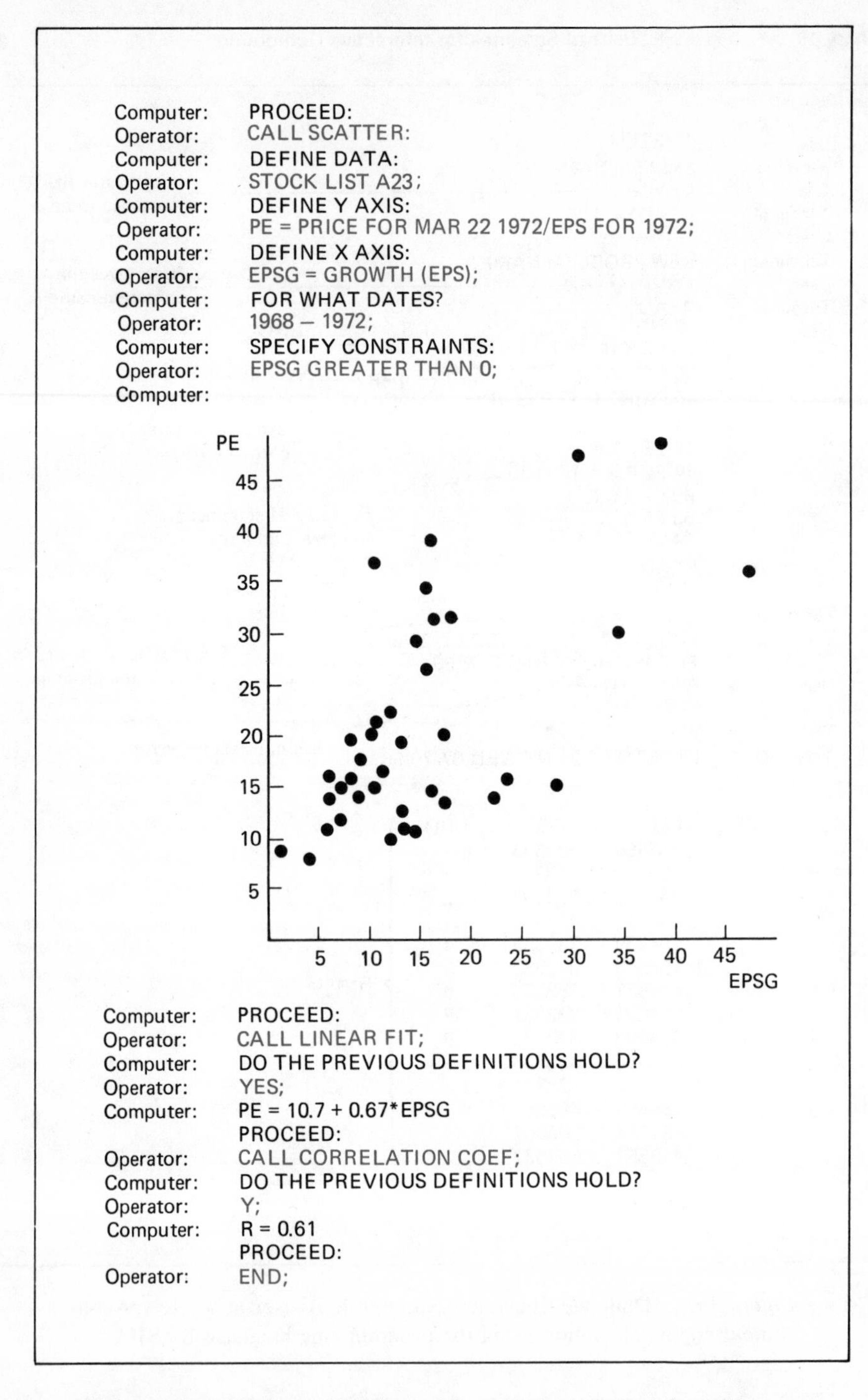

Computer: PROCEED:
Operator: CALL SCATTER:
Computer: DEFINE DATA:
Operator: STOCK LIST A23;
Computer: DEFINE Y AXIS:
Operator: PE = PRICE FOR MAR 22 1972/EPS FOR 1972;
Computer: DEFINE X AXIS:
Operator: EPSG = GROWTH (EPS);
Computer: FOR WHAT DATES?
Operator: 1968 – 1972;
Computer: SPECIFY CONSTRAINTS:
Operator: EPSG GREATER THAN 0;
Computer:
PE
45
40
35
30
25
20
15
10
5
5
10
15
20
25
30
35
40
45
EPSG
Computer: PROCEED:
Operator: CALL LINEAR FIT;
Computer: DO THE PREVIOUS DEFINITIONS HOLD?
Operator: YES;
Computer: PE = 10.7 + 0.67*EPSG
PROCEED:
Operator: CALL CORRELATION COEF;
Computer: DO THE PREVIOUS DEFINITIONS HOLD?
Operator: Y;
Computer: R = 0.61
PROCEED:
Operator: END;

Figure 19.2 Dialogue Specimen Number 2. A financial analyst at a typewriter-like terminal using a program designed for him to carry out stock analyses [2].

Dialogue Specimen 4 (Fig. 19.4)

Figure 19.4 shows a dialogue in which an operator is changing customer information as in the previous case, but here the operator has learned no special mnemonics or formats. Each screen tells the operator exactly how to respond. This form of dialogue is suitable for a casual operator who uses the terminal only occasionally and who is not highly trained in the dialogue usage, as in the previous case.

The increased ease of use of the dialogue is paid for by the increased number of characters it requires. If all the characters of Fig. 19.4 are transmitted from the central computer, this constitutes a much heavier communications load than with the brief dialogue structure of Fig. 19.3. However, if the terminal is "intelligent" and has a local microprocessor for generating the responses, a much smaller number of characters would be transmitted.

The terminal must respond with a data rate of several hundred characters per second; otherwise this form of dialogue would be very slow and frustrating for the operator. The responses of the operator are very brief, so if the response time and data rate are fast enough, a dialogue such as this can be speedy and effective.

Dialogue Specimen 5 (Fig. 19.5)

Much data have to be entered into computer systems, and this dialogue shows an operator "filling in a form" on a screen to enter data.

It could take place either *off-line* or *on-line* to a distant computer. If it is on-line, the computer can be programmed to detect operator errors, where possible, when the operator makes them. If it is off-line, interactive error control cannot take place, and errors must be corrected later. The on-line dialogue needs the data to be sent to the computer when they are entered. With the off-line dialogue they can be stored for transmission in a batch. The "form" which the operator fills in will not be transmitted if it can be stored at the terminal location (for example, in a terminal controller).

Dialogue Specimen 6 (Fig. 19.6)

Figure 19.6 shows a graphics terminal using a computer programmed for designing and testing logic circuits [3].

The user employs a *light pen,* which he can point to the screen to select items or to indicate where items should be positioned. He selects circuits from a circuit library and positions them on the screen. He moves the light pen to write them together. He can then test the operations of the circuit he has designed with the computer simulating its behavior. He modifies it at the screen until it behaves correctly. Then, using the last of the panels shown, he documents what he has designed.

A graphics user of this type may flip rapidly from one type of panel display to another, so he needs a fast data transmission rate if he is remote from

A Mr. Goldsmith telephones the airline to say that he wants to modify his reservation. The agent he speaks to instructs her terminal to display his record—Flight 21 on the 25th of March. She types:

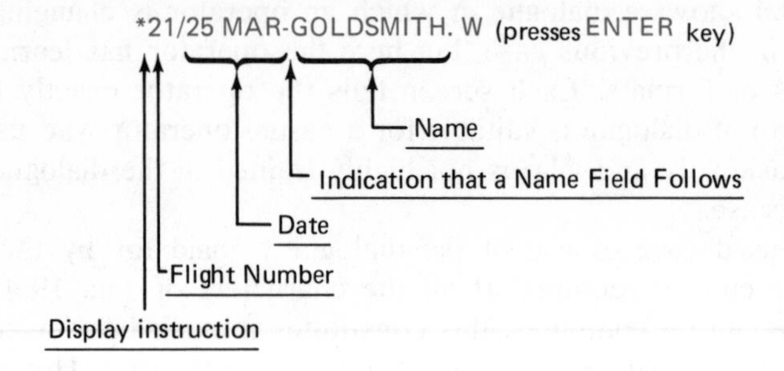

The computer types details of his record:

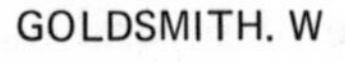

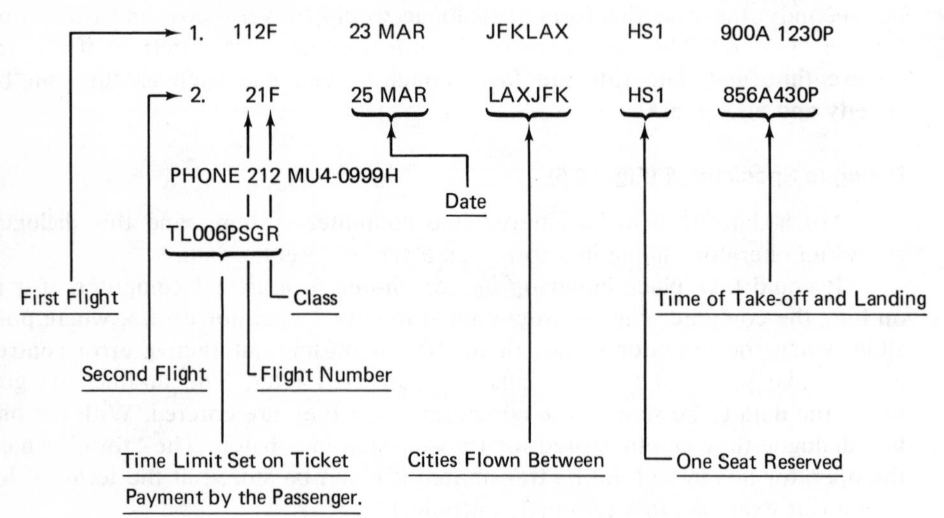

The operator checks that this is indeed the required record, and the passenger indicates that he wishes to change the second flight of his journey. He wants to fly back from Los Anglees (LAX) to New York's Kennedy Airport (JFK) on the 26th March rather than the 25th.

The agent informs the computer that the second flight of the journey is to be changed. She types:

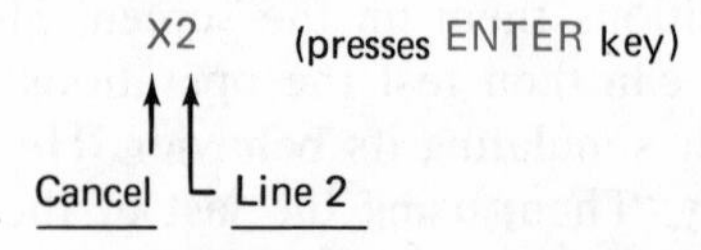

Figure 19.3 Dialogue Specimen Number 3. An airline reservation agent using a screen terminal designed for that application.

The computer responds:

NEXT SEG ENTRY REPLACES 2

She types:

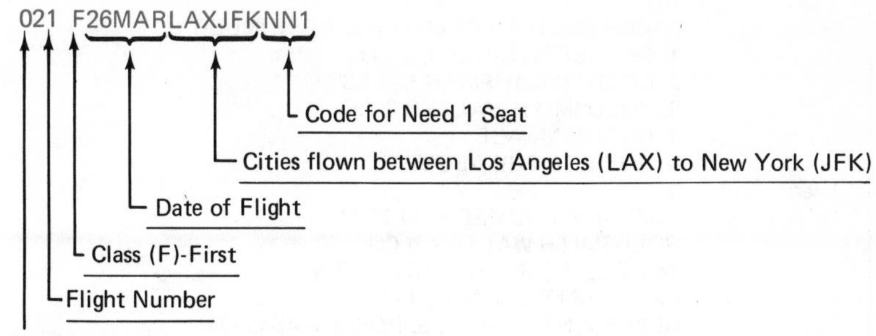

This statement is a request for a seat on the 26th of March rather than the 25th. "NN1" is a code meaning that the agent needs one seat. The computer replies with a replacement for line 2 above:

2. 21F 26MAR LAXJFK HS1 845A430P

The operator attempts to book this seat by pressing the "E" key meaning "END TRANSACTION":

E

The computer asks who requested this modification:

WHO MADE CHANGE

The operator replies that the passenger himself requested it. The "6" key means "change made by" . . . :

The computer indicates that this message has been received by placing an asterisk at the end of it.

6PSGR*

The operator again presses the "E" key, and this time the computer carries out the "END TRANSACTION" operation and updates the appropriate files. It responds:

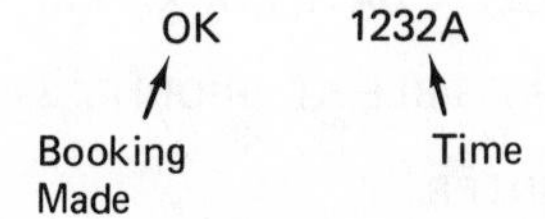

```
Screen:   ENTER CUSTOMER NUMBER IF KNOWN. IF NOT TYPE "N"
User:     N
Screen:   TYPE FIRST TWO LETTERS OF CUSTOMER'S NAME
User:     GO
Screen:   WHICH CUSTOMER? ENTER LINE NUMBER
          1. GOLDBERG SIDNEY & CO.
          2. GOLDEN KAUFMAN SALES CO.
          3. GOLDSMITH WALTER & CO.
          4. GOODMAN ABE
          5. NONE OF THE ABOVE
User:     3
Screen:   CUSTOMER NUMBER: 17-8341
          GOLDSMITH WALTER & CO.
          745, OLD STONE HIGHWAY, SPRINGS,
          EAST HAMPTON, N.Y., 11937
          REPRESENTATIVE E. S. ROPLEY #543271

          PRESS ENTER IF CORRECT
          PRESS 1 IF INCORRECT
User:     Presses ENTER key.
Screen:   WHICH INFORMATION IS TO BE CHANGED?
          ENTER LINE NUMBER
          1. NEW ORDER               9. INVOICE TO ADDRESS
          2. CANCEL ORDER           10. SHIP TO ADDRESS
          3. ORDER QUANTITY         11. CUSTOMER ADDRESS
          4. DELIVERY DATE          12. CUSTOMER NAME
          5. PRIORITY CODE          13. MISC. INSTRUCTIONS
          6. ORDER DETAILS          14. REPRESENTATIVE DETAILS
          7. DISCOUNT               15. BRANCH OFFICE
          8. TAX DETAILS            16. OTHER
User:     4
Screen:   WHICH ORDER IS TO BE CHANGED?
          ENTER LINE NUMBER
              ORDER #   ITEMS                       ORDER     DEL.
                                                    DATE      DATE
          1   573824    87289   00241   00029      4/1/77    4/1/79
                        00310   77139   04126
                        00900
          2   575931    00391   00700   00421     10/1/77    4/1/78
          3   575828    00750   00900             12/1/77   12/1/78
          4   579003    02107                      6/1/77    8/1/78
User:     2
Screen:   ORDER NUMBER 575931
          ITEM #        DESCRIPTION                       QTY.
          00391         MAGNETIC D EXTRACTOR 7 BR          5
          00700         LINK ADAPTOR ADDL                  5
          00421         CTC DIRECT CONTROL                 5
          ORDER DATE    10/1/77  DELIV DATE 4/1/79
          ENTER NEW REQUIRED DELIV DATE (XX/XX/XX)
User:     10/01/78
Screen:   EARLIEST DELIV DATE POSSIBLE AT PRIORITY 3 IS
          2/1/78
          IF THIS IS REQUESTED ENTER "Y"
User:     Y
```

Figure 19.4 Dialogue Specimen Number 4. A visual display unit user carrying out a clerical operation concerned with sales orders.

The terminal presents the operator with a "form" to fill in:

CUSTOMER DATA

ENTER THE FOLLOWING

▶-- LEGAL NAME
-- DEPARTMENT FOR ATTENTION
-- ADDRESS, STREET
-- CITY
-- STATE, ZIP
-- TELEPHONE
-- CONTACT

The operator types data into the form:

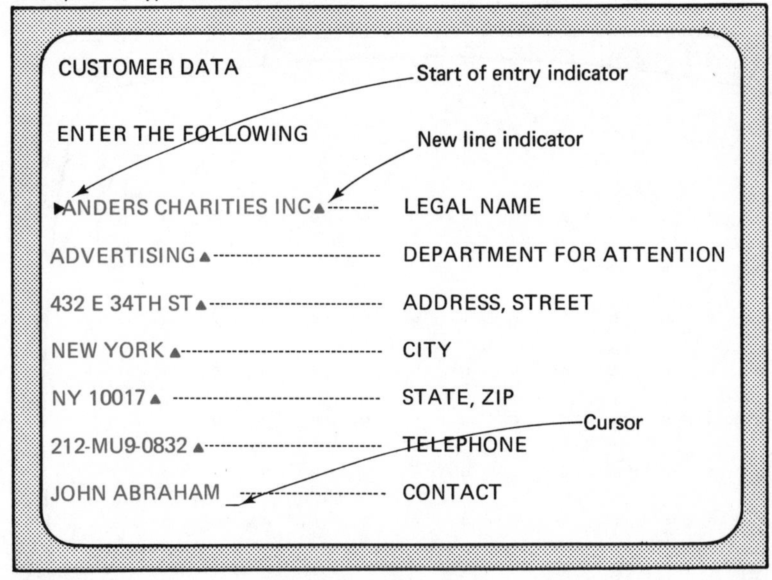

Figure 19.5 Dialogue Specimen Number 5. A data entry dialogue using a visual display unit. It could be on-line or off-line.

1

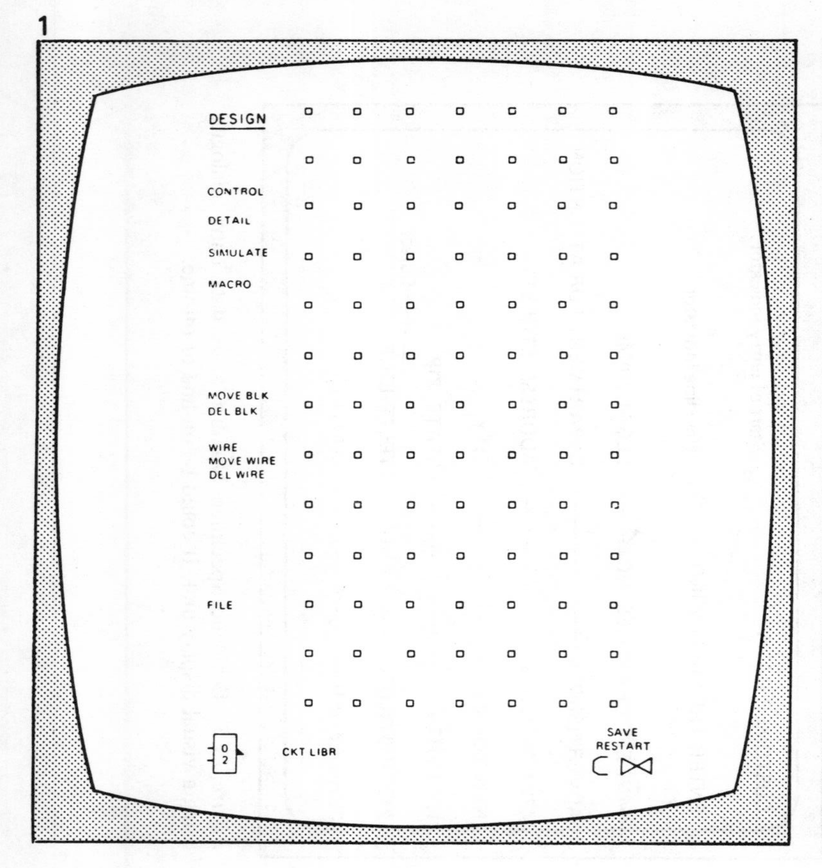
DESIGN
CONTROL
DETAIL
SIMULATE
MACRO
MOVE BLK
DEL BLK
WIRE
MOVE WIRE
DEL WIRE
FILE
O 2
CKT LIBR
SAVE
RESTART

2

DESIGN
O 2
CONTROL
DETAIL
SIMULATE
MACRO
MOVE BLK
DEL BLK
WIRE
MOVE WIRE
DEL WIRE
FILE
The user moves the components above the screen with a light pen.
N 1
CKT LIBR
SAVE
RESTART

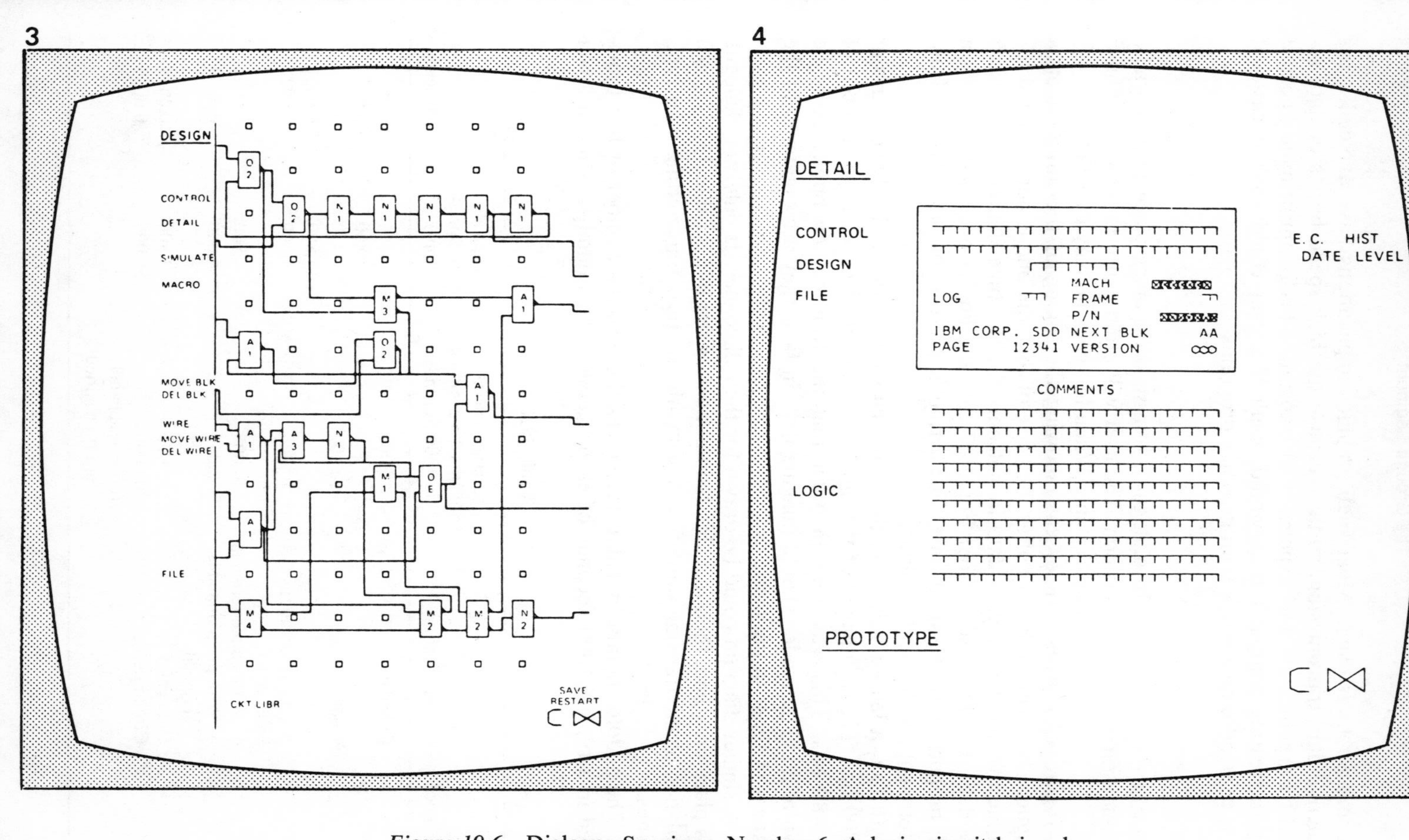

Figure 19.6 Dialogue Specimen Number 6. A logic circuit being designed with a graphics program. The user interacts with the screen with a keyboard and light-pen [3].

the computer he is using. Most such graphics applications have avoided using interactive data transmission, so far, because the high-speed bursts of data required by graphics are too expensive to transmit. This is unfortunate because many interesting computer uses would result if a user could have occasional access to remote computers with graphics programs.

SPORADIC TRANSMISSION

In dialogues, transmission takes place in a sporadic fashion. The user will often pause to read a response or to think. When he types he may do so slowly.

Figure 19.7 gives a time scale showing the timing of the airline agent's dialogue. The terminal has a buffer. Because of this, the line is not occupied for the time the agent is keying data in but only for the time it takes to transmit those data. It will be seen from Fig. 19.7 that the line is occupied for only a small proportion of the total time. Furthermore, there is no transmission from this terminal for the next 2 minutes.

In fact, a leased voice line to a terminal such as this is likely to transmit at 4800 bps. The occupancy of the line shown in Fig. 19.7 assumes a speed of only 480 bps. This was done in order that the transmission bursts could be drawn at all. The reader should imagine that, in fact, they are one-tenth of the thickness of those in the figure.

A similar diagram could be drawn for their dialogues. If only one dialogue takes place on a line, then that line is idle most of the time. Table 19.1 gives a calculation of the transmission efficiencies with one user transmitting over typical lines.

The middle column in Table 19.1 gives the *average* number of bits transmitted in a typical 1000 seconds of each dialogue. This number is much lower

Table 19.1

Specimen	Application	Average Quantity of Bits Transmitted in 1000 Seconds	Typical Line Speed (bps)	Efficiency
1	Calculation in BASIC	8,000	100	0.08
			4,800	0.0017
2	Stock analysis	10,000	150	0.067
			4,800	0.0021
3	Airline reservations	10,000	4,800	0.002
4	Sales order:			
	a. With simple terminal	40,000	4,800	0.0094
	b. With programmable terminal	2,000	4,800	0.0004
5	Data entry	15,000	4,800	0.003
6	Circuit design	200,000 (only new panels are transmitted)	56,000	0.0036

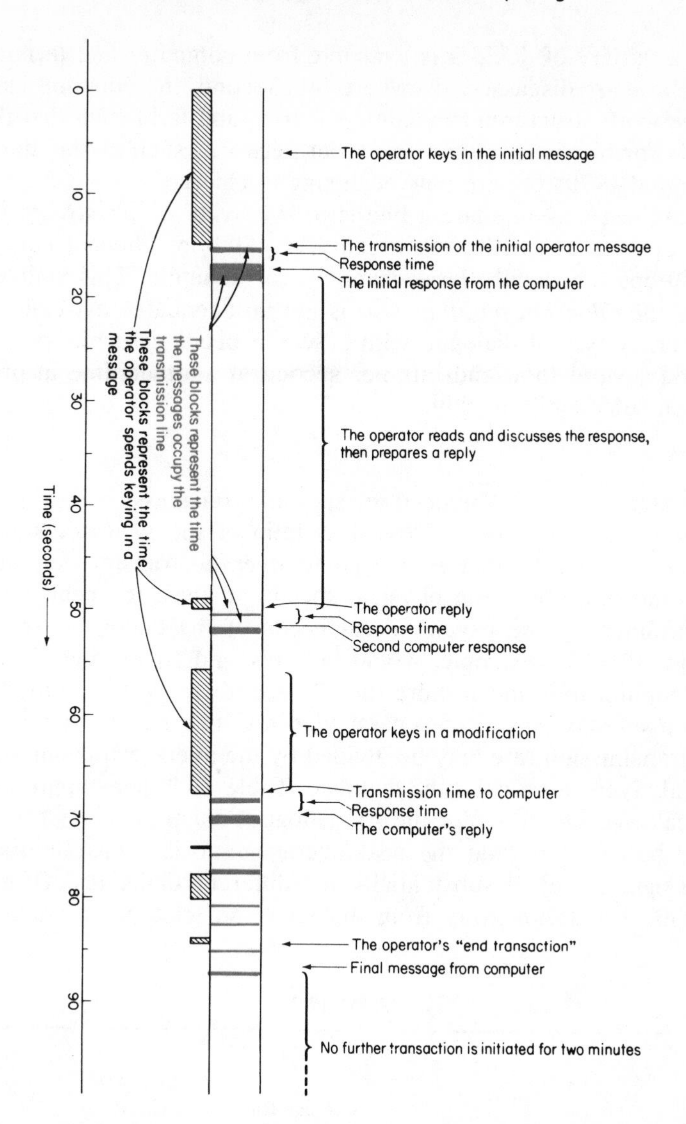

Figure 19.7 In man-computer conversations, transmission takes place sporadically.

than the number of bits that *could* have been transmitted because of the pauses in the transmission; 4800 bps have been used as the typical speed of a voice line.

SHARING

To improve the efficiency with which lines are utilized, many terminals must share the same line. There are two types of ways in which sharing can be accomplished.

First, a variety of devices is available from computer and terminal manufacturers. These are discussed elsewhere [4]. Second, the common carriers can provide a network structured to facilitate sharing and tariff such that the user is charged only for his share. Common carrier networks such as this did not exist prior to the mid-1970s but are now beginning to emerge.

The simplest way to share a line is to subdivide its bit stream into channels which operate at smaller bit streams. A 4800-bps channel could be split into ten 480-bps (or slightly less) channels, for example. This simple subdivision *(called time-division multiplexing)* is not an adequate answer to the problem, however. A typical dialogue with a screen needs *zero* bits per second at one time and several thousand bits per second at another time in order to fill up the screen sufficiently quickly.

PEAK/AVERAGE RATIOS

The need for sporadic transmission in a dialogue can be expressed as ratio of the *peak* transmission rate that is needed to *average* transmission rate. If we forget for a moment about the physical speeds of lines, the peak transmission rate is determined by the psychological needs of the dialogue user. The dialogue in Fig. 19.4, for example, would be slow and frustrating if the screens from the computer took much more than 2 seconds to be transmitted. In other words, a transmission speed of several thousand bits per second is desirable. This peak transmission rate may be divided by the average transmission rate to obtain a peak/average ratio for the dialogue. Table 19.2 gives approximate estimates of peak/average ratios for the six dialogues illustrated.

It will be observed that the peak/average ratio is a characteristic of the dialogue design. It varies substantially for different dialogues. Often it is as high as 1000. The trend away from dialogues on teletype machines to more

Table 19.2

Specimen	Application	Average Rate (bps)	Desirable Peak Rate (bps)	Peak/Average Ratio
1	Calculation in BASIC	8	100	12.5
2	Stock analysis	10	150	15
3	Airline reservations	10	2,000	200
4	Sales order:			
	a. With simple terminal	40	4,000	100
	b. With programmable terminal	2	4,000	2000
5	Data entry	15	1,500	100
6	Circuit design	200	200,000	1000

powerful dialogues on display units, graphics terminals, and distributed-intelligence systems often results in a higher peak/average ratio. This characteristic is very different from human telephone conversations, which have a peak/average ratio of 1 if the sound channel is kept continuously open between the speakers in both directions. (On some systems speech has been packetized with a peak-to-average ratio of 3.)

The higher the peak/average ratio, the greater the *inefficiency* of using a transmission channel *which transmits at a constant fixed rate.* Instead a channel is needed which transmits bursts of data when they are required with a suitably short response time.

RESPONSE TIME

Response time is a particularly important parameter in the design of user-computer dialogues. In many (but not all) dialogue situations the time interval between sending a message from the terminal and starting to receive a reply should not be more than about 2 seconds [1]. The mean one-way transmission delay should therefore be well under 1 second. Also, there shall not be too great a *variation* in response times. The standard deviation of one-way transmission time should also be well under 1 second.

TERRESTRIAL LINE CONTROL

A variety of line control procedures is used on earth for enabling terminals or data processing machines to share transmission lines [4]. The main types of procedures are variations on one or more of the following:

1. Frequency-division multiplexing.
2. Time-division multiplexing.
3. Polling.
4. Concentration.
5. Loop control.
6, Packet switching.

Most of these have disadvantages if employed without modification for interactive computer usage over satellite channels. We shall comment briefly on each of them.

1. *Multiplexing.* Frequency-division multiplexing and time-division multiplexing both subdivide a channel into smaller channels. They provide continuous channels, not the sporadic channels needed for interactive terminals. With the peak-to-average ratios of typical dialogues, *simple* multiplexing would give efficiencies similar to those in Table 19.1. Statistical (intelligent) multiplexors are much better.

2. *Polling.* Polling can give efficient usage of terrestrial telephone channels and is one of the most widely used line control procedures for interactive ter-

minals. In a typical polling scheme the computer addresses the terminals on the line in sequence and asks if they have a message to transmit:

COMPUTER TO TERMINAL A: Have you a message?
TERMINAL A TO COMPUTER: No.
COMPUTER TO TERMINAL B: Have you a message?
TERMINAL B TO COMPUTER: No.
COMPUTER TO TERMINAL C: Have you a message?
TERMINAL C TO COMPUTER: A message is transmitted.

Polling is efficient when the line polled has a fast turnaround time so that the polling messages have quick responses. Polling becomes less efficient as the number of locations polled increases. A satellite channel can link a very large number of locations, and the time to send a polling message backwards and forwards is at least 540 milliseconds. Polling is therefore inappropriate for interactive computing via satellites.

3. *Concentrators.* Concentrators and terminal cluster controllers are used to group together many transmissions at one point so that the link from that point can be used efficiently. The link from the concentrator to a computer is often a point-to-point connection, but sometimes multiple concentrators are connected by one line using polling or multiplexing. Concentrators interleave the sporadic dialogue messages from many terminals and cause some of them to wait in a queue briefly, waiting for channel time to be available. In this way a link can be used with high efficiency. Some intelligent multiplexors are equivalent in their efficiency to concentrators.

Concentrators can be used on point-to-point satellite channels just as on terrestrial links. They do not, by themselves, solve the geographical demand-assignment problem because they concentrate the traffic arriving at *one* geographical location. They can, however, be used on SPADE, SBS, or other demand-assignment systems.

Major computer manufacturers now have elaborate architectures for the use of intelligent concentrators and terminal cluster controllers (for example IBM's SNA, Univac's DCA, DEC's DECNET). The protocols permit concentrators, controllers, and computers to be interconnected, in some cases, with satellite point-to-point circuits. It is desirable that the protocols and frame sizes used should follow the principles discussed in the previous chapter.

4. *Loop control.* Using loop control techniques, a continuous stream of time slots travels in a loop like a train of box cars past the terminals. Each time slot can either carry data to a terminal, or be used if it is empty by the terminal to send data back to the computer. The time slots are allocated dynamically to whoever requires them. A satellite does not provide a suitable mechanism for building a loop because the propagation delay is too great.

5. *Packet switching.* Packet switching systems are constructed using a

mesh of point-to-point lines. Data to be transmitted are formed into *packets,* which are then passed from point to point in the network until they reach their destination. The packets, like envelopes being sent by the postal service, contain the address of their destination, the address of their sender, and some control information. Any form of data can be placed inside the packets, and lengthy data have to be sent in multiple packets.

A packet travels over many separate point-to-point links before it reaches its destination. If each of these were a satellite link, the total transmission time would be too long for low-response-time systems. A different protocol is therefore needed from that in terrestrial packet-switching. If packet transmission is used, some means of solving the geographical demand-assignment problem is needed, without multiple satellite hops.

PACKET BROADCASTING

In the next three chapters we shall discuss a group of techniques called packet broadcasting with which effective interactive systems can be built. The packet structure is somewhat similar to that in packet switching networks, but the mechanisms of packet broadcasting are entirely different. The reader is cautioned not to associate the two in his mind; they are entirely different techniques.

Packet broadcasting can give more efficient use of a satellite transponder than multiplexing, polling, or SPADE-like techniques when the following circumstances apply:

1. The transmissions have a high peak/average ratio.
2. Fast response times are needed.
3. There are many separate earth stations.

—in other words, when the transponder is used for remote interactive computing on a large scale.

REFERENCES

1. James Martin, *Design of Man-Computer Dialogues,* Prentice-Hall, Englewood Cliffs, N.J., 1973. Dialogues used with data transmission are discussed in detail.
2. J. J. Gal, "Man-Machine Interactive Systems and Their Application to Financial Analysis," *Financial Analysis.* May–June 1966. The dialogue illustration from this article is used in this chapter with minor modifications.

3. William H. Sass and Stephen P. Krosner, *An 1130-Based Logic Layout and Evaluation System,* IBM Internal Paper, IBM Corp., Kingston, N.Y., 1970. (See also the *GAZELLE Operators' Guide.)* More details of this dialogue are given in Ref. 1.

4. James Martin, *Teleprocessing Network Organization* and *Systems Analysis for Data Transmission,* Prentice-Hall, Englewood Cliffs, N.J., 1969 and 1972. These books include alternative methods of line organization.

20 ALOHA CHANNELS

An ALOHA channel is a form of demand-assignment time-division multiple access designed for interactive computer transmission. It does not provide continuous channels like the TDMA discussed in Chapter 16 but transmits the brief sporadic random bursts of data characteristic of interactive computing. It is suited to transmission users with a high peak-to-average ratio as discussed in Chapter 19.

The ALOHA system is a system implemented at the University of Hawaii for interconnecting terminals and computers via satellites and terrestrial radio links [1]. It provides a form of transmission discipline for interactive computing using broadcast channels (with or without satellites). Channels organized in this fashion are referred to as ALOHA channels.

If many thousands of independent interactive computer users are to share a satellite transponder, a variation of the ALOHA system would be one way to organize the transmission. ALOHA channels employ today's transponders and earth stations without modification. ALOHA techniques are appropriate for mobile radio and local wideband cable systems. They are attractive for future satellite systems if these permit large numbers of low-cost earth stations.

LINK CONTROL

ALOHA transmission control is, in essence, a very simple and apparently undisciplined TDMA scheme. The data to be transmitted are collected in a transmission control unit which forms them into a "packet" which it transmits in a high-speed burst as soon as the packet is ready.

On the original Hawaii system the packets were up to 704 bits in length. Each packet contained a transmission header of 32 bits containing the addresses of the receiving and originating locations and some control bits, 32 redundant bits employed by a powerful error-detection scheme, and up to 640 bits of data. The error detection is sufficiently powerful that less than one erroneous packet in a billion (10^9) would escape detection.

When the packet is complete the transmission control unit transmits it at the maximum speed of the link. These packet transmissions take place at random. All stations receiving on that frequency receive the packet. They all ignore it with the exception of the station (or stations) to which the packet is addressed. A station which receives a packet addressed to it transmits an acknowledgment if the packet appears to be free from error.

The sending station waits for the acknowledgment confirming correct receipt. It waits for a given period, and then if it has not received an acknowledgment, it retransmits the packet.

Packets are transmitted at random times, and so they occasionally collide. If two packets are transmitted at overlapping times, both will be damaged, and their error-detecting codes will indicate the damage. The transmission control units which sent the packets will therefore receive no acknowledgment. After a given time each transmission control unit retransmits its packet. It is important that they should not retransmit simultaneously or the packets will again collide. Each transmission control unit should wait a different time before retransmitting.

The different waiting time could be achieved by giving each transmission control unit a different built-in delay. However, there are many such control units, and the one with the longest delay would be at a disadvantage. Therefore the ALOHA scheme gives each control unit a randomizing circuit so that the time it waits before retransmission is a random variable. The retransmitted packets are unlikely to collide a second time, but there is a very low probability that this will happen and that a third retransmission will be made. A second or even a third collision does little harm if the delays before reattempting transmission are low compared to the desired terminal response time.

If a packet is damaged by other causes, such as radio noise, it will similarly be retransmitted.

The ALOHA protocol is illustrated in Fig. 20.1.

Packet broadcasting is a little like an unruly auction in which participants all shout their bids and the auctioneer acknowledges them. If a bid is not acknowledged, the person shouts it again. Occasionally two persons will shout a bid at once, and then both have to repeat their bid.

CHANNEL UTILIZATION

It is clear that the ALOHA protocol will work well if the number of transmissions is small compared with the total available channel time. A more interesting question is: What happens when the channel becomes busy?

Abramson's analysis of the ALOHA channel proceeds as follows [2]:

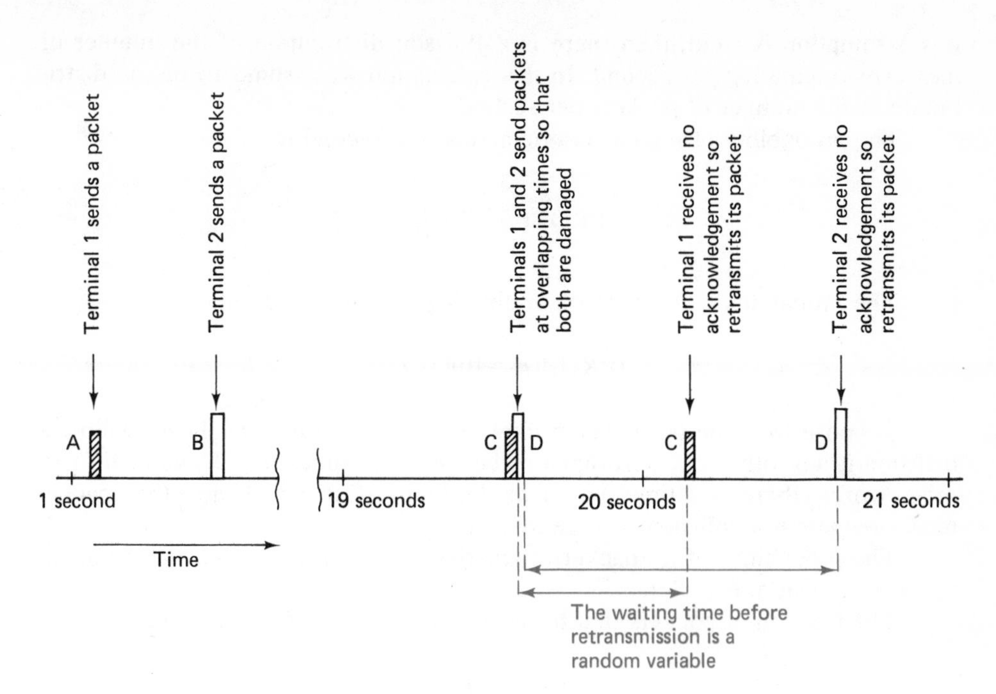

Figure 20.1 The "classical" ALOHA protocol.

Suppose that all the users of an ALOHA channel originate λ packets per second on average.

Suppose that the packets are of fixed length of duration T seconds.

The utilization of the channel as perceived by the users, which in conventional queuing theory is referred to as ρ, is then

$$\rho = \lambda T$$

$$\rho = \frac{\text{Total time the channel is in use for sending original packets}}{\text{Total time}}$$

In reality there will be no more than λ packets using the channels because some packets have to be transmitted more than once because of collisions. Let the total number of packets per second, including retransmitted packets, be λ'.

Because of the retransmitted packets, the actual channel utilization will be greater than that perceived by the users—greater than ρ. To understand how the channel behaves when it becomes busy we wish to know the relationship between ρ and λ'.

The assumption normally made in the analysis of interactive systems is that the users originate messages *independently* and *at random*. In other words, the probability of a message starting in a small time Δt is proportional to Δt. If this assumption is valid, then there is a Poisson distribution of the number of messages originating per second. In this calculation we assume a Poisson distribution of the number of packets per second.

The probability of n packets originating in a second is

$$\text{PROB}(n) = \frac{\lambda'^n \, e^{-\lambda'}}{n!}$$

The probability of no packets originating in a time of duration t is

$$[\text{PROB}(n = 0)]^t = e^{-\lambda' t}$$

Suppose that one particular packet originates at a time t_0. There will be a collision if any other packet originates between the times $t_0 - T$ and $t_0 + T$. In other words, there is a time period of duration $2T$ in which no other packet must originate if a collision is to be avoided.

The probability of no packet originating in the time $2T$, i.e., the probability of no collision, is $e^{-\lambda' \cdot 2T}$.

The fraction of packets that have to be retransmitted is therefore

$$R = 1 - e^{-\lambda' \cdot 2T} \tag{20.1}$$

The relation between the number of user-originated packets and the actual number of packets on the channel is therefore

$$\lambda = \lambda'[1 - R] = \lambda' e^{-\lambda' \cdot 2T}$$

Substituting into $\rho = \lambda T$,

$$\rho = \lambda' T e^{-\lambda' \cdot 2T} \tag{20.2}$$

Sometimes a retransmitted packet will collide a second time and have to be retransmitted again. The mean number of times a given data packet is retransmitted is

$$\begin{aligned} N &= 1 + R + R^2 + R^3 \\ &= \frac{1}{1 - R} \\ &= e^{\lambda' \cdot 2T} \end{aligned} \tag{20.3}$$

Substituting $\lambda' T$ into Eq. (20.2), we have

$$\rho = \frac{\log_e N}{2N} \tag{20.4}$$

In Fig. 20.2, ρ is plotted against N.

It will be seen that as the channel utilization (as perceived by the user) increases, the traffic which attempts to use the channel builds up at an increasing rate. A chain reaction develops, with the retransmitted packets themselves causing retransmissions so that above a certain throughput the channel becomes unstable.

Differentiating Eq. (20.4) shows that the maximum value of ρ is $\frac{1}{2e} = 0.184$.

The maximum utilization of a classical ALOHA channel is 18.4%.

It is clearly desirable to stop the users of an ALOHA channel transmitting when they reach the unstable part of the curve in Fig. 20.2. Interactive

Figure 20.2 Traffic on a "classical" ALOHA channel. The maximum usable channel utilization if $\frac{1}{2e} = 0.184$.

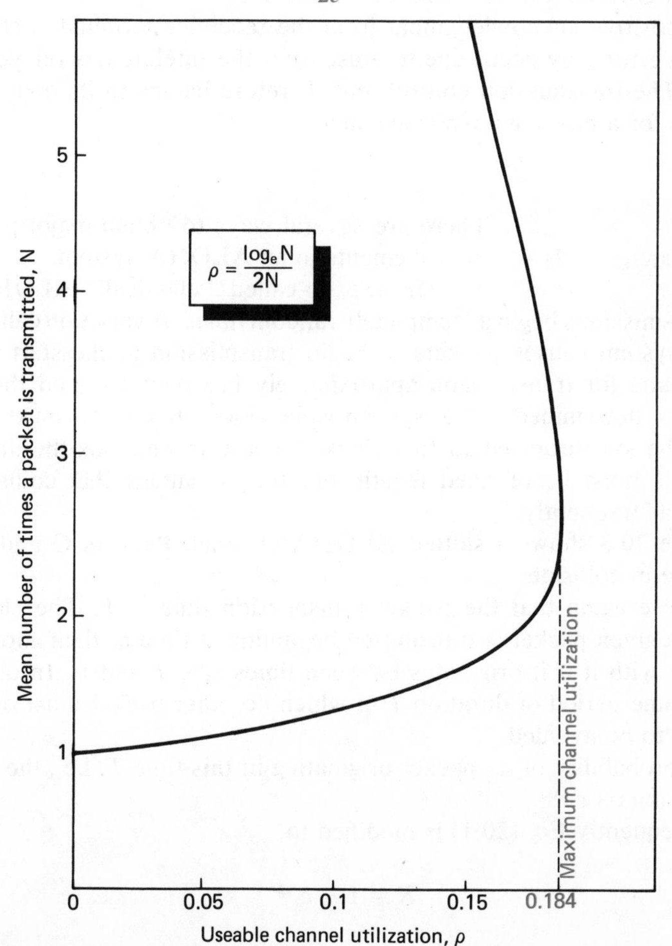

users might be slowed down either by giving them an artificially long response time or by giving them a visible warning when multiple transmissions occur.

It may be possible to achieve better channel control if the channel is used for a mixture of batch transmissions and interactive transmissions. The batch terminals could be made to pause for a while when any packet is retransmitted twice.

FEEDBACK

It will be at least 540 milliseconds before a transmitting station receives a positive acknowledgment from the receiving station. However, the transmitting station can detect a collision earlier than this because *it can listen to its own transmission being relayed by the satellite.*

The transmitting station can perform an error check on its own transmission approximately 270 milliseconds after it sends it. If it does not receive its own packet correctly, then it will retransmit it.

The positive acknowledgment from the receiving terminal is still required because an error may occur due to noise *after* the satellite has relayed a packet correctly. The transmission control unit therefore listens to its own packet and also listens for a positive acknowledgment.

SLOTTED ALOHA CHANNEL

There are several ways to obtain major performance improvements in an ALOHA system.

In the so-called "classical" ALOHA system packet transmissions begin at completely random times. A variation called a *slotted* ALOHA system causes packets to begin transmission at the start of a timing slot. The time for transmission approximately fills one slot, and the timing of the slots is determined by a system-wide clock. Each transmission control unit must be synchronized to this clock. Such a scheme has the disadvantage that packets must be of fixed length, but the advantage that collisions occur about half as frequently.

Figure 20.3 shows a slotted ALOHA channel. Packets C and D on this diagram are in collision.

Suppose again that the packet transmission time is T. The slot width is also T. If a given packet is transmitted beginning at time t_0, then another packet will collide with it if it originates between times $t_0 - T$ and t_0. In other words, there is a time period of duration T in which no other packet must originate if a collision is to be avoided.

The probability of no packet originating in this time T, i.e., the probability of no collision, is $e^{-\lambda' T}$.

Consequently Eq. (20.1) is modified to

$$R = 1 - e^{-\lambda' \cdot T} \tag{20.5}$$

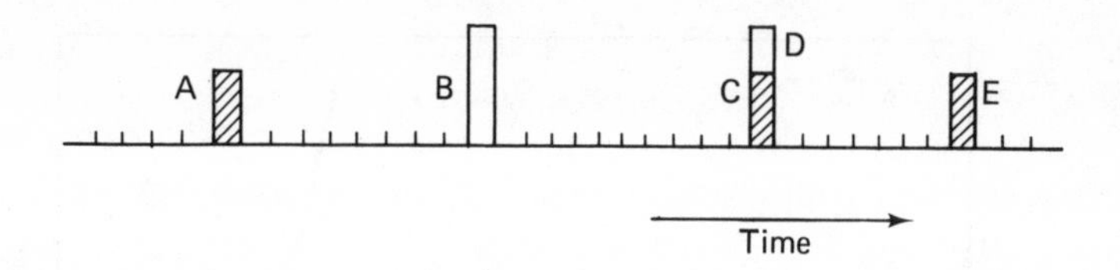

Figure 20.3 A slotted ALOHA channel. Each packet must begin on one of the time divisions.

Equation (20.2) is modified to

$$\rho = \lambda' T e^{-\lambda' T} \tag{20.6}$$

Equation (20.3) is modified to

$$N = e^{-\lambda' T} \tag{20.7}$$

Substituting, we then have

$$\rho = \frac{\log_e N}{N} \tag{20.8}$$

This is plotted in Fig. 20.4. The channel utilizations for a given retransmission rate are double those for a classical ALOHA channel.

The maximum utilization of a slotted ALOHA channel is 36.8%.

The transmission control unit is slightly more complicated because it must maintain the slot timing. A transmitter will send out periodic timing pulses for the entire system, and each control unit must synchronize its activity to these pulses. The synchronization needs careful adjustment as the transmission paths from the satellite to the control units differ slightly in length.

TRANSMISSION PROBABILITIES

From the above equations the probability that a packet will have 1, 2, . . . , n collisions can be calculated.

The probability that a collision will occur on any particular transmission is R.

For a classical ALOHA channel

$$R = 1 - e^{-2\lambda' T} \tag{20.1}$$

For a slotted ALOHA channel

$$R = 1 - e^{\lambda' T} \tag{20.5}$$

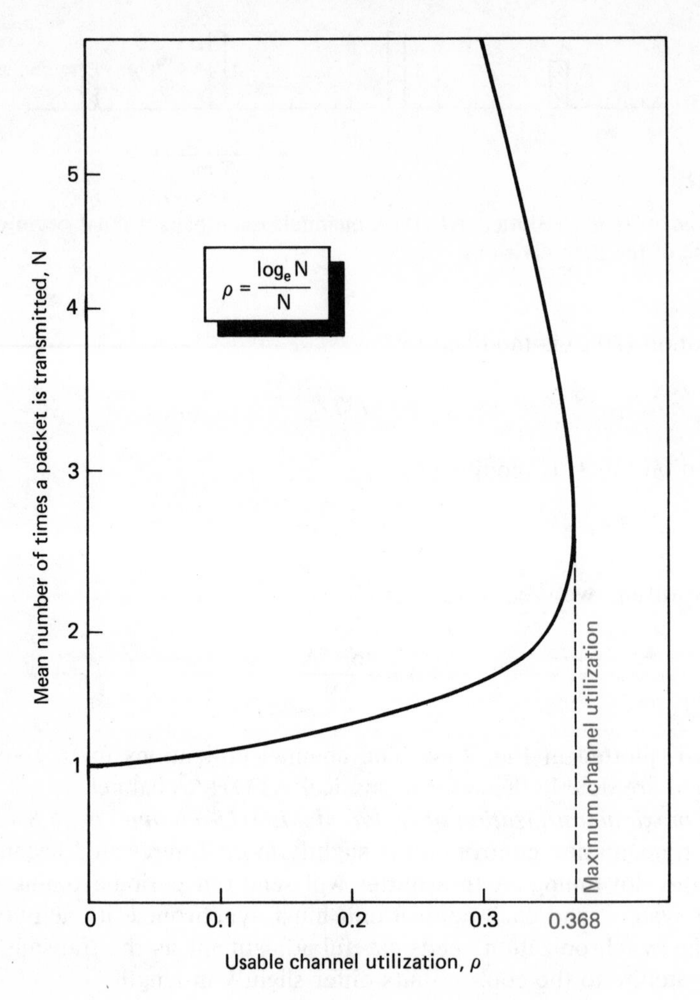

Figure 20.4 Traffic on a slotted ALOHA channel. The channel utilization is twice that of the classical ALOHA channel of Fig. 20.3, but only fixed-length packets are sent.

Let P_n be the probability that a given packet of data is transmitted n times in total (i.e., it suffers $n - 1$ collisions). Then,

$$
\begin{aligned}
P_1 &= 1 - R \\
P_2 &= R(1 - R) \\
P_3 &= R^2(1 - R) \\
&\vdots \\
P_n &= R^{n-1}(1 - R)
\end{aligned}
\tag{20.9}
$$

These probabilities are plotted in Fig. 20.5.

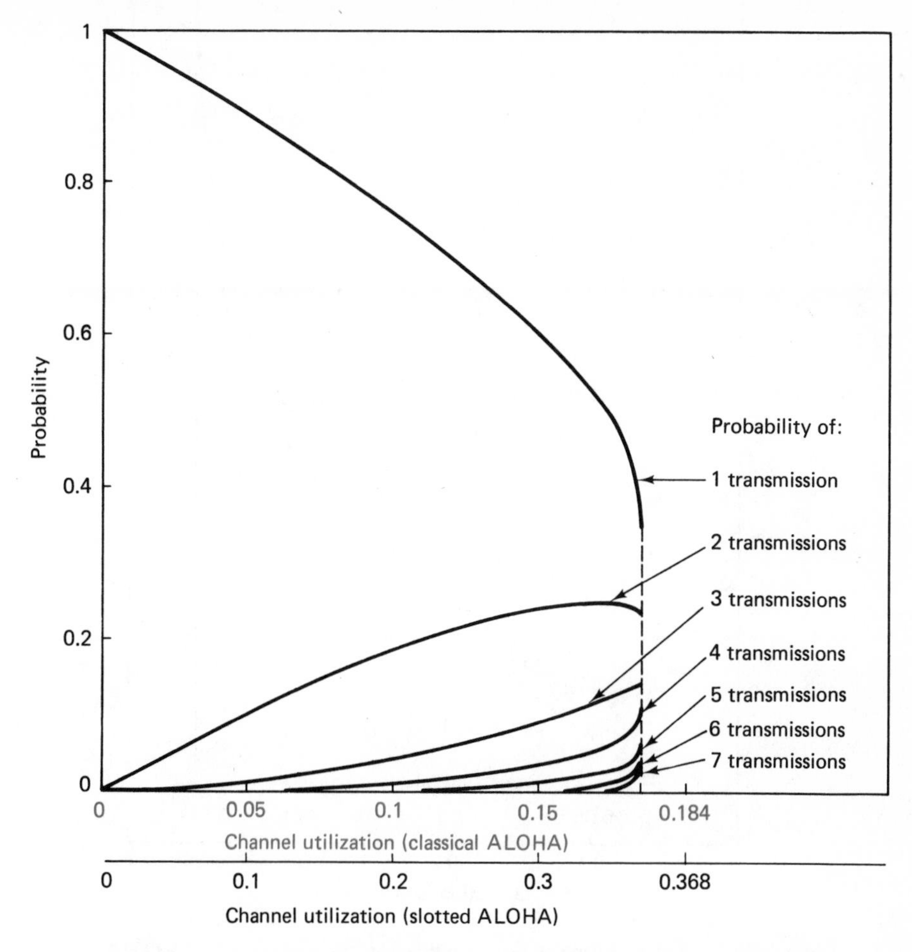

Figure 20.5 Probabilities of retransmission of a packet on ALOHA channels.

MEAN TRANSMISSION DELAY

Using the probabilities in Eq. (20.9), the mean and standard deviations of the transmission delay can be calculated.

The plot in Fig. 20.6 assumes that the satellite transmission time is 270 milliseconds, that a transmitting station listens to its own packets and hence detects a collision 270 milliseconds after transmitting the packet, and that it then pauses 200 milliseconds, on average, before retransmitting.

It will be seen that the mean delay time is low enough for interactive computing providing that the channel utilization is below about 16% for a classical ALOHA channel or 32% for a slotted channel.

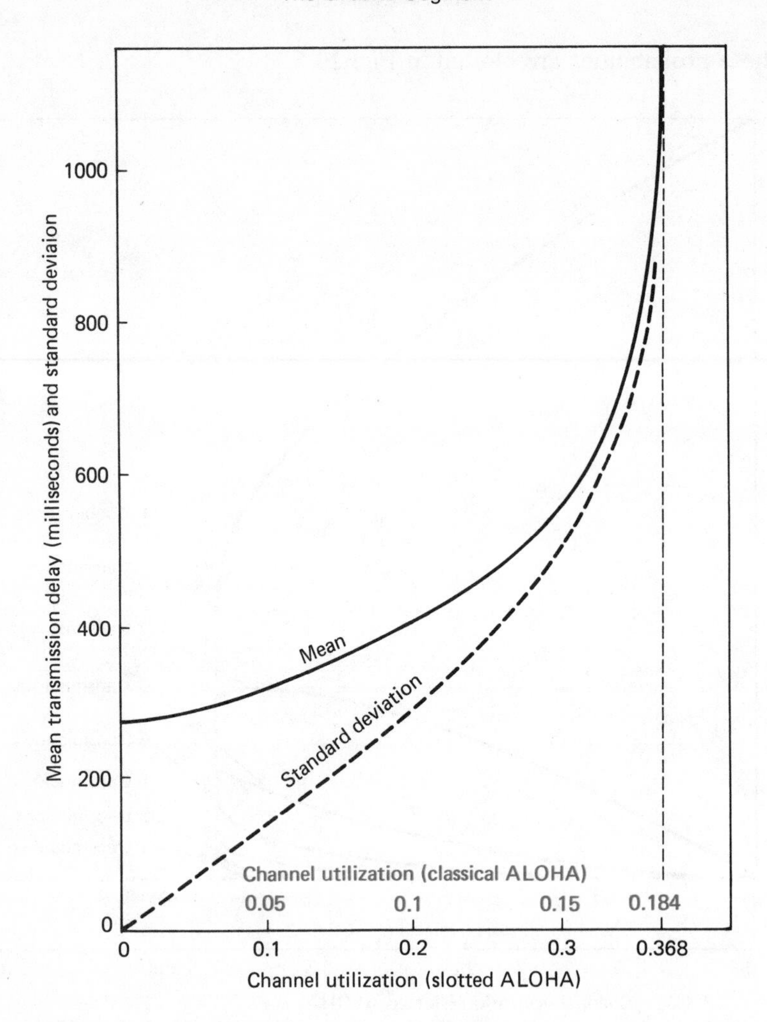

Figure 20.6 Typical one-way transmission delays using ALOHA control on a satellite channel.

EARLY WARNING Any technique which prevents the separate transmission control units from transmitting at the same instant will improve the utilization of a packet radio channel. One possibility is to use a reservation protocol which permits the control units to reserve certain time slots. Another technique is to use an early-warning system.

An early-warning system may divide the terminals up by priority. If there are two priority classes, the high-priority terminals send out a signal at a fixed interval about 270 milliseconds before they transmit. All transmission control units hear this warning signal. The low-priority ones avoid the time slot in question, leaving it free for the high-priority use. If two high-priority terminals warn that they want the same slot, both must decide whether or not to reallocate the time of their transmission. This action lowers the probability of collision, but could still result in either a collision or a slot being left empty. A multiple-priority early warning system could be used.

DISCRIMINATION

When frequency modulation is used a radio receiver circuit can be built to discriminate between weak and strong signals, and to reject the weak ones. Such a technique can greatly increase the efficiency of a packet radio system.

The transmission control units can be organized so that they vary substantially in power. There is then a high probability that two colliding packets will differ in power sufficiently greatly that one will be received correctly. The time slot is then not wasted, although the weaker packet would have to be retransmitted. It is interesting to note that if one of the packets could survive in *all* collisions, then 100% channel utilization could be achieved on a slotted channel. In practice the discrimination could never be this good, but it might be a reasonable objective that one packet should survive in half of the collisions. The maximum channel utilization would then be much higher than 36.8%.

Unless a transmission control unit had the ability to change its transmitting power, discrimination would have the effect that some terminals always had low priority. This may or may not be a disadvantage. It would be advantageous, for example, to give batch terminals a lower priority than interactive ones. Interactive terminals with a dialogue structure needing a fast response should be given priority over those for which a larger response time is acceptable.

AN ALOHA TRANSPONDER

ALOHA systems work very well over existing transponders. Amazingly, they have worked with satellites regarded as "dead" in that they no longer have enough power to relay telephone channels adequately. The Hawaii system worked with the early NASA satellite ATS-1, for example. The reason is that the transponder switches off its transmitter after transmitting each packet and hence conserves enough power for the brief packet transmission. It is possible that other dead satellites will remain valuable for packet transmission.

If one of the many transponders on a modern satellite were *designed specifically for ALOHA operation,* it could be made to function more efficiently. First, a circuit in the transponder could carry out an error check on the packet. The satellite need then only relay good packets, thus conserving power. Power, rather than bandwidth, is the limiting factor in today's transponders, and the relatively low channel utilization of the ALOHA protocol would help preserve this power.

Second, the transponder could be designed to use discrimination with frequency modulation, attempting to isolate the strongest packet when a collision occurs and relay it correctly.

It is, perhaps, unlikely that a specially designed ALOHA transponder will be included in a satellite unless a large organization commits itself to the marketing of ALOHA transmission. It is interesting to estimate the capability of an ALOHA transponder.

Given today's transponder bandwidth, such a transponder might relay 10 million bits per second with FM modulation designed for fairly low cost earth stations. Suppose we assume the use of a slotted channel operating at a maximum channel utilization of 35%, and assume that only 40% of the bits in the packets are data bits—the others are overhead or wasted. If we accept a figure of 10 bps per terminal on average for terminals using interactive dialogues (see Table 20.2), then the number of terminals that could be interactive at one time is

$$\frac{10 \text{ million} \times 0.35 \times 0.4}{10} = 140{,}000$$

If at any one time only an eighth of the terminals are in active use, such a transponder might support about a million terminals.

REFERENCES

1. N. Abramson, "Packet Switching with Satellites," National Computer Conference, *AFIPS Conf. Proc.*, Vol. 42, 1973. The ALOHA system is a research system at the University of Hawaii, supported by ARPA under NASA Contract No. NAS2-6700 and by the National Science Foundation under NSF Grant No. GJ-33220.
2. N. Abramson, "The ALOHA System—Another Alternative for Computer Communications," Fall Joint Computer Conference, *AFIPS Conf. Proc.*, Vol. 37, 1970.
3. N. Abramson, "Packet Switching with Satellites," National Computer Conference, *AFIPS Conf. Proc.*, Vol. 42, 1973.
4. L. Kleinrock and S. Lam, "Packet Switching in a Slotted Satellite Channel," National Computer Conference, *AFIPS Conf. Proc.*, Vol. 42, 1973.

21 PACKET RADIO TERMINALS

Where ALOHA techniques are used they can be extended beyond the satellite link and earth stations to the computers and terminals which employ the link.

Today, the earth stations are usually some distance from the end users, and the gap is normally bridged with leased or dial land lines. An appealing alternative is to transmit the packets by radio from the terminal or computer to the earth station and then onward over the space link. To do this there would be a transmission control unit at each data processing location, with a small radio transceiver replacing today's modem at about the same cost.

The University of Hawaii system operates with an ALOHA transmission control unit at each terminal location, transmitting to an earth station in Oahu. The system employs two 100-kHz channels in the UHF band, at 407.350 MHz and 413.475 MHz, one for transmitting *to* the terminals and one for transmitting *from* them. Each channel carries data at a rate of 24,000 bps. The transmission control units and transceivers cost $3000 to build in 1973 (less than the cost of many modems) and Abramson, who directed the project, estimates they would cost less than $500 to produce in quantity [1].

A packet radio system can use one frequency or two. If a separate channel is used for transmitting from an earth-station controller to the terminals, there is no problem with packet collisions on this channel; the controller can determine all the packet transmission times. Packet collisions would occur on the channel *to* the station. On the other hand, if a single frequency is used for transmission in both directions, some bandwidth is saved.

DEVICES USED

Figure 21.1 shows the types of devices interconnected by the University of Hawaii ALOHA system. In 1974, two satellites were in use, INTELSAT IV and NASA's ATS-1 satellite. The connection to the latter uses a very inexpensive earth sta-

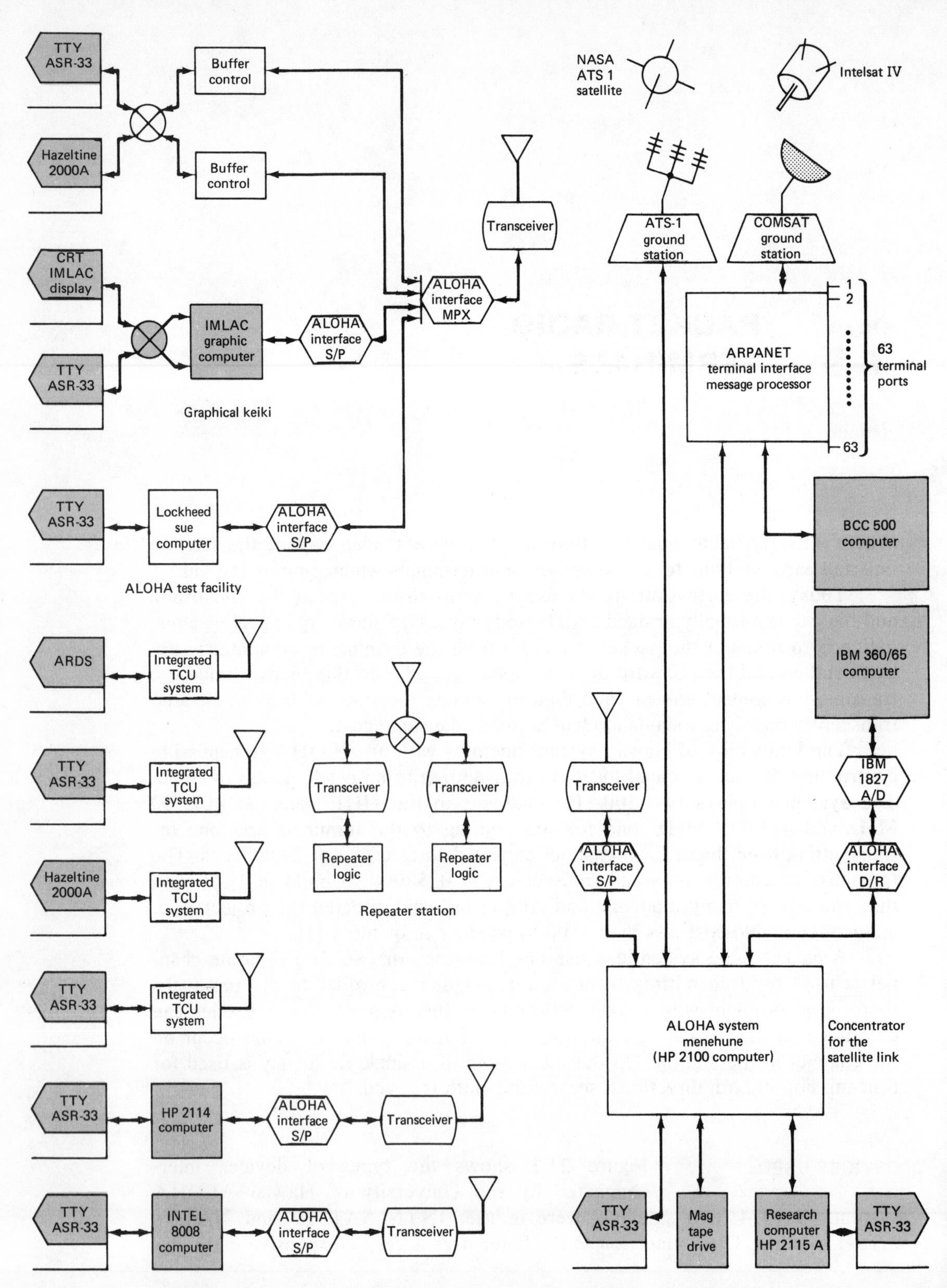

Figure 21.1 Types of devices interconnected in the University of Hawaii ALOHA System, 1974.

tion. The INTELSAT IV channel provides a connection to the ARPA nationwide packet-switching network.

Both types of satellite channel are connected to a minicomputer which collects and queues the packets and is called a MENEHUNE (Hawaiian word for "IMP"). Packets reach the MENEHUNE both from machines nearby which are connected to it by cables, and from distant machines which transmit to it over an ALOHA broadcast link. There are thus two types of packet broadcasting links: satellite links and terrestrial links. The latter are important for connecting the many scattered terminals to the earth stations.

At the locations remote from the earth stations a variety of data processing devices is connected to ALOHA transmission control units. In some locations single teletype machines or other terminals are connected. In other locations groups of terminals share a transmission control unit.

When a user employs his terminal he may be communicating with machines via the satellite link or he may be using a computer attached to the MENEHUNE by cable. Packet broadcasting is not useful only for satellite links. The reader should think of it as an inexpensive means of attaching any form of terminal to remote computers without using the local telephone loops.

LOW DUTY CYCLE

A packet radio transmitter is different from a conventional radio transmitter in that it does not transmit continuously; it transmits occasional very brief bursts of data. Because of this, a relatively small power source can be used—a source capable of occasional brief bursts of high power but not continuous transmission. A small hand-held transmitter is therefore practicable.

It is indeed desirable that the power source should *not* be capable of powering continuous transmission because one type of failure that would play havoc with packet radio is a transmitter being stuck ON.

POTENTIAL USES

Packet radio, given the spectrum allocation it needs, has the potential of revolutionizing some aspects of interactive computing. The following are potential uses of packet radio:

1. Radio units built into terminal controllers to permit them to continue operating when failures occur on their telephone connections.
2. A large population of portable hand-held terminals which fit into the user's pocket—perhaps somewhat larger than an HP 67 pocket computer.
3. The burglar and fire alarms of a community capable of radioing immediate details of violations to police or fire headquarters.
4. Weather, seismic, and other monitors parachuted into forest areas.
5. Radio devices on every vehicle which identify the vehicle for automatically paying

tolls and parking fees, for opening garage doors, for security control, and for enabling police to detect traffic violations or stolen cars.

6. Terminals in vehicles for sending messages, making inquiries, or fleet scheduling.
7. Terminals on boats for navigation and other purposes.
8. Portable terminals for children to act as a super-toy and give them access to libraries and computer-assisted instruction.
9. Graphics terminals connected to remote computers and requiring bursts of transmission at a speed too great for connection by telephone lines.
10. Computer terminals in developing nations and other areas where the local telephone lines are inadequate.
11. Portable terminals for military field operations in areas where other forms of communication are poor.

HAND-HELD TERMINALS

Particularly intriguing, perhaps, is the prospect of a hand-held terminal, somewhat larger than an HP 67 pocket computer—small enough to fit in one's pocket. Today's technology could provide a highly reliable, secure, pocket terminal which would enable individuals to send and receive messages. They could communicate with other individuals, computers, and data libraries, possibly on the other side of the world via local concentrators, data networks, and satellite relays.

Such a device could have an alphanumeric keyboard; the HP 67 has enough space on its keyboard for 40 keys. To be generally capable of man-computer dialogues it would need a screen which could display short text messages. A gas panel, LED, or liquid crystal screen would probably be used. The screen could either be designed to display alphanumeric characters or else could be a rectangular dot matrix.

Roberts [2] visualized such a terminal with a dot matrix screen with 80-dot-per-inch resolution and 2.8 inches × 1 inch in size, i.e., 244 × 80 dots. This could display 8 lines of 32 characters or small diagrams. The messages to the screen would be sent in character blocks, each block being a 7 × 10 dot matrix. If, instead, the machine handled 7-bit characters, an alphanumeric message would require one-tenth as many bits, but the machine would need character generation logic. There is a trade-off between bandwidth and logic requirements. The logic required for character generation would be less than that in an HP 67.

Such devices might be used on a 50,000-bps UHF channel. In a terminal dialogue using the 256-character screen, the average transmission rate to the machine over a long period of dialogue might typically be 10 characters per second. If these are transmitted as 7-bit characters, the channel might accommodate 500 active terminals. A single local broadcast system might therefore have thousands of users of whom not more than 500 would be likely to be active at once.

The potential applications of such a device are as diverse as the computer industry itself.

A Collins Radio study designed a pocket data terminal. ARPA, the U.S. Department of Defense Advanced Research Projects Agency, has operated a packet radio network with substantially larger portable terminals connected to the ARPANET network since 1977.

REPEATERS

In a packet radio system some of the terminals might be too far away from the controlling location or satellite earth station for an inexpensive transmitter to reach it. In such a case radio repeaters are used. The transmitters used in the Hawaii system have a range of many miles, but repeaters are used for interisland transmission and to reach terminals in the shadow of mountains.

Conventional radio repeaters cannot receive and transmit on the same frequency at the same time because the strong transmitted signal blots out the weak incoming signal. They receive on one frequency band, translate the signal to a different frequency band, and then transmit it. A packet repeater can, however, operate on a single frequency. It switches off its receiver momentarily while it transmits a packet. Operating at a single frequency saves the expense of the frequency translation equipment. The Hawaii repeaters use a single frequency for relaying packets to the central control location (MENEHUNE) and a different frequency for relaying packets back to the terminals. If two frequencies are used in this way the repeater antennas pointing toward the central station can be highly directional.

When terminals are distributed over a wide geographic area, many repeater stations may be used (Fig. 21.2). Such may be the case in areas of low population density and large distances, where the alternatives to radio would be expensive. In areas of high population density such as a city, repeaters may not be used as all terminals are close enough to the central station. Repeaters may be used in a city to avoid problems with high-rise buildings.

When large numbers of terminals are used it is desirable to select the design options such that the complexity resides in the central location and such that the repeaters are inexpensive and the terminals as inexpensive as possible.

When multiple repeaters are used, a problem that must be solved is the cascading of packets. A packet from certain terminals may reach more than one repeater or reach the central station as well as a repeater. Unless the repeaters have directional antennas, packets from one repeater may reach other repeaters. Some packets might travel endlessly in loops. The unnecessary cascading of packets could substantially lower the effective channel capacity. It is necessary to devise a protocol that prevents packets multiplying themselves.

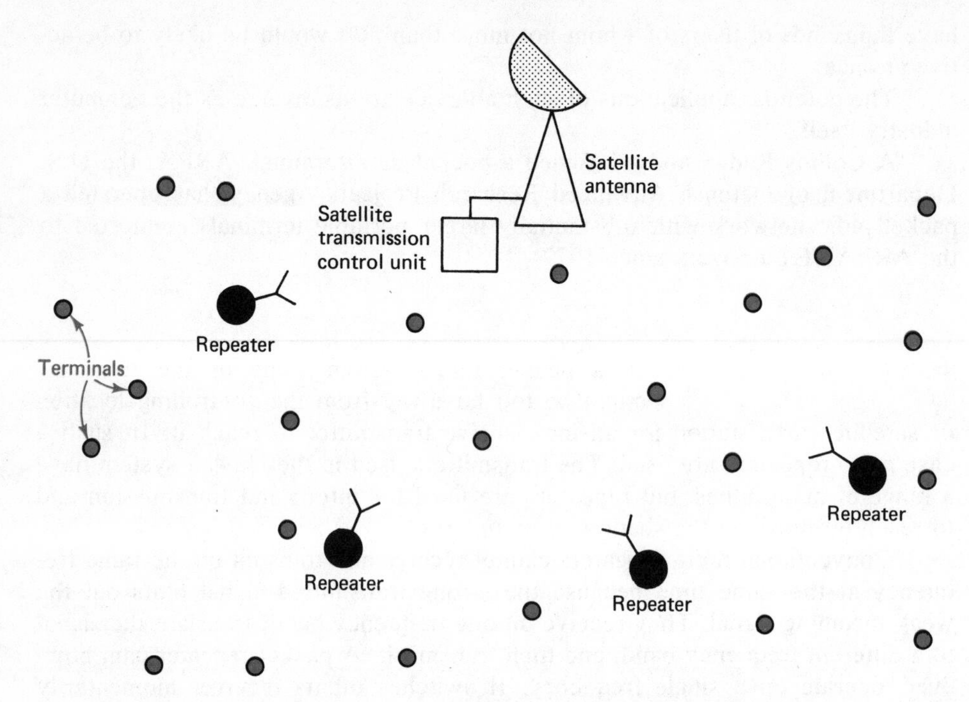

Figure 21.2 Terminals scattered over a wide geographic area may be served by multiple packet radio repeaters relaying transmission to a central control station.

REFERENCES

1. N. Abramson, "Digital Broadcasting in Hawaii—The Aloha System," paper presented at Canadian International Seminars, Ottawa, 1974.
2. L. Roberts, "Extension of Packet Switching to a Hand Held Personal Terminal," *AFIPS Conference Proceedings, SJCC 72,* pp. 295–298.

22 BURST RESERVATION SYSTEMS

ROUND-ROBIN PROTOCOL

An ALOHA protocol is only one of several alternatives for transmitting sporadic bursts of data from small satellite earth stations. Perhaps the simplest is a protocol in which each earth station is allowed to transmit in turn in a round-robin fashion (rather like the commutator drawn in Fig. 10.6). Each station transmits a burst the length of which was predetermined. All stations listen to the bursts, listening for messages addressed to themselves.

The disadvantage of a round-robin system is that the traffic load at a station may vary second by second. The station therefore wants a way to vary the amount of channel space it uses. It cannot vary its usage of channel space without informing the other earth stations that it is going to do so, or confusion will result. A variety of schemes is possible with which a station could *make a reservation* for time slots before it uses them.

RESERVATIONS

One difference between an ALOHA system and a conventional demand assignment system is that in the conventional system some form of *reservation* is made for a portion of the transmission capacity; in an ALOHA system it is not.

A system which takes channel space at random without reservations may be appropriate for brief sporadic transactions which need quick service—like individuals catching taxis at random in New York. Transactions from computer terminal users conform to this random pattern. A system *with* reservations can normally achieve higher channel utilization; however, the users have to wait longer on average before (successful) transmission. The taxis of New York could be fuller on average if all passengers had to make a reservation (there would be fewer taxis); however, the passengers would usually *have to wait longer* to obtain a taxi. Nonreservation systems tend to degrade more seriously when the load swings up to high levels, like trying to find a taxi in New York in a thunderstorm.

PACKET RESERVATIONS

It is possible to design a reservation system for interactive data transmission that will have the advantage of an ALOHA system in handling sporadic burst traffic but which will give better utilization of the satellite channel capacity. It will not give so fast a mean response time as an ALOHA system, but even so will complete a one-way trip in well under 1 second and hence is fast enough for interactive computing.

Such a system has been described as a *burst reservation system* or *packet reservation system*. It differs fundamentally from the FDMA or TDMA systems described in Chapters 15 and 16 in that it does not reserve *continuous* channels, such as telephone channels. It reserves time slots in which a block of data bits can be sent. It is thus more akin to packet switching than circuit switching, whereas systems such as SPADE and MATE (Chapters 15 and 16) are more akin to circuit switching. Burst reservation satellite channels will provide a natural extension to today's packet switching networks. Furthermore, because of the high capacity of satellite transponders, burst reservation systems could be used for applications entirely different from those which employ terrestrial packet switching networks.

SINGLE QUEUE

Suppose that every station wishing to send a burst could request a time slot for that burst on a single satellite channel, and the first free time slot is reserved for it. We then have what is, in effect, a single-server queue. All bursts queue for the use of the satellite channel. The queuing time (after the burst is allocated a slot in the queue) is low because each burst is transmitted very quickly when its time comes—in 1 millisecond, say. If the channel is heavily loaded so that there are 10 bursts queuing for the channel, a burst has to wait for 10 milliseconds after being allocated a slot before it is transmitted. This is short compared to the propagation delay.

Before the burst can join the queue, the earth station sending it must have a valid reservation. It will take about 270 milliseconds to send a request for a reservation and possibly another 270 milliseconds to receive a reply telling it when to transmit. This second delay can be avoided with an ingenious control mechanism. The transmission itself will take another 270 milliseconds, and there may be a few milliseconds of "queuing" before the transmission.

Curve 2 in Fig. 22.1 shows the mean transmission time for a burst reservation system on which the subchannel for reservations and other protocol messages occupies one-sixth of the channel capacity. The maximum channel

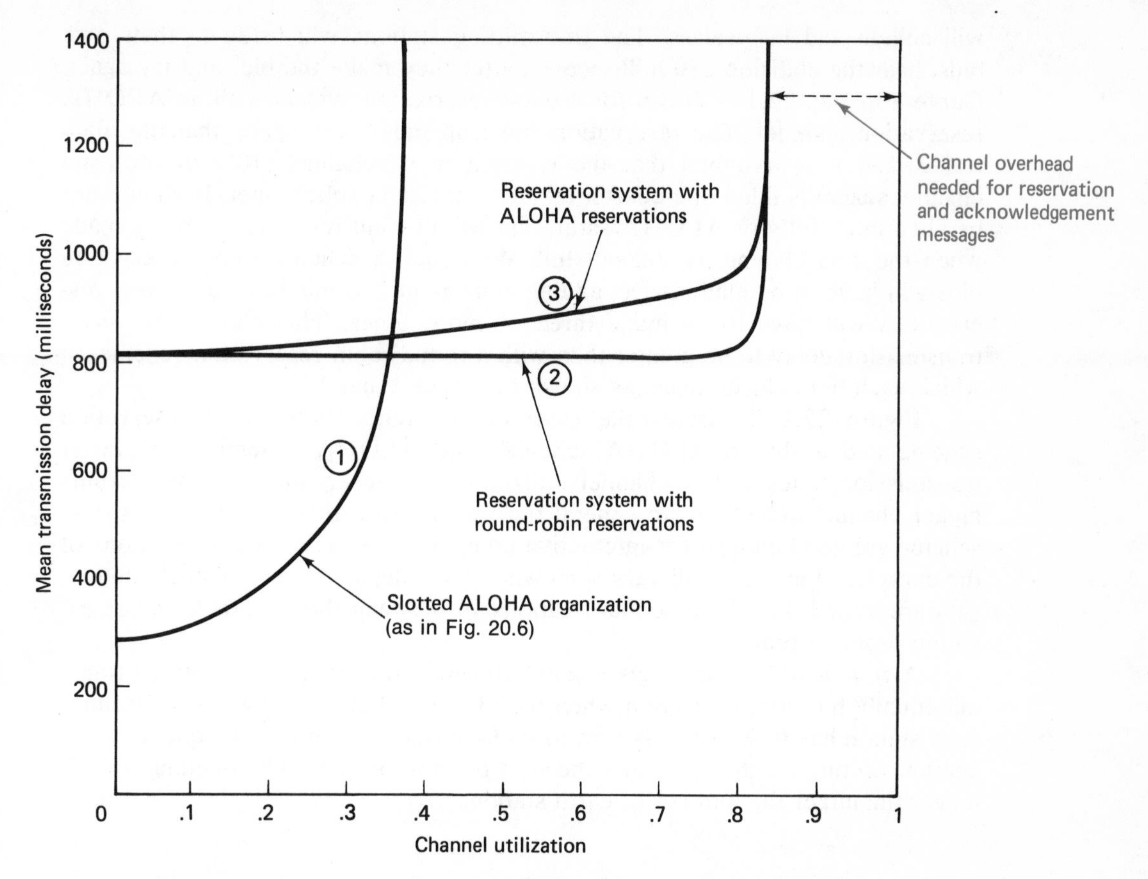

Figure 22.1 Mean transmission delay and channel utilization on ALOHA systems and reservation systems in which the protocol messages occur one sixth of the channel capacity.

utilization is thus 83.3%. The mean transmission time is close to 3×270 milliseconds until the queue begins to build up. The additional queuing time will be barely noticeable until the channel has almost reached its saturation level. The curve assumes that each earth station has a reservation subchannel which always succeeds in transmitting a reservation when required—part of the 83.3% overhead.

For contrast, curve 1 in Fig. 22.1 shows a classical ALOHA system.

If there are many earth stations, it may be difficult to provide a reservation subchannel which does not itself cause delay. For a small number of stations, each station may be allocated a fixed slot, round-robin fashion, in which it can transmit its reservation. If there are many hundreds of earth stations, one approach is to give the reservation channel an ALOHA protocol. Each station can then make bids at random, as in Fig. 20.3. Occasionally two bids will collide and be useless. The transmitting stations will listen to their own bids, hear the collision 270 milliseconds after they make the bid, and try again. Curve 3 in Fig. 22.1 is drawn for a burst reservation system with an ALOHA reservation channel. The reservations take up much less space than the data traffic, and it is assumed that the reservation subchannel (10% of the total channel space) is filled one-quarter as full as the data subchannel. It should not be filled more fully or ALOHA saturation will prevent reservations being made when the data channel is still not full. With such a scheme some reservation bids will have to be made twice, adding more than 270 milliseconds; a few reservations will have to be made three or more times. This causes the *mean* transmission delay to be greater than with a round-robin reservation scheme, in which each bid is made once, as shown in curves 2 and 3.

Figure 22.1 illustrates the essential difference between a reservation scheme and a slotted ALOHA scheme. The ALOHA scheme gives faster transmission times at low channel utilization; the reservation scheme permits higher channel utilization. In general the transmission times of the reservation scheme are good enough for interactive computer use. The precise positions of the curves in Fig. 22.1 will vary somewhat when details of the channel subdivision are varied, but the essential relationship between the ALOHA and reservation protocol remains.

A *round-robin* scheme gives good channel utilization if all stations transmit equally but poor utilization when they have widely differing loads. Because each station has to wait for its turn to come around, the messages queue at the station, waiting for the messages ahead of them to be sent. The queuing time is longer the larger the number of earth stations.

SMALL MESSAGES

The advantages of a reservation system apparent in Fig. 22.1 apply when the data messages are not too short. With very short messages the channel capacity needed to make the reservation may be greater than the channel capacity needed to send the message. Figure 22.1 is drawn for a situation in which one-sixth of the channel capacity is for protocol messages. For *very* short messages, then, the simple operation of a slotted ALOHA channel can give better throughput than a reservation system.

Many terminal dialogues do indeed employ very short messages—so short in many cases that they would be shorter than reservation messages.

Rather than sending a reservation message the system might as well send the data message itself.

There is much to be said, therefore, for a dual protocol which makes reservations for long messages and which sends very short messages without reservations. The channel capacity would be divided into large time slots which can be reserved and small time slots which cannot. Figure 22.2 shows one way

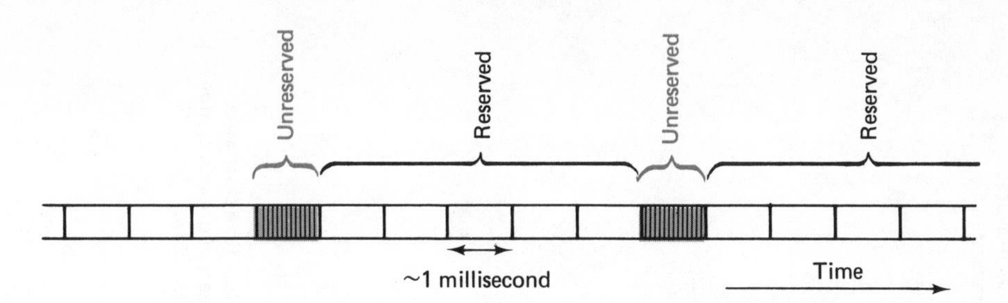

Figure 22.2 One method of interleaving reserved and unreserved time slots.

of organizing such a channel. The reservation and other protocol messages would travel in the unreserved time slots.

PROTOCOL

To establish a *single-queue* reservation system there are some important details which must exist in the protocol. First, there must be an acknowledgment mechanism. The sending station must retain each packet until it receives positive confirmation that it has been received correctly. If it does not receive this confirmation, it will resend the packet. This will be a rare event (unlike on an ALOHA system) normally caused by transmission errors. A very thorough error-detecting code is necessary.

The channel subdivision must be organized so that the acknowledgment messages can be sent without incurring high overhead. Like the reservation messages, they are short. Several reservation or acknowledgment messages can share the same transmission from one station, if necessary. If there are not too many earth stations, these protocol transmissions may be sent at predetermined times. Each earth station is allocated a time slot in which it sends its protocol messages. Alternatively, the acknowledgment messages, reservation messages, and short data messages could all travel together in an ALOHA subchannel.

AN ARPA SATELLITE CHANNEL

L. G. Roberts [1] proposed a reservation system which could be used as an extension of terrestrial packet switching networks such as ARPANET, using satellite channels of 50,000 bps or more. In this system all protocol messages are handled in an ALOHA fashion. The system is always in either a *reservation state* or an *ALOHA state.* In the reservation state reserved blocks of data are sent; in the ALOHA state, protocol or short data messages are sent.

Figure 22.3 illustrates the operation of the system. This illustration could relate to a 50,000-bps channel. The time allocated for reserved bursts is 25 mil-

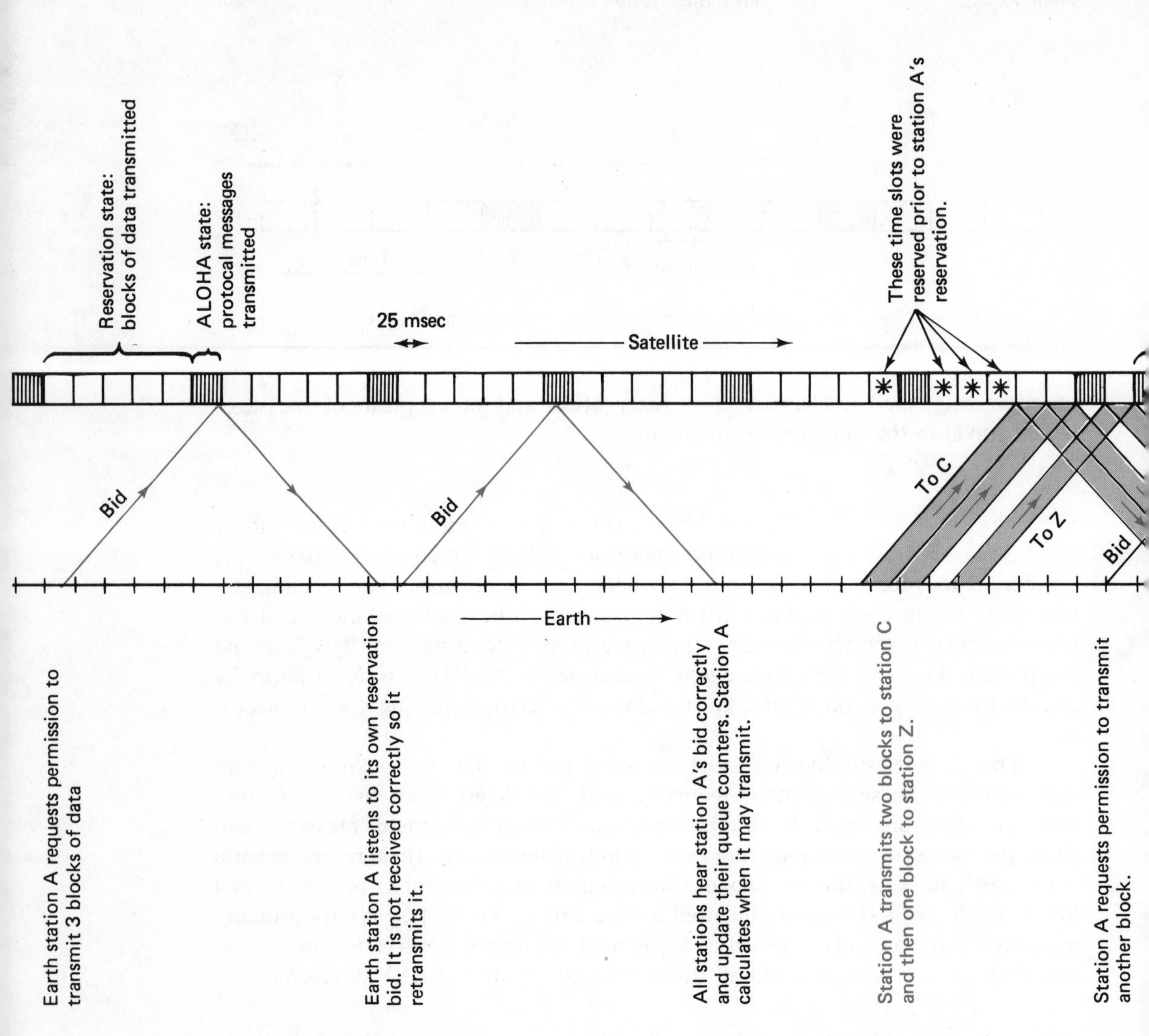

Reservation state: blocks of data transmitted
ALOHA state: protocal messages transmitted
25 msec
Satellite
These time slots were reserved prior to station A's reservation.
Bid
Bid
To C
To Z
Bid
Earth
Earth station A requests permission to transmit 3 blocks of data
Earth station A listens to its own reservation bid. It is not received correctly so it retransmits it.
All stations hear station A's bid correctly and update their queue counters. Station A calculates when it may transmit.
Station A transmits two blocks to station C and then one block to station Z.
Station A requests permission to transmit another block.

Figure 22.3 A burst reservation system with an ALOHA protocol mechanism [1].

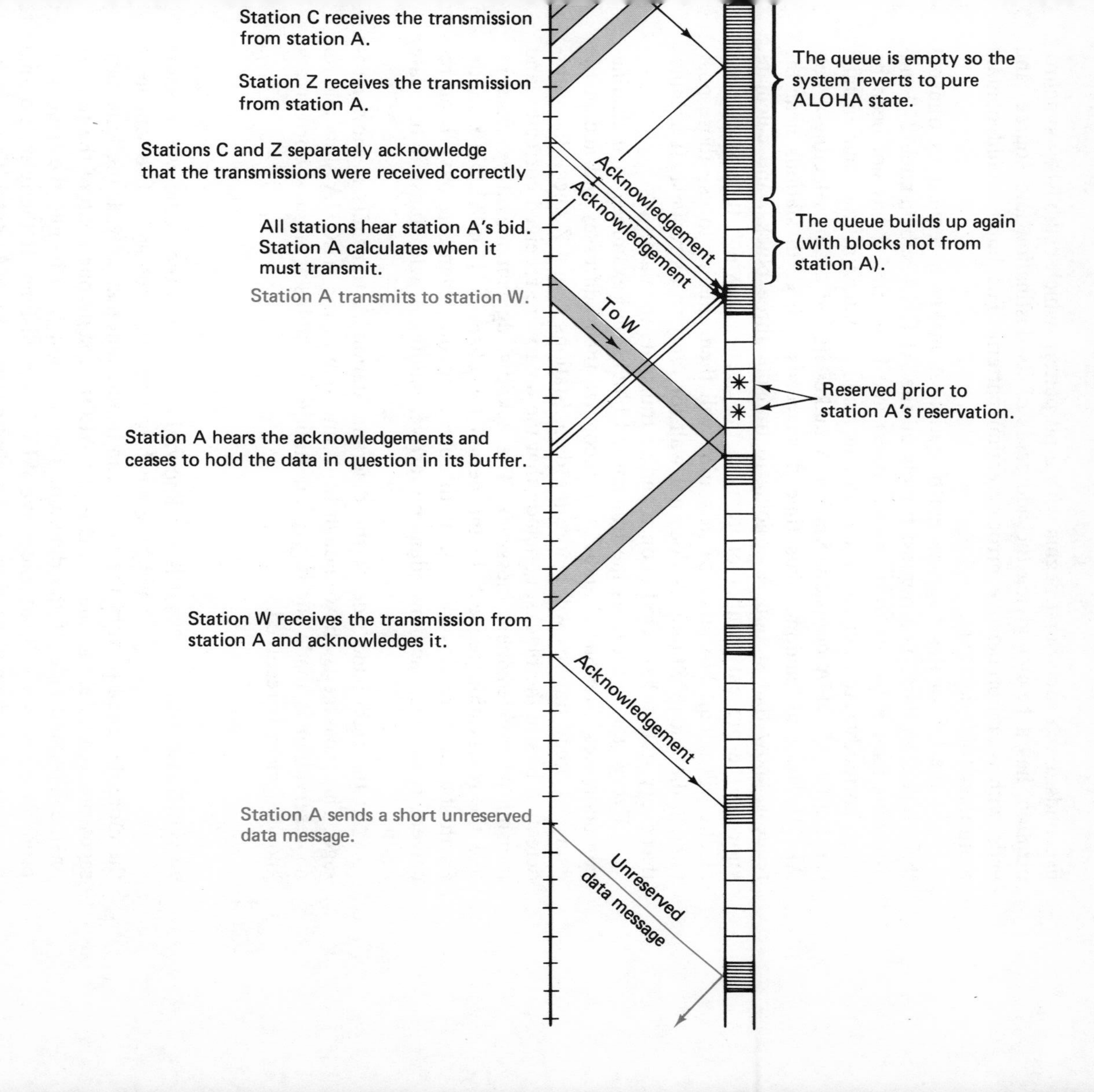

Station C receives the transmission from station A.
Station Z receives the transmission from station A.
Stations C and Z separately acknowledge that the transmissions were received correctly
All stations hear station A's bid. Station A calculates when it must transmit.
Station A transmits to station W.
Station A hears the acknowledgements and ceases to hold the data in question in its buffer.
Station W receives the transmission from station A and acknowledges it.
Station A sends a short unreserved data message.
The queue is empty so the system reverts to pure ALOHA state.
The queue builds up again (with blocks not from station A).
Reserved prior to station A's reservation.
Acknowledgement
Acknowledgement
To W
Acknowledgement
Unreserved data message

liseconds. Each such burst begins with a bit pattern which establishes synchronization, has a header giving the addresses of its destination and source, and ends with a comprehensive error-detection pattern. The burst would contain more than 1000 data bits.

At the left of the diagram, earth station A makes a request to transmit three blocks of data. The request travels on the ALOHA subchannel. 270 milliseconds later it listens to its own request and hears that it has not been relayed correctly. Another earth station had made a request at the same instant, causing an ALOHA collision. Station A makes its bid again, selecting a new ALOHA slot at random. This time it succeeds. All earth stations hear the request, know that station A is going to transmit three blocks, and calculate, knowing the present number of items in the queue, at which time slots these blocks will occur. No other earth station will then attempt to use those positions in the queue. Station A does not wait for reply to its request. It assumes that every station has acted upon it and so transmits the data.

Every station receives the transmitted blocks and examines their destination addresses. Stations C and Z will recognize their addresses and accept the data. All other stations will ignore the data. Stations C and Z check the error-detection bits in the blocks, conclude that the data were received correctly, and transmit acknowledgment messages. The acknowledgment messages are sent, like the reservation request, in the period the channel is in ALOHA mode. Earth station A retains the data until correct acknowledgments have been received. If no such acknowledgments arrive, station A will transmit the data again.

At the right-hand side of the diagram, station A transmits a short message. This message is short enough to fit into one of the ALOHA slots and so no reservation is made for it. Again, station A retains it until a correct acknowledgment is received.

VARYING THE MIX

It is not known how much of the traffic will be short and how much will be long messages. The scheme is therefore designed to vary the mix it can handle automatically. If the long messages predominate, the channel is in the ALOHA state one-sixth of the time as on the left-hand side of the diagram. This is enough to handle the necessary protocol messages without excessive ALOHA collisions. If there are no long messages, the channel reverts to the continuous ALOHA state. This can be seen happening in the center of the diagram.

If the channel is in the continuous ALOHA state, the first reservation request will cause it to allocate a time slot in reservation mode. As soon as the reservation queue goes to zero the system reverts to continuous ALOHA mode.

In Roberts' proposal [1] the reserved time slots were 1350 bits in length and the nonreserved slots were 224 bits. Both of these include the overhead of synchronization, error detection, and addressing. Other slot sizes may be appropriate to channels of different speed.

PROTOCOL ACCURACY

To make a burst reservation system work, the protocol messages must be relayed accurately. Errors in some of the reservation messages would cause chaos. This is especially so when there are a large number of ground stations competing. The protocol messages should therefore have exceptionally safe error-detection facilities and possibly error-correction coding also. It may be desirable to send more than one copy of each reservation message. Whatever technique is used, careful attention is needed to all possible types of protocol failure.

COMPARISON

It is complex to compare one protocol with another because the comparison must make assumptions about the traffic mix that is handled. For different situations, different techniques appear preferable, and there are so many different potential uses of satellite channels.

Roberts [1] makes an interesting comparison of a packet reservation system like that in Fig. 20.3 with other systems, on the assumption that the traffic would be broadly similar to that on ARPANET, a multipurpose network of computers used mainly in universities. He assumes, for purposes of comparison, that all earth stations send and receive an equal volume of messages; half of these are single packets of 1270 bits (including the header) and half are multipacket blocks of 8 packets; both the numbers of single packets and multipacket blocks arrive with a Poisson distribution. This assumption biases the results slightly in favor of a reservation system in that it does not have very short messages, which would favor an ALOHA protocol, or large batches of data, which would favor a *round-robin* approach somewhat more. Figure 22.4 gives curves for such a traffic pattern. The *round-robin* curves relate to a system in which each earth station in strict rotation has a time slot allocated to it in which one packet can be transmitted.

If each earth station handles a *large* volume of interactive data from a concentrator or packet switch, then the unevenness of high peak-to-average ratios is statistically leveled, and a variable TDMA system with reservations, like that in Fig. 16.4, becomes efficient.

REFERENCES

1. L. G. Roberts, "Dynamic Allocation of Satellite Capacity through Packet Reservation," National Computer Conference, *AFIPS Conf. Proc.,* Vol. 42, 1973.

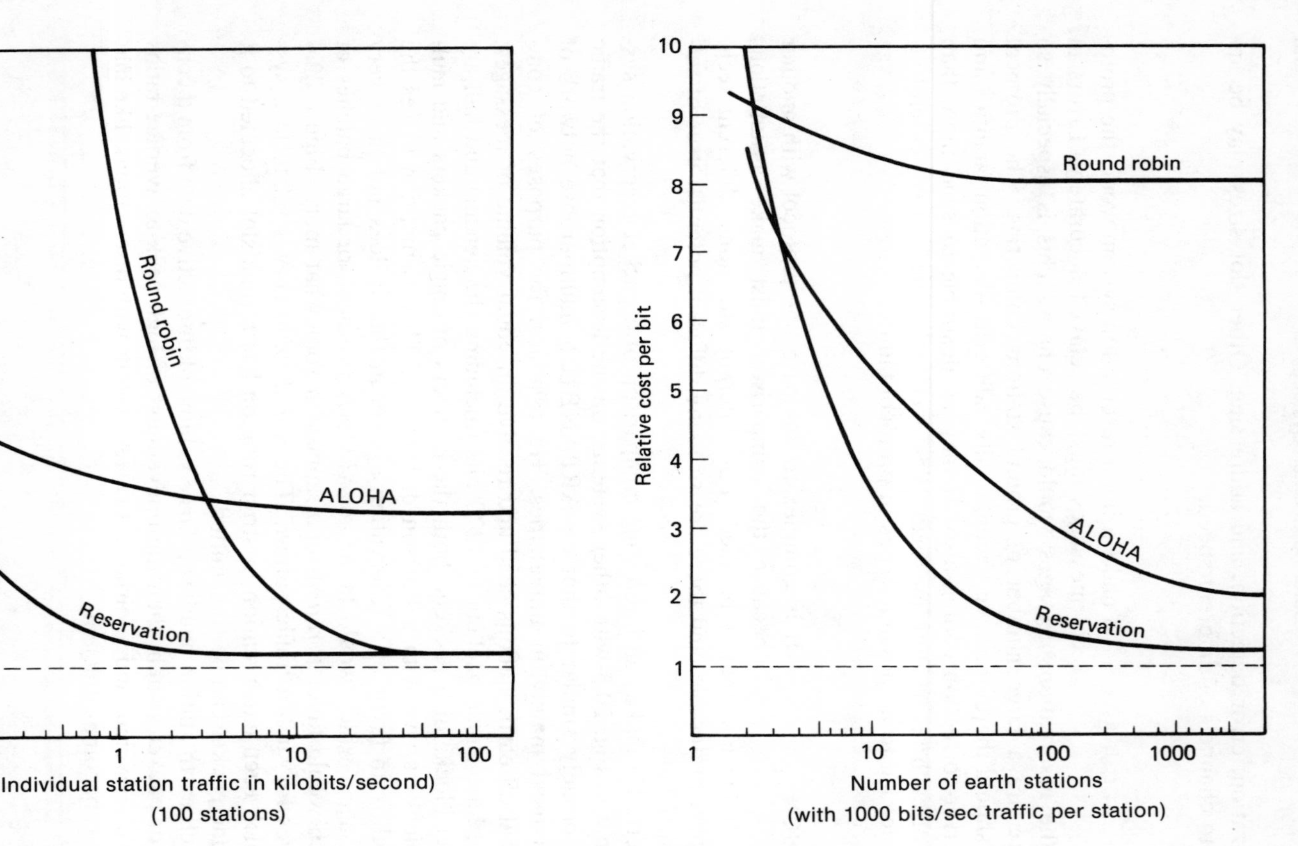

Figure 22.4 Comparison of a round robin, ALOHA, and burst reservation protocol.

23 AVAILABILITY

Satellite channels have a number of negative aspects which are worse than on equivalent terrestrial links:

1. A quarter-second propagation time.
2. In some designs a poor signal-to-noise ratio such that forward error correction is needed on data channels.
3. Occasional severe increases in noise and error rate caused by storms on satellites using frequencies above 10 GHz.
4. A potential privacy risk in that a transmission can be picked up by earth stations everywhere in a satellite beam.
5. Occasional transmission outages caused by eclipses.
6. The small, but finite, possibility of satellite failure.
7. The possibility of radio-jamming of satellites used by the military.

The satellite systems designer needs to appreciate these disadvantages and design around them. We have already discussed errors and response times. In this chapter and the next we shall discuss availability, security, and privacy.

SPACE SEGMENT AVAILABILITY

The space segment exhibits a different pattern of availability from the ground segment. Satellite failures cannot be fixed by repair men but are very rare. Most satellite systems have at least two satellites in orbit, so that if one fails the users can switch to the other. A spare satellite is usually kept on earth, so that it can be launched quickly if necessary to replace a failed satellite.

The satellite is designed with enough redundancy in its components to enable it to remain useful after component failures. A user may be confronted with the failure of a transponder allocated to him, but the system can switch to a different transponder at a different frequency. Figure 23.1 shows the redundant components designed into an SBS satellite. Each pair of traveling wave tubes has one backup, and the receiving electronics is duplicated. In general, anything which is likely to fail has a backup.

In addition, the engineering of the satellite components is done so as to maximize their probable lifetime in space. It is economic to use costly materials and processes to achieve this end.

A satellite's availability is related to its age and orbital lifetime. When a satellite has been operating for some years some of the redundant electronics in Fig. 23.1 may have replaced failed components. The probability of a serious failure is then somewhat higher. If failure probability is calculated from the *mean time between failure* of components, then the variation of availability with orbital lifetime can be estimated. If a satellite is to have a longer lifetime, it will need more redundant components. SBS in its FCC filing quotes a satellite availability of 0.9999 for a lifetime of 7 years, using the redundant components shown in Fig. 23.1. During the satellite lifetime it is estimated that there will be a few component failures, taking about 15 minutes to correct. A failure of the attitude control will probably not occur during the 7 years. If it does, it may take 3 hours to correct, and one such failure is included in the estimate of 0.9999 availability.

So far, commercial (as opposed to experimental) satellites have had a remarkably good history of failure-free operation, once they are operating correctly in orbit. The failures have mostly occurred during the launch or checkout period. No successfully operating commercial satellite has failed completely (at the time of writing), though there have been occasional transponder failures and misfirings of the attitude control jets.

ECLIPSE AND OUTAGES

As we discussed in Chapter 2, the satellite undergoes eclipses and sun outages.

A domestic satellite can be positioned so that the eclipses occur shortly after midnight when they do little harm. Storage batteries can keep some (or all) of the transponders operating during the eclipse.

Sun outages, caused when the sun passes behind the satellite and shines directly into the antenna, occur at less convenient times. Transmission may be blotted out for up to a quarter of an hour.

The satellite becomes unavailable for certain predictable periods during eclipses. The systems designer has to decide what the system will do during eclipses. The satellite may be designed to keep functioning when the earth's shadow passes across it (Fig. 2.12) by using on-board storage batteries. The batteries may operate some but not all transponders. During the less frequent

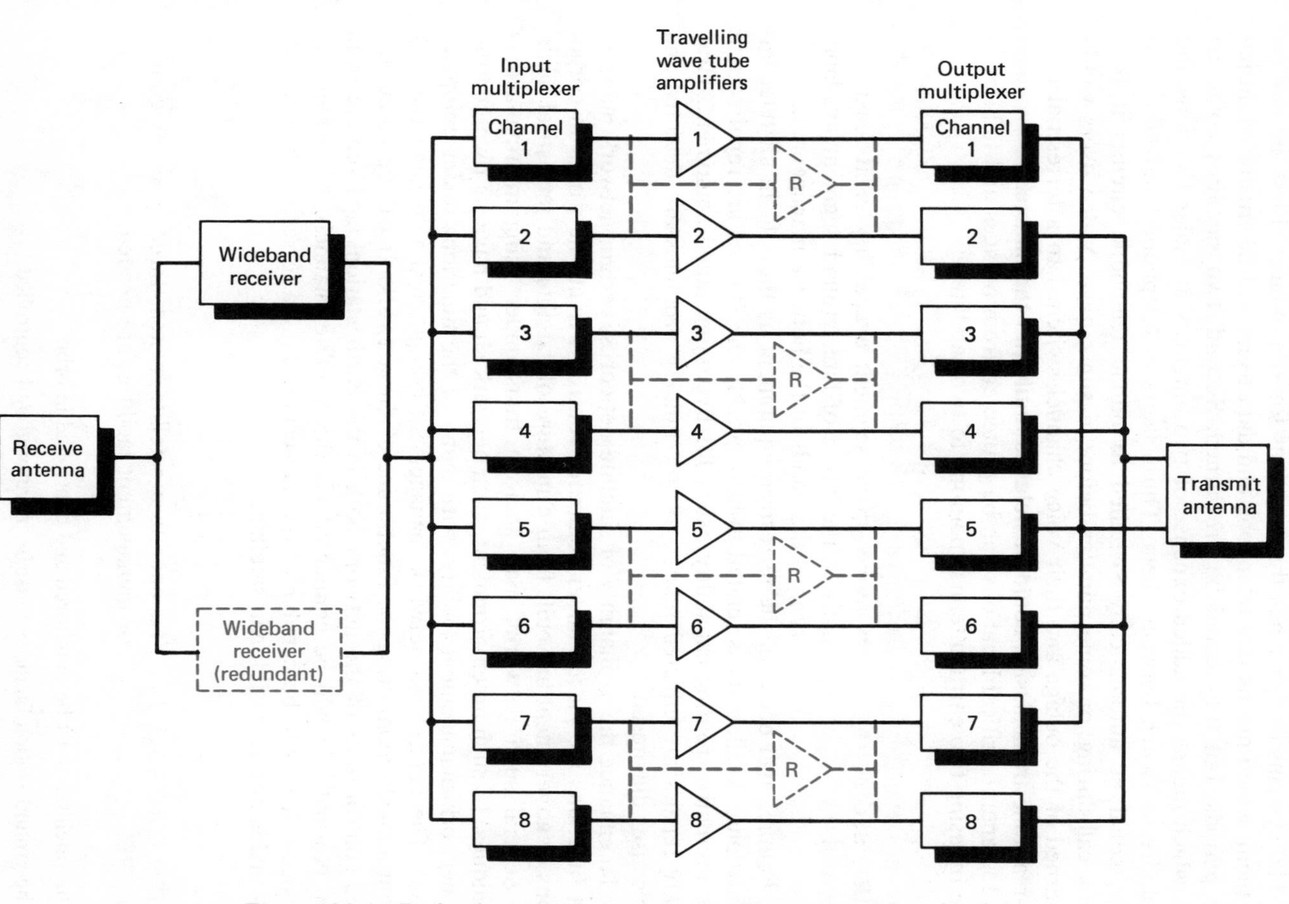

Figure 23.1 Redundant receiver and travelling wave tubes in the SBS satellite, designed for system dependability. The system now has ten active travelling wave tube amplifiers with six spares.

occasions when the sun passes behind the satellite, the transmission is blotted out for up to a quarter of an hour.

The systems designer has three options for sun outages. First, he can ignore them, which means that telephone and data users will be frustrated during these periods, but they could be prewarned. Second, two satellites could be used, which means the added expense of two antennas at major locations and the ability to switch between them. Third, users of telephone bandwidth, and below, could be automatically switched to public telephone circuits if they make a call during, or some minutes before, a sun outage. Video users would be warned of the outage and their voice channel switched to a terrestrial link. Non-real-time transmission could be delayed until after the outage.

In general, critical locations can be protected from outages and failures in space by using two earth antennas pointing to two satellites.

GROUND SEGMENT AVAILABILITY

Whereas space segment availability is at least partially in the hands of fate, ground segment availability is in the hands of the network management.

Failures will occur in the terrestrial equipment as they do in all telecommunications. Availability is related to how quickly the failures are repaired and what alternative routing capability exists for bypassing failed equipment. How quickly repairs are effected relates to how quickly problems are detected and their causes diagnosed.

To enhance the availability of satellite networks, certain network management functions are placed under computer control. Private networks in corporations or government are built from a mixture of satellite and terrestrial facilities. Several such networks may share a transponder using multiple-access techniques. In such an environment, each network should have its own centralized network management facility, with many of the functions under computer control. The centralized network management computer should have the task of continuously monitoring the functioning of the network. It will be connected to the control units of the network both in the earth stations and remote from them. Normally, it will be connected via the satellite channels, but when failures occur it should be capable of quickly establishing a terrestrial line to network nodes not accessible via satellite.

SYSTEM CONTROL FACILITIES

To operate a satellite system three types of control or management facilities are needed:

1. The satellites must be monitored and controlled in orbit.
2. The ground station facilities must be monitored and controlled.
3. Where separate telecommunications networks are derived from the same satellite system, each separate network needs to be monitored and managed.

SATELLITE CONTROL FACILITY

The *satellite control facility* is a location at which the position and performance of the satellite is monitored. Adjustments to the orbital position and attitude of the satellite are made. When it is necessary to switch to alternative equipment on board the satellite this can be done. The satellite control facility will have an exacting task to play during the launch and initial positioning of the satellite.

One, or probably more than one, earth station will have the task of tracking the satellite, receiving telemetry signals from it, and sending commands to it. Figure 23.2 shows the Western Union center which performs these tasks for the WESTAR satellites and acts as the satellite control facility. The SBS system will have two *telemetry, tracking, and command* earth stations (TT&C) on opposite sides of the United States. These stations have a large antenna (10 meters) and special equipment for transmission and reception. They can operate both in the 4/6-GHz and 12/14-GHz bands. 4/6 GHz are used during the initial transfer orbit maneuvers when positioning the satellite, because it is necessary to make use of the existing worldwide tracking network which employs these frequencies. 12/14 GHz will be used for telemetry and control when the satellite is operational, but the 4/6-GHz telemetry and control equipment could be used in an emergency.

The *satellite control facility* will be connected to the *telemetry, tracking, and control* earth stations by terrestrial leased lines. It may be co-located with one of them. The system can operate if one of the TT&C stations has failed.

The telemetry equipment in satellites is normally duplicated. Satellite operating parameters are sampled constantly, encoded, and transmitted to earth. To track the satellite, tones are sent to it, which it receives and retransmits on a telemetry down-link. The difference between the transmitted and received phases of the tones can be used to determine the range of the satellite, and the range from more than one point is measured. The range and telemetry signals are fed into a minicomputer (duplicated for reliability) at the satellite control facility. They are compared with preselected value ranges, and if any parameter falls outside the preset limits for a number of consecutive times, an alarm is sounded, and the faulty parameter is displayed.

When commands are to be sent to the satellite from the *satellite control facility* they are processed and displayed at the facility before transmission. The satellite receives the command, uses a security procedure to verify that it is valid, decodes the command, and transmits it back to earth before executing it.

SYSTEM MANAGEMENT FACILITY

While the *satellite control facility* monitors the space segment, a *system management facility* monitors the ground segment of the system.

The earth stations on many systems are connected to a central *system management facility* by leased terrestrial lines as well as by the satellite link it-

Figure 23.2 Western Union TT&C center.

self. Operating parameters for the earth stations are transmitted periodically to the system management facility so that the health of the ground segment is monitored. This form of monitoring is important for domestic satellite systems which have unattended earth stations.

Earth stations designed to be part of a reliable communications system have redundant components. Intelsat earth stations routinely achieve 0.9999 availability (less than one hour outage in a year). This gives continuity of service for the entire Intelsat global system of better than 0.999. This is achieved by duplicating items and minimizing the need for repairmen.

The FCC filing of SBS [1] estimates that the availability of its unmanned earth stations will be 0.9991. The SBS *system management facility* automatically detects failures and is on-line to the maintenance centers (Fig. 23.3). It has equipment which enables it to use all transponder channels so that it has a satellite link as well as a terrestrial link to every earth station.

Like the satellite control facility, the SBS system management facility contains duplexed computers. Its computers and personnel perform several functions:

1. Monitoring the performance of the earth stations.
2. Collection and analysis of earth-station performance data.
3. Collection and analysis of network utilization data.
4. Diagnosis and recommendation of corrective procedures when failures occur.
5. Management and control of system services.
6. Initialization and reinitialization of the earth-station TDMA equipment.
7. Change in system configuration.
8. Billing for various services.

The earth-station control computer will contain diagnostic software which can be brought into operation by the system management facility.

The system management facility is quite separate, in the SBS system, from the multiple-access control. The system management facility could be out of action and the customer earth stations would function normally. All the functions shown in Fig. 16.6, needed for normal TDMA operation, are performed elsewhere.

There is one exception to this. The system management facility can control a pool of spare capacity for each transponder. A customer network may be given the capability to request the use of this pooled capacity for a brief period for some temporary high-speed digital operation. The use is limited to demands for more than 448 kbps. A customer who needs a video channel, or a very high-speed connection to a data base for a brief period, can request it.

SUBNETWORK MANAGEMENT

When multiple customer networks are derived from a shared satellite, a customer wants to be able to manage his own network independently of any other users. He needs a console which will tell him the status of his network and earth stations, and give him usage statistics so that he can see how effectively his network meets his traffic requirements.

On the SBS system such a console is a remote terminal connected to the system management facility. It enables customers to access a portion of the

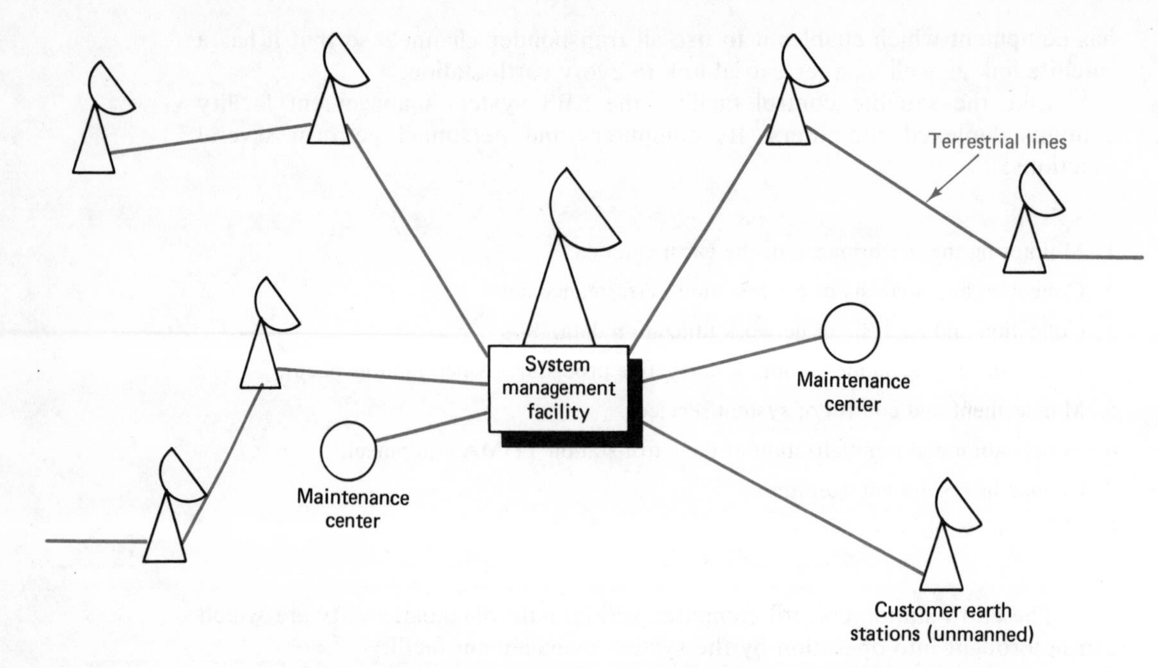

Figure 23.3 The SBS System Management Facility continually monitors the health of the SBS ground segment. It is connected by terrestrial links (as well as via the satellite) to all customer earth stations. Failures are automatically reported to a local maintenance center, which is also connected by terrestrial lines to the System Management Facility.

system management facility's data base to obtain information about their own network.

AVAILABILITY

For system availability as seen by its end user, the following components must all be operating:

1. Earth station transmitting equipment, including the modems and multiple-access equipment.
2. The up-link.
3. The satellite.
4. The down-link.
5. Earth station receiving equipment, including the modems and multiple-access equipment.

If the availability of the above items are a_1, a_2, a_3, a_4, and a_5, respectively then the system availability a is

$$a_s = a_1 \bullet a_2 \bullet a_3 \bullet a_4 \bullet a_5$$

The availability of the satellite and earth stations can be made very high and depends on how much is spent on redundancy and quality components. The availability of the link, especially at frequencies above 10 GHz, is affected by the weather, and to overcome the effects of weather large antennas and powerful amplifiers may be used.

In the SBS system [1] the objective for end-to-end link availability is 0.995. The satellite portion of this is estimated to have an availability of 0.9999, and the earth stations together, 0.9991.

If the up-link and down-link together have an availability a_L, then

$$0.995 \leqslant 0.9999 \times 0.9991 \times a_L$$

$$a_L \geqslant \frac{0.995}{0.9999 \times 0.9991} = 0.996$$

This figure sets a limit on the percentage of time the links may be out due to bad weather. One link, up or down, should have an availability of 0.998 if the overall objective is to be met. 0.2% of the time it may be unavailable because of bad weather.

Figure 23.4 shows the effects of bad weather on transmissions at 12 and 14 GHz, measured by Comsat Laboratories at Clarksburg, Maryland. It will be seen that 0.2% of the time rain attenuation will be about 3 decibels or more worse than average on the 14-GHz link and about 2 decibels on the 12-GHz link. Three decibels of rain attenuation may be assumed in the link equation (Chapter 7) in order to meet the availability objective.

To give a link availability comparable to that of the earth station, the signal should penetrate the rain 0.05% of the time. The curves indicate that this would need approximately an extra 4 decibels engineering into the link. This could be done by making the earth antennas 1.58 times larger in diameter or by using more expensive amplifiers and preamplifiers. Many SBS customers will have 5-meter antennas; some will have 7-meter antennas. Larger antennas could be used in special cases.

BIT ERROR RATE

Availability of the space link needs to be defined in terms of the bit error rate. A storm does not simply switch the transmission off; it increases the error rate. Availability of the link will therefore differ depending on what is defined to be the maximum acceptable bit error rate.

The above quote of an availability of 0.995 is based on an error rate of 1 bit error in 10^7, but this would be achieved only with the aid of error-correcting codes. As we have commented digitized speech transmission can sound good

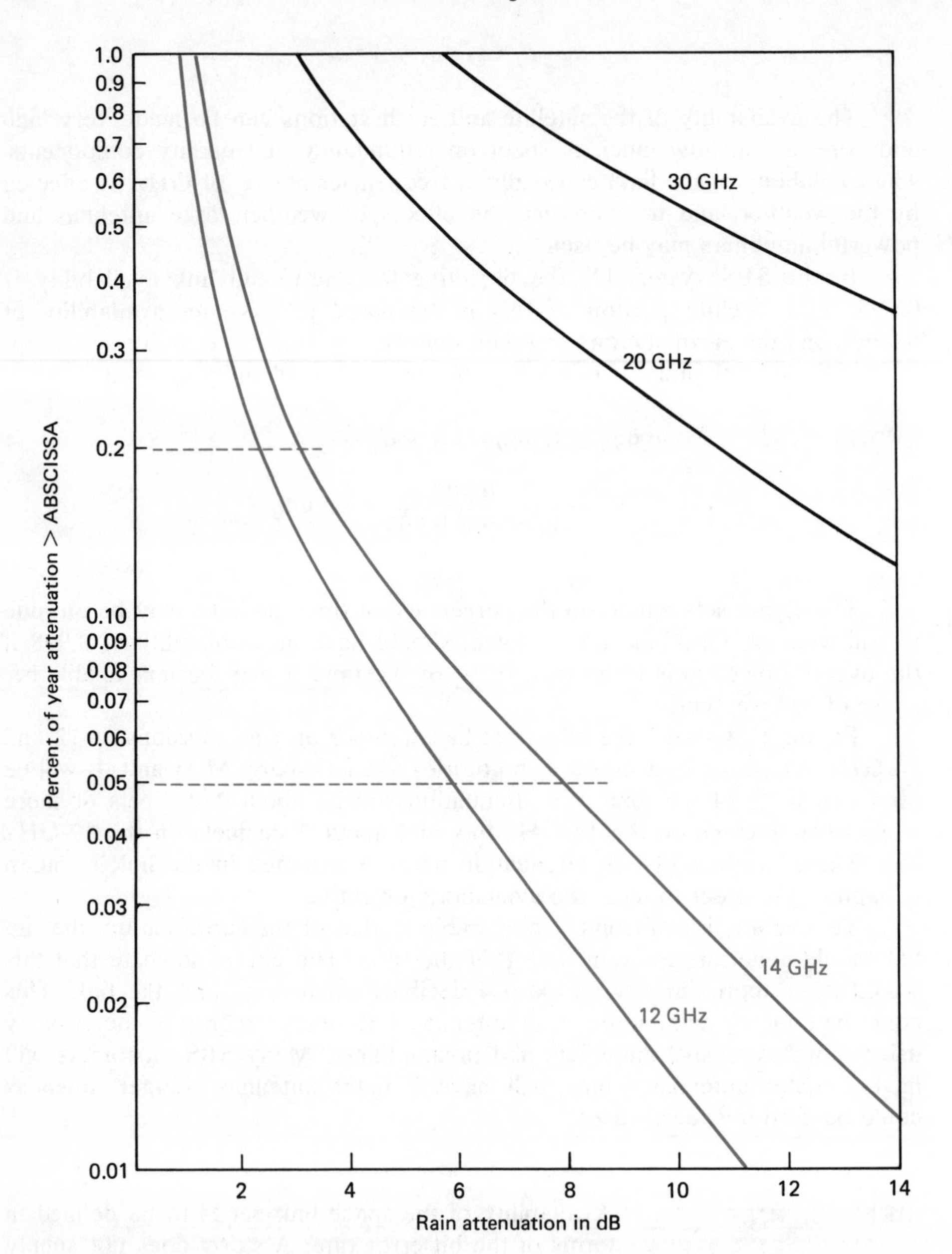

Figure 23.4 The effects of rain must be built into the availability equations. These curves show typical attenuation caused by rain. The 12/14 GHz curves are from Comsat Laboratories data collected at Clarksburg, Maryland, over a 3-year period [1]. Compare with Fig. 25.7.

even with a high bit error rate. With some codecs speech of good quality can be obtained with a random bit error rate as high as 1 in 10^3 or even higher. Such speech channels survive worse rain storms than channels requiring a lower error rate.

SYSTEM MANAGEMENT FUNCTIONS

Box 23.1 summarizes the basic system management functions which are needed for enhancing availability and for other purposes. The system management facility would normally reside at one of the earth stations. A combination of monitoring computers and staff would keep the system functioning well. The signal control units also need to have some functions for network management built into them (not included in the diagrams earlier in the book such as Figs. 15.2, 15.4, and 16.6) which permit them to interface with the central network management facility.

The control staff, or authorized managers, will be able to display status reports when they request them. By repeatedly sending messages through the network, the management computer will rapidly become aware of any failure. It is necessary to determine exactly where the failure has occurred. Parts of a corporate network are likely to come from different vendors—some from telephone companies, some from satellite equipment vendors, some from computer vendors, and so on. The management facilities should be designed to determine which organization to call for repair work. Diagnostic messages will be sent with the purpose of isolating the failure.

The network should be designed with automatic alternate routing. When

BOX 23.1 Basic system management functions

Space Segment

• Telemetry	Monitoring the telemetry signals from the satellite
• Diagnosis	Sending commands to the satellite to check its health and diagnose problems
• Reconfiguration	Sending commands to the satellite to switch in redundant equipment
• Tracking	Obtaining knowledge of the exact orbital position

BOX 23.1 Continued

• Orbital adjustment	Firing the satellites jets to adjust its position
• Attitude control	Making the satellite point to the correct position on earth
Earth Segment	
• Monitoring	Continual checks that the earth stations are functioning correctly
• Diagnosis	Determining the cause of any failures
• Maintenance	Dispatching maintenance engineers when needed
• Alternative routing	Switching traffic to alternative channels, including public telephone channels, to circumvent failures
• Measurement	Collecting measurements of traffic and system usage
• Traffic analysis	Analysis of the measurements to determine overloads and needs for reconfiguration
• Reconfiguring	Changing the allocation of channels and other facilities to meet changes in demand
• Initialization	Initializing and reinitializing earth stations
• Program loading	Loading and changing programs used in the various earth stations
• Operator assistance	Providing human assistance to users when necessary
• Security	Prohibiting the interconnection of facilities which are not specified
• Status reports	Providing customer management with displays showing the status of their network, when they request them
• Billing	Sending bills for system usage

segments of the network are unavailable, users' calls should be switched automatically to different paths. They may be routed via the telephone network to the nearest operational earth station. The alternate routing facility may be programmed into a private branch exchange, or other separate switches may be used.

The objective of both alternate routing and automatic fault isolation is to maximize the availability to the users, i.e., give them an appropriate good *grade* of service. The alternate routing capability may be used to bypass the satellite during sun outages.

REFERENCE

1. *FCC Application of Satellite Business Systems for a Domestic Communications Satellite System,* Federal Communications Commission, Washington, D.C., 1976.

24 PRIVACY AND SECURITY

A question commonly asked about satellites relates to privacy. The signals from the satellite can be received anywhere in the satellite beam. How can we ensure that they are not picked up by unauthorized persons?

The same question could be asked of much terrestrial transmission. Telephone lines can be tapped very easily on users' premises and a terrestrial microwave beam can be picked up about as easily as a satellite beam with a private antenna. The interception of a satellite beam needs expensive equipment to extract the wanted signal from the processes of encoding and multiple-access which we have described. To extract a voice or data signal from the SBS bit stream of 41 mbps would be very complex. Nevertheless, it is possible, and one organization might be tempted to extract another organization's signals in certain rare situations.

The best and probably the only safe way to protect telecommunications from wiretapping is to use cryptography. As speech and other signals are sent *digitally* on satellite channels, cryptography can be applied to the digital bit streams. The modules which do the speech processing can be associated with modules for encryption and decryption.

Since World War I, much has been written about cryptography and code breaking. The subject, however, has drastically changed its nature with the advent of computers. When a computer or computer-like logic can be used to do the coding, an immeasurably more complex form of enciphering can be used. On the other hand, the computer will now be used to aid in the decrypting and to search at high speed through very large numbers of possible transformations. In balance, if both sides act prudently, the sender is generally better off as a result of computing technology. When the transmission is between computers, the machines can scramble the data in a truly formidable fashion.

Most of the ciphers that were used prior to the development of computers can be broken by using computers [1]. More complex enciphering is essential.

VALUATION OF SECRECY SYSTEMS

A number of different criteria can be used to judge systems for encryption. Claude E. Shannon, the communications theory pioneer, developed a mathematical treatment of secrecy systems and in a classic paper [2] lists the following five criteria as being the most important. As in many aspects of systems design, there is some incompatibility among the most desirable features.

1. *High degree of secrecy.* Some systems, for example, those employing a key used once only, have absolute theoretical secrecy. Other systems can yield a diligent cryptanalyst some information but do not yield a unique "solution." Those that do yield a unique solution differ widely in the amount of work needed to obtain it and in the amount of material that must be intercepted. Good enciphering techniques require a very high work factor for deciphering.

2. *Smallness of key.* The key must be sent to the transmitting and receiving locations by some method that cannot be intercepted. In military and espionage circles the key has often been memorized. In the past it has generally been desirable that the key should be small. Now, with disk and magnetic tape storage, a scheme with a very large key can be practical. It is still convenient, however, to use a relatively small key that can be easily and frequently changed.

3. *Simplicity of enciphering.* Complex mechanisms for enciphering and deciphering can be expensive. In the past it has been desirable that the mechanisms should be simple. Now elaborate mechanisms can be built with programming and with microelectronic circuitry. It is still desirable that they should be fast and neither delay the transmission process nor consume an excessive amount of computing time or main memory space.

4. *Low propagation of errors.* In some types of cipher an error occurring in the transmission of one character leads to a large number of errors in subsequent characters. This is the case, for example, when the result of deciphering one element of a message is used as the key for deciphering the next element. It is desirable to minimize this error propagation.

5. *Little expansion of message size.* In some secrecy systems the size of the message is substantially increased. This is sometimes done to invalidate the use of statistical techniques in deciphering. It is desirable to minimize message expansion.

Shannon showed that when encrypting natural language these five criteria cannot be met simultaneously. If any one of them is dropped, the other four can be met fairly well. For satellite usage we are likely to drop criterion 2 or 3, either employing storage capability to give a very long key or or else employing computing capability to achieve a complex enciphering process. Because computers are available to the enemy, the work factor for deciphering must be extremely high.

TIME AVAILABLE FOR BREAKING THE CODE

If an intruder has a very long time available for breaking a code, he is more likely to succeed. Again, he is more likely to succeed if he has a very large amount of text to work on. The computer system should be designed so that whenever possible it minimizes the time available to the code breaker. This can be done by designing the enciphering program so that it can be changed at suitably frequent intervals and employing a key for deciphering that is also changed frequently.

On some systems the time available for cracking the code can be made very short. The intruder may be trying to break into a computer system. However, the key is changed sufficiently frequently that if he takes several days to break the code the result will still not enable him to gain access to the system.

On the other hand, some commercial data retain their value for a very long time. Data concerning oil drillings or mineral prospecting, for example, could be of great value to a thief and in some cases may retain their value for years. The thief has plenty of time to break the code. In such cases an enciphering method might be used that cannot be cracked by computing techniques, for example, the use of a *once-only key.*

ONCE-ONLY KEYS

One of the simplest and most effective techniques is to use a key sufficiently long that it is only employed once. A Vernam system [3] employing a key that is never repeated has been in use for military communications for many decades.

The Cipher-Printing Telegraph System used by the U.S. Army Signal Corps reads two five-channel paper tapes. One contains the information to be sent, coded conventionally, and the other is a tape of similar form having characters punched in it at random and with every tenth character numbered so that the tape can be set to any designated starting position. This is referred to as the key tape. It is prepared in advance, and both the transmitting station and the receiving station need to have identical copies of it. A simple electromechanical device then combines the bits on the two tapes and perforates a new tape containing the message so encrypted. The cipher-message tape prepared in this way is sometimes transmitted over telegraph lines and sometimes delivered by messenger or mail. If the key tape is short and repeatedly used, it is possible to break the code or messages transmitted. However, if the key tape is used only once, the code is regarded as unbreakable. Sometimes the key tape is on a roll 8 inches in diameter. Over telegraph lines operating at 45 words per minute, this takes 7 hours to transmit. With a key tape this long, it is almost impossible to break the code used, even if the key tape is used a number of times. On some machines two key tapes are used that can operate in different positions relative to one another.

With computer-to-computer transmission, such a technique can give the highest level of security with little programming. The sequence of characters in the key can be very long—many million. The equivalent of the telegraphic key tape might now be an entire disk pack filled with random numbers. Both the sending installation and the receiving installation have the same disk pack. The message is transmitted in fixed-length blocks of, for example, 100 characters. The bits in these are binary added (or EXCLUSIVE-ORed) with blocks of 100 characters taken sequentially from the key disk pack. This is simple, quick, and virtually uncrackable if the contents of the key disk pack are changed frequently.

With speech the number of bits transmitted is so great that a once-only key would be excessively long. Some means of making the key go further are necessary.

KEY LEVERAGE

In the preceding example the key is used in a simple manner. It is possible to make a given quantity of key encode a larger quantity of data by utilizing it in a more complex way. This is referred to as *key leverage*. The key may be made to go further by many different methods. Suppose that instead of adding one segment of key to the data, two or several segments are added or subtracted. These segments are selected by some addressing mechanism that is reproducible at the receiving machine. The addresses may themselves come from a segment of the key.

Cryptographic hardware can use a relatively small key and scramble its association with the transmitted information in formidable ways. Figure 24.1 shows the enciphering technique employed by an IBM device called Lucifer [4]. Lucifer can be used in conjunction with any conventional terminal. The device enciphers and deciphers messages of any length in groups of 128 bits; this is done under the control of a cipher key of 128 arbitrarily chosen bits. The key can be furnished from either a 16-byte plug-in read-only store module which would be changed at suitably frequent intervals or else from a magnetic-stripe card like a credit card which authorized users could carry. If the latter method is used, the computer that receives the enciphered transmission must have a means of retrieving the identical key from its files. To accomplish this, the user identification may be transmitted from the terminal before enciphering begins.

In the deciphering process, the steps are performed in the reverse sequence. The device contains a circuit that can distinguish between plain text and cipher text, so it can receive a mixture of the two.

If the cryptanalyst had to proceed by trial and error testing each possible

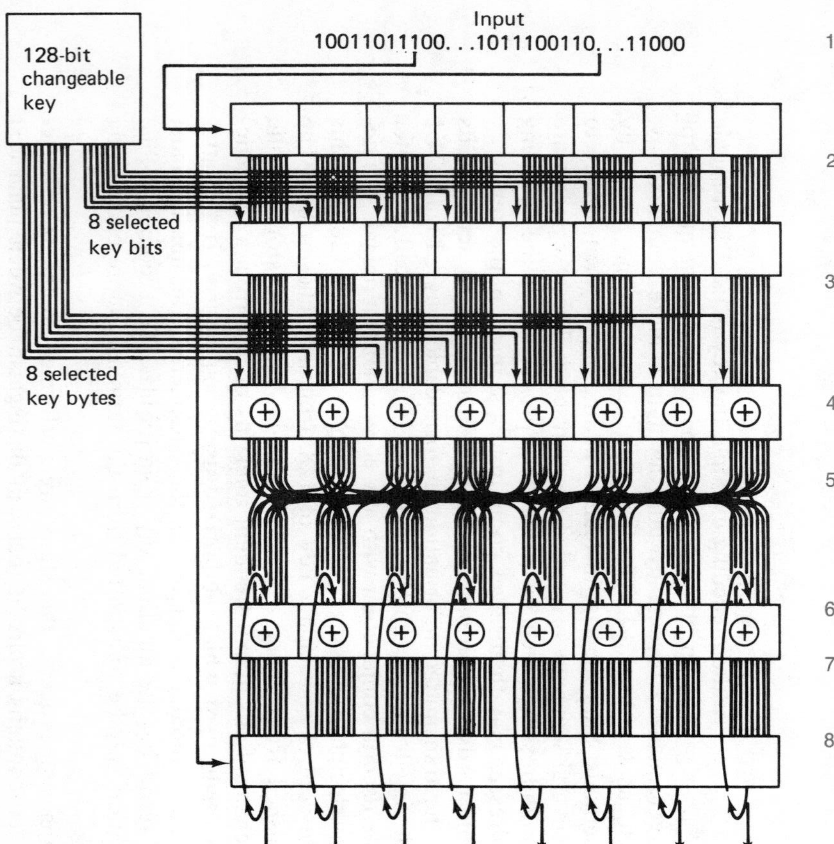
128-bit
changeable
key
Input
10011011100. . .1011100110. . .11000
8 selected
key bits
8 selected
key bytes
1. The message to be enciphered is read in blocks of 128 bits (16 bytes).
2. It is split into two halves of 8 bytes each.
3. According to the value of each of 8 selected bits from the key, one of two different nonlinear transformations is performed on each of the 8 bytes in the top half.
4. The 8 bytes in the top half are then added to 8 selected bytes from the key with modulo 2 addition.
5. The 8 bytes are scrambled.
6. The 8 bytes in the bottom half of the block are then added to the top half.
7. The above six steps are then repeated with the two halves interchanged and with different selected key bits.
8. After 16 such rounds, alternated with 15 interchanges, the encipherment of this block of message is complete.

Figure 24.1 An IBM device, called LUCIFER, which enciphers and deciphers data using a 128-bit key. A U.S. Federal Standard for cryptography, DES (Data Encryption Standard) uses an algorithm with a 64-bit key, also originating from IBM [5].

key, and he used a computer which could test the keys at the rate of 1 per microsecond, he would take an average time of

$$\frac{2^{128}}{2} \text{ microseconds} = \frac{2^{127}}{10^6 \times 60^2 \times 24 \times 365 \times 100} \text{ centuries}$$

$$= 5.4 \times 10^{22} \text{ centuries}$$

(ten million million times the age of the earth).

Complete trial and error will clearly never be used in solving cryptograms other than the simplest. The art of the cryptanalyst must be to find shortcuts that avoid full trial and error. Consequently, the encoding technique must do everything possible to prevent shortcuts from working.

An encoding technique designed to force the cryptanalyst to use a high proportion of trial and error will be different for voice encryption and text encryption or data encryption. The separate voice and data encoding units, illustrated in Fig. 16.5, could each contain different enciphering modules.

However, the most powerful way to crack a cipher is to obtain specimens of the same information enciphered and unenciphered. If the enciphering is done at the earth-station equipment, it may be possible to do that by tapping telephone lines before the information reaches the earth station or after it leaves. When cryptography is used with vital data, therefore, it is desirable that the enciphering and deciphering should take place at the *terminal* location. The same is true of highly secure voice transmission. For most telephone conversations and the majority of data, encryption at the earth station would suffice, making the satellite link about as secure as terrestrial links. It should be realized, however, that encryption at the earth station does not give the highest level of security unless the lines to the earth station are also secure.

U.S. FEDERAL STANDARD

The U.S. National Bureau of Standards has a Federal Standard for data encryption, the Data Encryption Standard (DES). This is employed in a variety of data processing hardware and software packages. It is implemented on mass-produced LSI chips, and is appropriate for end-to-end use on satellite links [5].

DEFENSE IN DEPTH

Certain persons in an organization must be given the cipher key. Like a key that locks up a bank vault, it should only be given to persons who can be trusted. The possibility is always present, however, of the key falling into the wrong hands or being misused.

A principle that should be employed in computer cryptography is that a person who has misappropriated the key should not be able to crack the system if he does not know the enciphering algorithm. Conversely, a person who knows the enciphering algorithm but does not have the key should not be able to decipher the data. *The management responsible for security must then ensure that no person can obtain both the algorithm and the key.* Both the key and the algorithm will be closely guarded, and both may be changed at suitably frequent intervals.

Enciphering modules can be designed so that although they are mass producible, part of the enciphering process differs from module to module. Step 5 in Fig. 24.1, for example, could be manually and easily changeable. Other algorithms are modifiable by changing a random-number generator or seed. Different organizations sharing the same satellite but not using it to communicate with one another would use different enciphering modules. The algorithms employed in these modules can be changed frequently.

JAMMING

The possibility of jamming is not normally taken into consideration in the design of commercial satellite systems. To jam a satellite would be a major criminal act which could quickly be stopped, or else an act of major international hostility which seems feasible only in time of war. It is not impossible that minor acts of war in space will occur in the not too distant future. One can imagine, for example, the Arab satellite being jammed by highly directional transmissions from a huge cheap chickenwire antenna carved like a bowl in the Israeli desert.

If a satellite seems particularly vulnerable to an act of war, it is, perhaps, worthwhile to reflect that other systems which form the infrastructure of modern society are also vulnerable to attack at certain critical nodes.

A satellite could be jammed either in the up-link or the down-link. Jamming the down-link would require the positioning of a hostile satellite in the region of the satellite to be jammed. This would take time, and countries with sophisticated space monitoring facilities would detect what was happening in time to plan evasive action, such as switching to an alternate satellite.

More practicable and much less expensive would be the jamming of the up-link with large earth antennas. Here the element of surprise could be achieved.

It is possible to design techniques for military satellites which attempt to evade the effects of jamming. Frequency-hopping can be used to make a signal hop continuously from one frequency to another. The receivers hop in the same pattern. Frequency-hopping forces the jammer to spread his signal across the entire usable bandwidth and hence reduce the jamming noise density on any one channel. The most effective antijamming technique is called *spread-spectrum modulation.* With this technique, modulation by a pseudorandom bit stream of high speed is superimposed upon the data modulation. This has the effect of spreading it across a wider spectrum. The same modulated bit stream is subtracted from the received signal. The technique can be used both for encrypting the signal and for making it less susceptible to jamming because the jamming noise must occupy a larger spectrum. The jammer cannot tell whether his jamming is effective.

Any antijamming technique is *highly* wasteful of satellite bandwidth and hence only likely to be used on military satellites.

DESTRUCTION

More serious than jamming would be the destruction of a satellite by enemy action. This destruction of a geosynchronous satellite would be expensive and would need high-quality space equipment. It would be far more difficult than, say, the destruction of the London Post Office tower through which most British trunk calls and television are routed. It could only be regarded as a major act of war.

If a satellite is destroyed, it could be replaced fairly quickly using the standby facilities which are normally available.

Multiple war scenarios are possible involving space activities. There has been much discussion of the possibility of Russian destruction of American satellites, and new space-age means of protecting them.

Satellites, like other technologies, are vulnerable to certain types of harm. It would not be prudent for a society to depend exclusively on satellite links. There should be alternative facilities. Survivability in an immensely complex technological society depends on diversity.

REFERENCES

1. B. Tuckerman, "A Study of the Vigenère-Vernam Single and Multiple Loop Enciphering Systems," *IBM Report No. RC2879,* Thomas J. Watson Research Center, Yorktown Heights, N.Y., 1970. A fascinating illustration of how ciphers which used to be commonly used can be broken at a computer terminal.

2. C. E. Shannon, "The Communication Theory of Secrecy Systems," *Bell System Tech. J.,* Vol. 28, Oct. 1949.

3. A. S. Vernam, "Cipher Printing Telegraph Systems for Secret Wire and Radio Telegraphic Communications," *J. AJEE,* Vol. 45, 1926, pp. 109–115.

4. *IBM Research Reports,* Vol. 7, No. 4, 1971. Published by IBM Research, Yorktown Heights, N.Y.; an issue on cryptography.

5. U.S. Federal Information Processing Standard #46. *Data Encryption Standard.* National Bureau of Standards, Washington, D.C.

25 TRADE-OFFS IN SYSTEMS DESIGN

In this chapter we summarize the major trade-offs in the ground segment. The chapter is complementary to Chapter 9, which summarizes the trade-offs in the space segment. Ideally the ground segment and space segment should be designed jointly. Often, however, the designer of the ground equipment is confronted with a *fait accompli;* the satellite is already in place. Satellite designers, on the other hand, have to try to imagine what uses will be made of their satellites during the satellite lifetime.

Figure 25.1 summarizes the major trade-offs affecting the ground segment and may be compared with Fig. 9.1. As with Fig. 9.1, it is a broad summary and does not show details, for example what form of multiple-access control is used.

As we stressed in Chapter 9, there are major economies of scale in the satellite itself. It is economically attractive for some applications to make the satellite heavy and powerful, thus permitting a high traffic volume or cheap earth stations. It is a function of ground segment design to permit the satellite to be shared among as many users as possible in a manner which gives efficient use of the transponders.

TRADITIONAL COMMON CARRIERS

The trade-offs perceived by the traditional common carriers are likely to be different from those perceived by new entrants to the field. To a traditional common carrier a satellite is a means to *augment* the existing network. To a new entrant, a satellite is a means to *bypass* the existing network.

AT&T would use satellites to dramatically increase the quantity of long-distance channels, to make Picturephone trunking more feasible, and to lower the investment cost of expanding the long-distance network. Western Union uses satellites to lower the cost of leasing long-distance channels from AT&T.

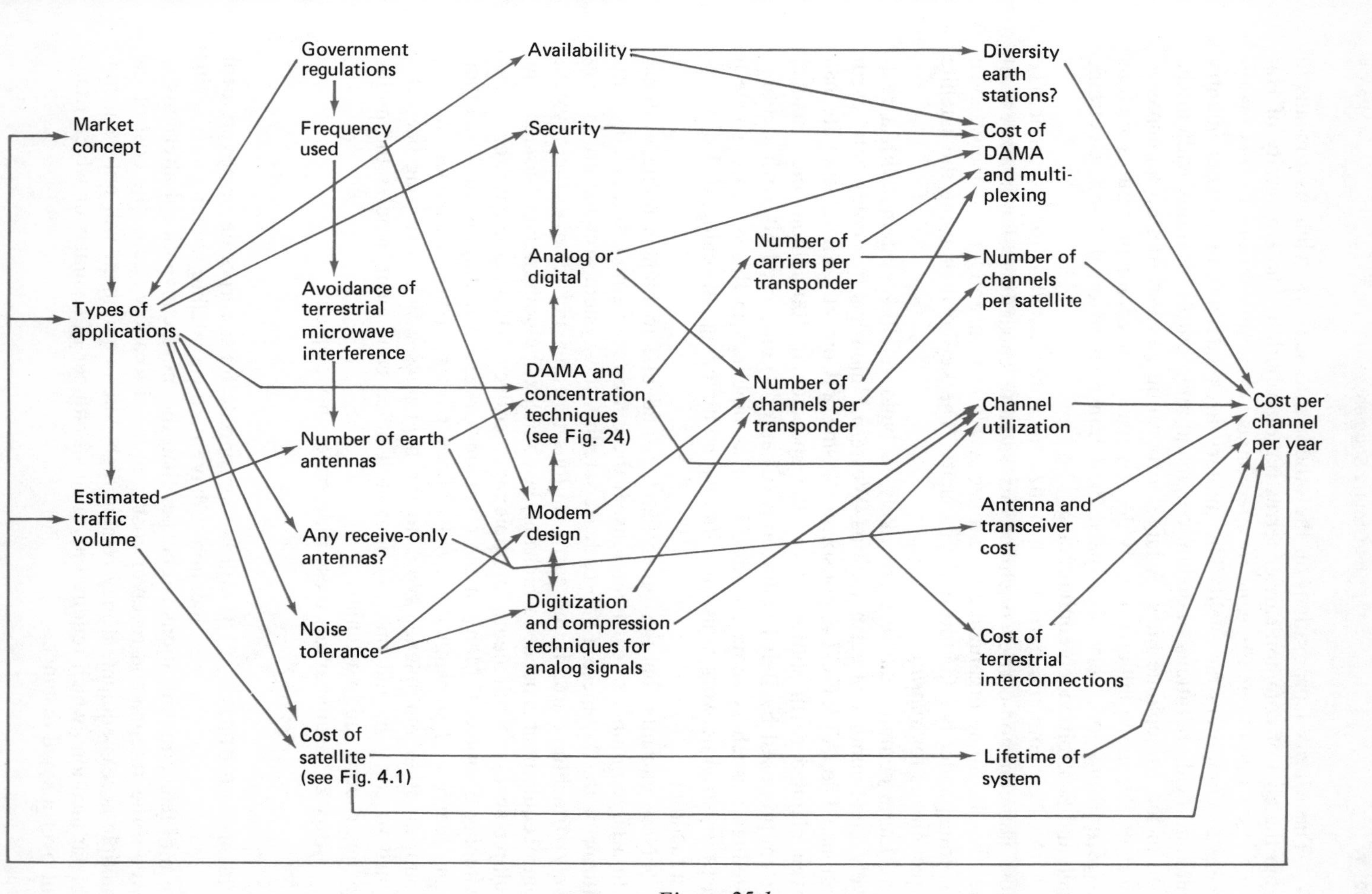

Market concept
Types of applications
Estimated traffic volume
Government regulations
Frequency used
Avoidance of terrestrial microwave interference
Number of earth antennas
Any receive-only antennas?
Noise tolerance
Cost of satellite (see Fig. 4.1)
Availability
Security
Analog or digital
DAMA and concentration techniques (see Fig. 24)
Modem design
Digitization and compression techniques for analog signals
Number of carriers per transponder
Number of channels per transponder
Diversity earth stations?
Cost of DAMA and multi-plexing
Number of channels per satellite
Channel utilization
Antenna and transceiver cost
Cost of terrestrial interconnections
Lifetime of system
Cost per channel per year

Figure 25.1

The Muzak Corporation, on the other hand, sees a satellite as a means to avoid the cost of common carrier terrestrial distribution. The University of Hawaii sees a satellite as a means to interconnect computer terminal users everywhere to computers far afield. The military uses satellites to connect infantry units and field operations to distant command and control locations without the need of local telephone lines. A large corporation can use rooftop antennas to lower its telephone bill and to provide new uses of telecommunications such as image transmission, very high-speed data transmission on demand, electronic mail, and possibly video conferencing.

A large common carrier can utilize an entire satellite for its traffic. To most non-common-carrier organizations satellite capacity is far in excess of their needs. Some organizations can use a whole transponder. To many, even one transponder has far too much capacity. The key to their using the satellite is techniques for *sharing* it.

Large common carriers are likely to build satellite systems with a relatively small number of earth stations. Western Union has five continental earth stations. The AT&T-GTE consortium using a Comsat General satellite has seven planned earth stations. Even the spectacular 100-million-voice-channel system proposed by Bell Laboratories (Chapter 9) only used 50 earth stations. Satellites in such systems are thought of as pipelines in the sky to enhance a vast terrestrial network which has its own multiplexing, switching, and distribution facilities.

New entrants, on the other hand, would like to build satellite systems with a large number of earth stations. If these are receive-only stations, as with Muzak or the Japanese TV broadcast satellite, the system architecture is relatively straightforward. If the earth stations transmit and receive, then system control equipment is needed which may be complex, including demand-assignment multiple-access equipment, multiplexors, codecs, data concentrators, and switching equipment. Many of the common carrier earth stations do little other than receive and transmit the signals. They pass a television channel, mastergroup, or other multiplexed block to the land lines which have long been designed to carry it, and the traditional toll office equipment handles the multiplexing, routing, and switching.

Box 25.1 summarizes these differences in viewpoint.

COST TRADE-OFFS

It will be appropriate for a corporate or government location to have its own satellite antenna *only* if this form of telecommunications is cheaper than the more conventional alternatives. Even if the antenna and receiver cost is reduced to a low figure, the cost of the multiple-access equipment may remain high. Such equipment will not be justifiable at locations which require only a small number of channels or which transmit over a small distance.

BOX 25.1 Differences in viewpoint of traditional common carriers and other organizations concerning the use of satellites

Motivations of Independent Organizations in Establishing Satellite Channels	Motivations of Traditional Common Carriers in Establishing Satellite Channels
• Establish minimum-cost channels	• Respond to rate-of-return regulation (in USA), which encourages high capital investment
• Provide new business opportunities	• Preserve the existing plant, much of which has a 40-year write-off
• Introduce competition to established common carriers	• Preserve existing corporate control
• Create facilities which carry all types of traffic	• Augment the telephone network
• Permit many organizations without very high traffic volumes to share the satellite	• Use the satellite for their own traffic
• Provide multiple-access demand-assignment of individual channels	• Carry existing mastergroups and other groups
• Provide burst multiplexing to give flexible demand-assignment of all types of signals	• Extend terrestrial circuit-switched system
• Bypass the local loop bottleneck so that wideband video and data signals can be carried	• Preserve today's local distribution
• Have satellite antennas at all major premises	• Use a small number of earth stations in regions of highest traffic density
• Use small, cheap, antennas	• Use large, expensive, earth stations which concentrate high traffic volumes
• Minimize the total system cost	• Maximize the transponder utilization
• Use a public telephone network to bypass the effects of storms (above 11 GHz)	• Use space diversity to bypass the effects of storms (about 11 GHz)
• Bypass today's terrestrial facilities	• Augment today's terrestrial facilities.

Evaluating the break-even point between satellite and land-line usage is complex, and the break-even point will change substantially for the next decade or two. The cost of terrestrial links is roughly proportional to distance, and that of satellite links is independent of distance, as shown in Fig. 25.2.

Where the curves of Fig. 25.2 cross depends on the number of channels the facility is designed for. The investment cost of terrestrial links is highly dependent on channel capacity. Below about 100 voice circuits, wire cable is used. From 100 to 10,000 voice circuits, microwave links are the lowest in cost. For many thousands of voice circuits, coaxial cable links are the cheapest [1]. Figure 25.3 shows approximately how the cost per voice channel per mile varies with the number of voice channels.

A given satellite handles a fixed number of channels irrespective of distance. $2500 satellite investment cost per channel is typical of domestic satellites at the time of writing, and this cost will fall dramatically if satellites of higher capacity are launched. To this we must add the cost of the earth-station equipment. Let us suppose that a system is designed with small antennas and multiple-access equipment operating with voice channels. The cost of one such installation on user premises is, say, $1 million. If that location has N voice channels, we shall assume that the total investment cost per voice channel is $\$2500 + \$1$ million$/N$. Thus for 10 channels going to a building the investment cost is $102,500 per channel; for 1000 channels it is $3500 per channel. These figures are plotted in the upper dotted curve in Fig. 25.4, which compares them with the terrestrial cost for links of 50, 100, and 200 miles. It will be seen that investment cost for a satellite system is lower than a 200-mile terrestrial link for up to about 2000 channels. A 50-mile terrestrial link is lower than satellite

Figure 25.2

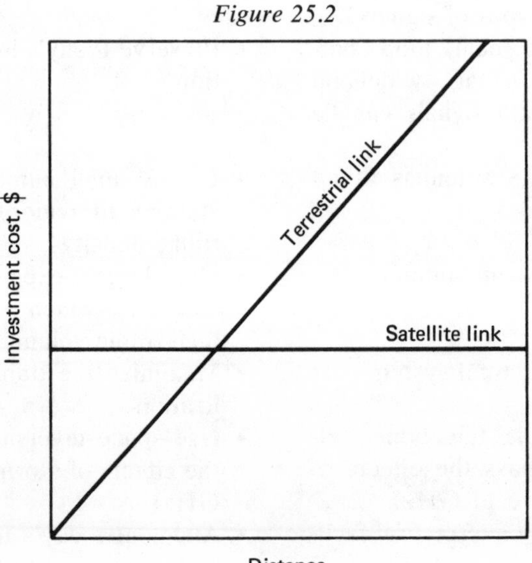

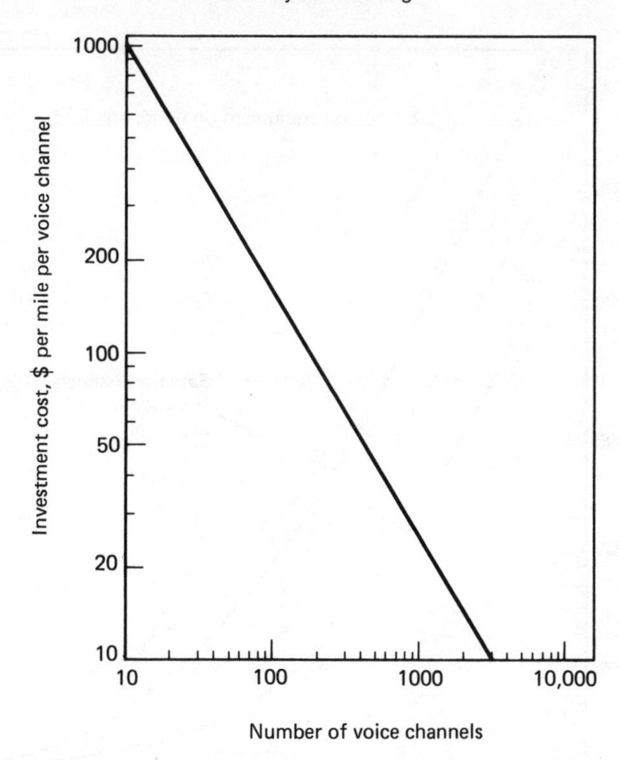

Figure 25.3 Economies of scale in terrestrial circuits [1].

costs. The lower dotted curve of Fig. 25.4 does a similar calculation for equipment on the user's premises costing $ 1/4 million.

These figures form a *very rough* illustration of the trade-offs and in practice *will vary greatly as the technology changes.* A user will not be interested in *investment costs* but in how much the common carriers charge him—which sometimes has little relation to the investment costs. At the time of writing a user may be charged about $1000 per month for a satellite voice channel, or about $2000 per month for a 2000-mile leased terrestrial voice channel. He may in the near future be able to lease a roof antenna and demand assignment equipment for 100 channels for less than $5000 per month.

LOW EARTH-STATION COST

The key to corporate use of satellite systems is the achievement of sufficiently low-cost earth stations. Among the factors that permit lower transmit/receive earth-station costs are the following:

1. *Small antenna.*
2. *Fixed (nontracking) antenna.*

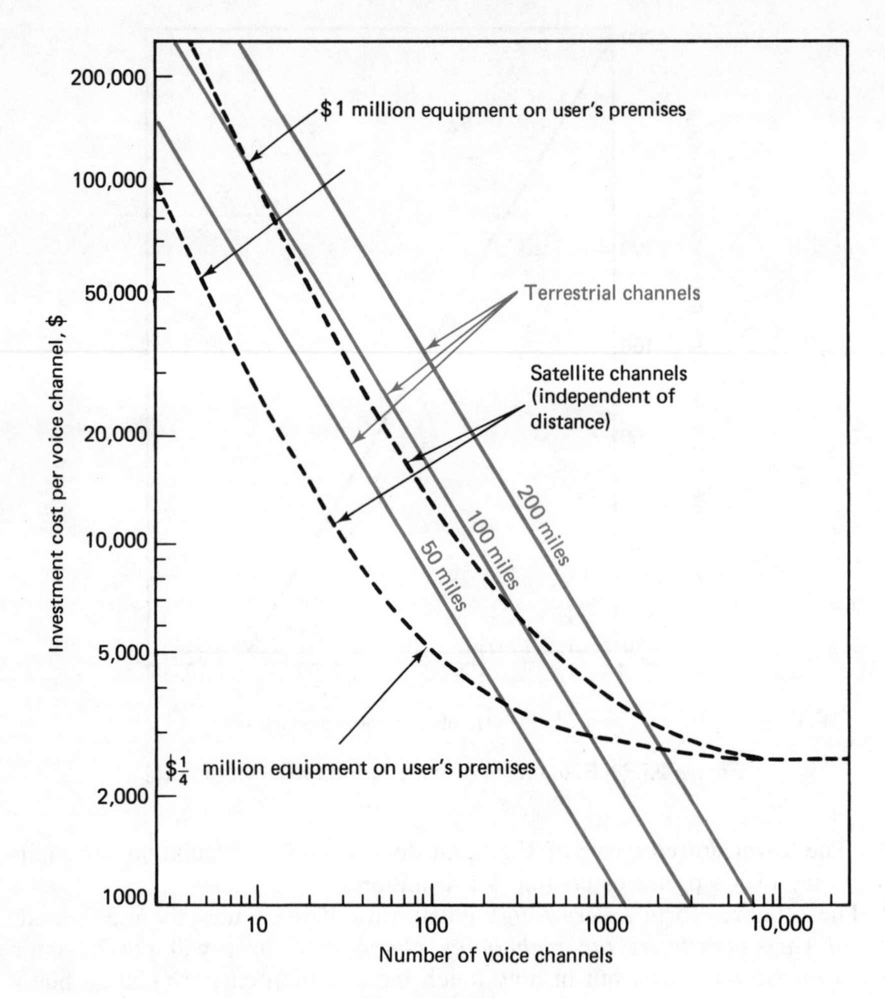

Figure 25.4

3. *Uncooled receiver.*
4. *Demand-assignment technique* which is not overly sophisticated.
5. *Low-bandwidth transponder* on satellite.
6. *Modulation technique* which can tolerate a substantial level of noise and interference.
7. *High modulation index.*
8. *Low information rate per channel,* permitting unsophisticated modulation and concentration techniques.
9. *Frequency* which gives freedom from common carrier interference.
10. *Site location* which permits freedom from common carrier interference.

11. *Mass production,* i.e., a large community of similar earth-station users.
12. *Fabrication technique* for earth stations which is susceptible to mass production, as with the Japanese TV broadcast satellite receivers.

The move to frequencies higher than the common carrier band of 4/6 GHz is important, first to avoid common carrier interference and second because it permits smaller antennas.

BAD WEATHER

12/14-GHz and 20/30-GHz links have excellent properties when the weather is good. Figure 25.5 repeats a curve from Chapter 8 which illustrates the combined effects of loss, attenuation, and noise when the skies are clear. The high frequencies have better properties than 4/6 GHz because the gain of antennas of the same diameter is

Figure 25.5 Composite effect of link loss, atmospheric attenuation and atmospheric noise.

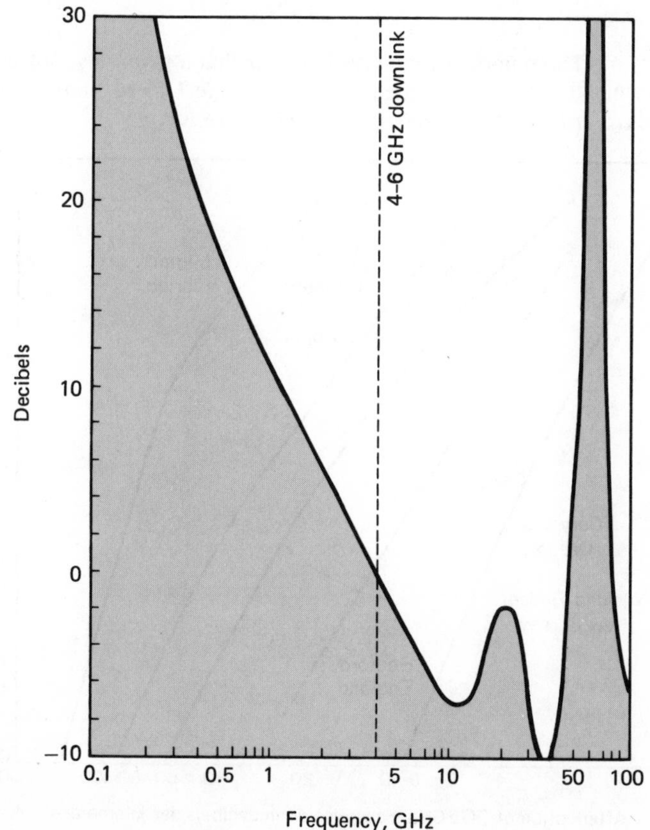

higher at the higher frequencies. Figure 25.5 is drawn for a ground angle of elevation of 15°. For higher angles the curve favors the higher frequencies still more (see Fig. 3.7). As illustrated in Chapter 3, most domestic satellite systems can avoid angles of elevation less than 35°.

The trouble with the higher frequencies is that their properties are much worse when the weather is very bad, as illustrated in Figs. 8.4 and 8.5. Moderate rain and cloud cover does not do much harm, but the link deteriorates badly in very heavy storms.

The amount of very heavy rain varies widely from location to location. Figure 25.6 shows the number of minutes per year that the attenuation on a 30-GHz radio path exceeds certain limits. The result is not directly related to the amount of time it rains because it depends on the intensity of the rain. England, where one rarely goes out without a raincoat, does better than Miami, because when it does rain in Miami it rains cats and dogs.

Figure 25.7 shows the effects of an intense rainstorm on satellite transmissions at 12/14 and 20/30 GHz. Storms like that in Fig. 25.7 are rare. Figure 23.4 showed the percentage of time that various attenuations occur.

Figure 25.6 The number of minutes per year that a given level of attenuation at 30 GHz is exceeded for parts of the United States and England. (*Reproduced from reference 4, with permission.*)

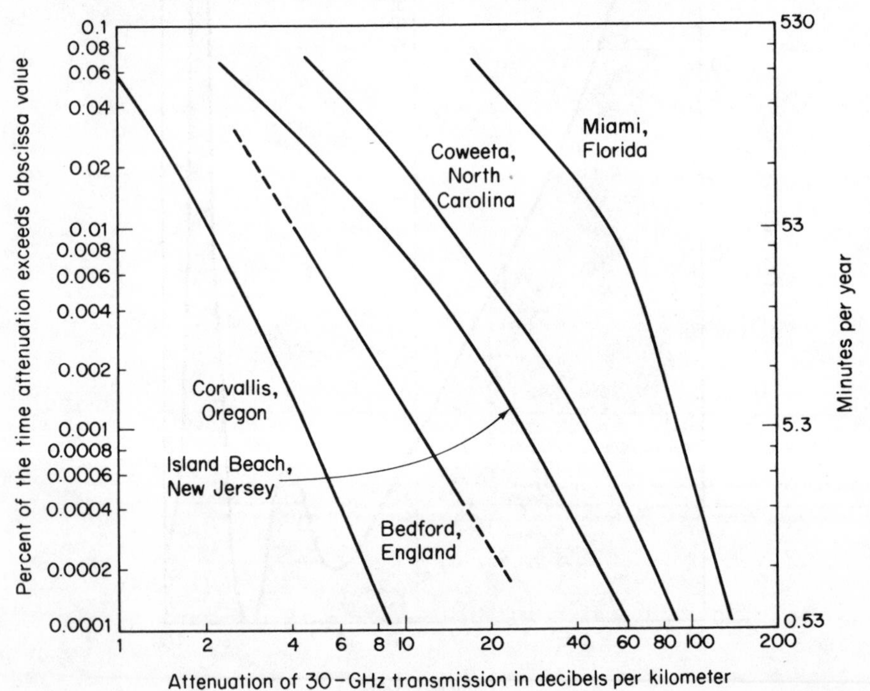

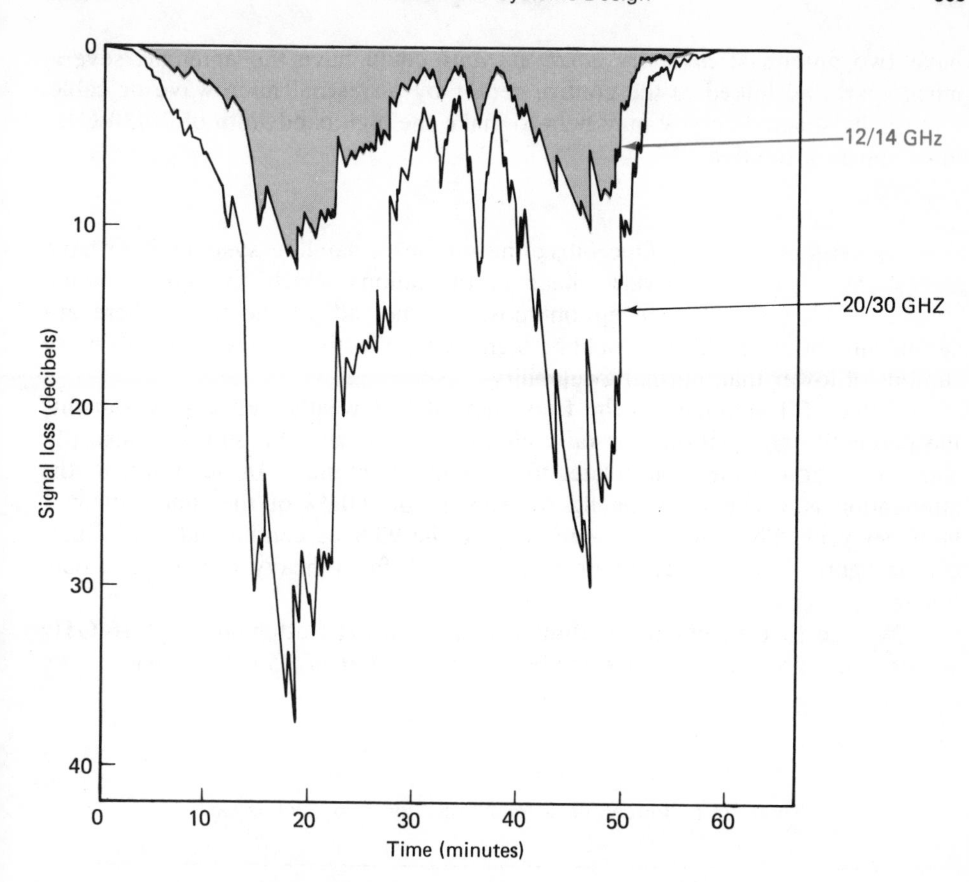

Figure 25.7 The profile of a very intense rainstorm (exceeding 100 millimeters per hour). Rain this intense lasts for a relatively short time. Compare with Fig. 23.4.

SPACE DIVERSITY The most intense storms are in fact the most limited in area. Consequently, a trade-off called *space diversity* will be used for common carrier systems. Two earth stations operating at the higher frequencies would serve an area and would be separated by several miles. Because very intense storms are limited in diameter, when one station operating at 12/14 GHz was suffering a loss of 12 decibels due to a storm the other would normally be experiencing less intense rain with less than 3 decibels of loss. The same concept could be more valuable at 20/30 GHz when the station in a heavy storm could be suffering 40 decibels of loss and its neighbor less than 10. Space diversity requires the ability to switch channels to avoid the most intense storms. Such an ability could be made an integral part of systems such as the Bell System. Many common carrier earth stations are likely to have two antennas; *diversity earth stations* could have the antennas several miles apart and linked to the control center by terrestrial microwave or cable. Eventually space diversity may help to make the high bandwidth of 20/30-GHz links appear attractive.

FAIR WEATHER ANTENNAS

One intriguing option of satellite systems is to have very cheap earth stations which give good enough reception most, but not all, of the time. There are certain situations in which it could be highly cost-effective to have low-cost earth stations of lower than normal availability.

Table 25.1 summarizes the frequency of bad weather effects by showing the percentile attenuations for rain, clouds, and air and the sum of these [2]. The top section of the table relates to the 99.99 percentile. In other words, the attenuation is worse than the figures shown for 0.01% of the time—about 1 hour per year. The other two sections give the 99.9 percentile and the 97 percentile figures. The figures relate to a site with fairly heavy rainfall, 100 centimeters per year.

We see that except for 10 hours in the year, the attenuation of 16-GHz waves is not greater than 10.9 decibels and that that of 35-GHz waves is not

Table 25.1 Percentile attenuations for rain, clouds, and air†

		Attenuation			
		Frequency (GHz)			
Percentile	Element	*4*	*16*	*35*	*90*
99.99	Rain (dB)	0.5	26.4	105	320
	Clouds (2 g/m³, 0°C) (dB)	0.5	5.6	24	150
	Oxygen and water vapor (dB)	–	0.3	0.8	3.5
	Total (dB)	1.0	32.3	129.8	573.5
99.9	Rain (dB)	0.1	6.6	26.4	80
	Clouds (1.5 g/m³, 0°C) (dB)	0.3	4.0	17	120
	Oxygen and water vapor (dB)	–	0.3	0.8	3.5
	Total (dB)	0.4	10.9	44.2	203.5
97	Rain (dB)	–	0.4	1.5	4.7
	Clouds (1.0 g/m³, 0°C) (dB)	0.2	2.7	12	82
	Oxygen and water vapor (dB)	–	0.3	0.8	3.5
	Total (dB)	0.2	3.4	14.3	90.2

† Source: Adapted from Ref. 2.

more than 44.2 decibels. The attenuation of 90-GHz waves is high, however. These frequencies have been selected because they do not coincide with atmospheric absorption peaks. The attenuation levels shown suggest that with large and powerful satellites the 20/30 GHz could be used successfully.

Let us suppose that a 20/30-GHz satellite is used with 15-meter ground antennas and that this provides good transmission 99.9% of the time when the attenuation in atmosphere, clouds, and rain is no worse than 30 decibels. 97% of the time it is no worse than 2 decibels, and so an antenna with 28 decibels less gain can be used. A 1-meter antenna would give good transmission 97% of the time.

Certain important locations, then, in the corporation could be equipped with a 15-meter antenna. Other locations would have an inexpensive 1- or 2-meter unit. Let us assume that the effect of this would be that the locations with the 15-meter antenna have an availability of 0.998. Locations with the 2-meter antennas have bad reception 3% of the time, i.e., 262.8 hours in total but only 60 *working* hours per year.

What would be the effect of this?

The 0.998 availability system would be better than today's telephone system, which experiences a low number of local loop and other failures. The locations with the 2-meter antennas may be provided with a private branch exchange which automatically places calls between corporate locations on the satellite network if possible, but if not, it places them on the toll telephone system. A priority system may be used so that only certain telephone extensions or computer terminals are permitted to place nonsatellite calls, in order to keep costs down.

If video conferencing is used, there may be no alternative to satellite channels. There would rarely be problems with video conferences at the important locations. The locations with 2-meter antennas would have good transmission 97% of the time. For the other 3%, video pictures with varying degrees of "snow" on the screen might be tolerated *providing that the sound quality is good* so that people's voices are clear. Good sound quality could be achieved either by setting up toll telephone voice links to operate in parallel with the snowy picture transmission or by allocating several times the normal bit rate to the sound channel and using powerful error-correcting codes.

If music is transmitted to the 2-meter-antenna locations, it may be tolerable to do without it 3% of the time and perhaps use local music tapes as backup.

Great use will no doubt be made one day of satellites in education. Schools will have access to televised material and computer-assisted instruction, both of much higher quality than most that have been demonstrated to date. The antennas serving the schools will need to be inexpensive, and a compromise of 3% unavailability may be acceptable because teachers can keep the students occupied with alternative material.

NON-REAL-TIME TRAFFIC

Perhaps the major appeal of antennas which only work in fair weather is for non-real-time traffic. The sending of electronic mail, facsimile documents, monetary transactions, cables, or batches of computer data can wait until a storm has passed. As we indicated in Chapter 13, the potential volume of non-real-time satellite traffic is gigantic.

Given the right satellite, electronic mail could be handled with very low-cost earth stations.

DAMA TECHNIQUE

An important item in the center of Fig. 25.1 is DAMA *(demand-assignment multiple-access)* technique. The multiple-access technique has a major effect on the cost per channel and the accessibility of the satellite from users' premises. Box 25.2 summarizes multiple-access techniques.

Multiple-access can be accomplished in a highly elaborate and flexible manner, requiring expensive control equipment at the earth station, or it can be accomplished in a simple inexpensive manner which gives poorer utilization of the satellite. One of the simplest forms of multiple access is the fixed allocation of a transmission frequency to an earth station. The earth station can be allocated one or more of the SPADE frequency bands in Fig. 15.1, and no other station is permitted to use those bands. This is rather like allocating a station a *leased* telephone line rather than a dial-up line, except that signals transmitted on it can be received by any other earth station.

Demand assignment equipment, which varies the number of channels allocated to a station on a demand basis, is generally more expensive than equipment for fixed assignment but gives better utilization of the expensive channels. In some cases a fixed bandwidth assignment may be used and that bandwidth allocated to users on a demand basis. The ALOHA technique is a simple way to allow computer users to share a channel. A burst reservation system is generally a more efficient way but requires more expensive control equipment.

Substantial savings can be achieved with equipment which intermixes burst traffic, such as interactive data, and continuous signals, such as speech; or which intermixes continuous channels of different capacity; or which interleaves real-time and non-real-time traffic. But again, the more elaborate the control equipment, the greater its cost. Large users can afford elaborate control equipment.

SIZE OF TRANSPONDER

If all earth stations have very high volumes of traffic to send, a large transponder is economical, as with the Bell System study described in Chapter 9 which proposed a transponder of throughput of 630 million bps. If all earth stations have a small traffic volume, transponders of smaller bandwidth than today's are

BOX 25.2 Summary of multiple-access techniques

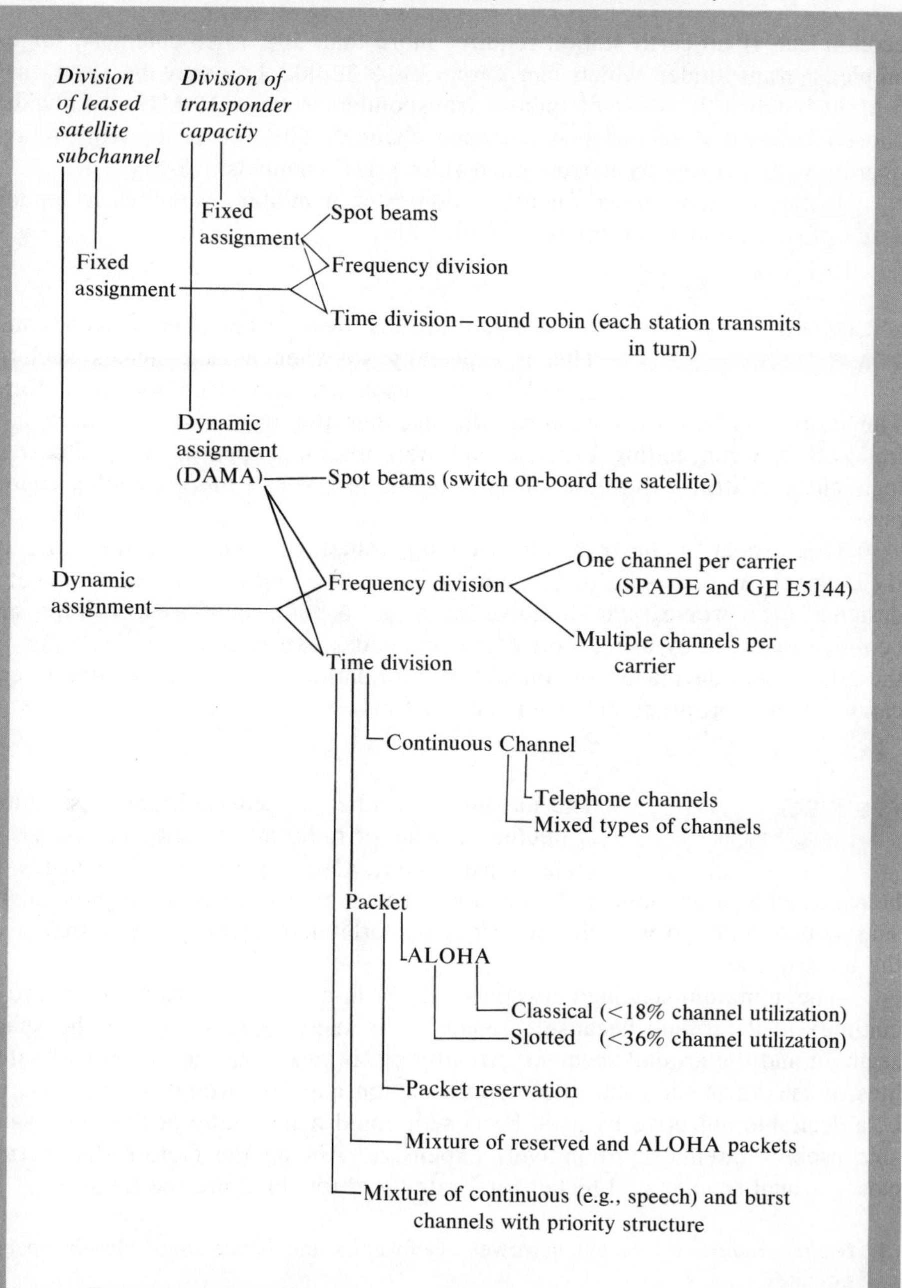

Division of leased satellite subchannel
Division of transponder capacity
Fixed assignment
Spot beams
Fixed assignment
Frequency division
Time division – round robin (each station transmits in turn)
Dynamic assignment (DAMA)
Spot beams (switch on-board the satellite)
Dynamic assignment
Frequency division
One channel per carrier (SPADE and GE E5144)
Multiple channels per carrier
Time division
Continuous Channel
Telephone channels
Mixed types of channels
Packet
ALOHA
Classical (<18% channel utilization)
Slotted (<36% channel utilization)
Packet reservation
Mixture of reserved and ALOHA packets
Mixture of continuous (e.g., speech) and burst channels with priority structure

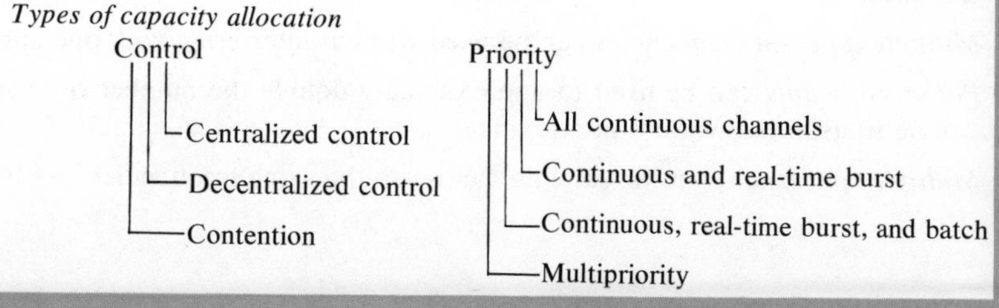

Types of capacity allocation
Control
Centralized control
Decentralized control
Contention
Priority
All continuous channels
Continuous and real-time burst
Continuous, real-time burst, and batch
Multipriority

economical. If no earth station requires more than 300 voice channels, for example, a transponder which can carry 300 × 32,000 bps may be used—say, 6-MHz bandwidth. Most of today's transponders have a 36-MHz bandwidth, chosen because it carries one televised channel. The cheapest earth stations operate with a relatively narrow-bandwidth UHF channels.

It may be economical for future domestic or military satellites to contain a mix of transponders of different bandwidths.

ENCODING COMPLEXITY

Most digital signals yield to compaction techniques. This is especially so when analog signals such as speech or television are converted to digital form. The more complex the encoding, the cheaper the transmission. There is a trade-off between coding expense and transmission expense. As the cost of logic circuitry drops, the trade-off is swinging in favor of more complex encoding.

One aspect of the trade-off in coding complexity relates to the noise on the channel. As discussed in Chapter 17, a satellite link can be lower in cost if designed for a worse signal-to-noise ratio, and a poor signal-to-noise ratio can be made tolerable by using error-correcting codes. An error-correcting code, on the other hand, decreases the amount of information that a given bit stream can carry, so an appropriate balance must be struck.

RESOURCE CONSERVATION

The maximum number of geosynchronous satellites is limited because of radio interference between satellite transmissions. Orbit space therefore needs to be regarded as a precious and limited resource which must be used judiciously. This is especially so with the more popular orbital segments such as that over the United States.

The minimum spacing between satellites in a given geosynchronous arc is complex to determine because it depends on many factors both in the space segment and the ground segment. An attempt to maximize the number of satellites or maximize the total satellite information capacity would not necessarily be a desirable objective by itself because it could make many of the most valuable uses of satellites prohibitively expensive. Among the factors that permit closer orbital spacing and higher total satellite throughput are the following:

1. *Higher frequencies* permit narrower beamwidths and hence more closely spaced satellites.
2. *Multiple different frequencies* can be used without interfering with one another.
3. *Polarized beams* can be used to approximately double the number of channels that can be transmitted without interference.
4. *Multiple spot beams* on the satellite can reuse the same frequencies.

5. *Large satellite antennas* give narrower beams, and hence more spot beams can reuse the same frequencies.
6. *Large earth antennas* transmit and receive over a narrow angle, permitting closer satellite spacing.
7. *Frequent orbit adjustments* give more accurately stationed satellites and hence permit closer spacing.
8. *Accurate attitude control* prevents narrow spot beams drifting away from their target.
9. *Accurate earth antenna* pointing reduces the possibility of interference. A rooftop antenna carelessly aligned could cause trouble.
10. *Choice of modulation method* can give protection from interference.
11. *High modulation index* helps give immunity from interference but lowers the throughput of a given bandwidth.
12. *Single carrier per transponder* gives the highest information throughout per satellite.
13. *Low noise temperature of earth-station receiver* increases tolerance to interference.

Conserving the orbital resource is directly at odds with lowering the cost of satellite channels. Most of the above measures increase the cost of either the satellite or the earth stations. The argument against conservation is that it is desirable to make satellites as useful as possible to man in this decade, and that needs cheap earth stations and antennas on school and hospital rooftops. The argument might say that the next decade will take care of itself by moving up to frequencies of 20/30 GHz and these, as indicated in Chapter 9, can present man with a truly prodigious information capacity.

REFERENCES

1. Eugene V. Rostow, "A Survey of Telecommunications Technology," the U.S. President's Task Force on Communications Policy, *Staff Paper No. 1*, Washington, D.C., 1969.
2. "Future Communications Systems via Satellites Utilizing Low-Cost Earth Stations," prepared by the Electronic Industries Association for the President's Task Force on Communications Policy, Washington, D.C., 1969.
3. Burton I. Edelson and Andrew M. Werth, "SPADE System Progress and Application," *COMSAT Tech. Rev.*, Vol. 2, No. 1, Spring 1972.
4. D. C. Hogg, "Millimeter-wave Communication Through the Atmosphere," *Science*, No. 3810, Jan. 5, 1968.
5. Leroy C. Tillotson, "A Model of a Domestic Satellite Communication System," *Bell System Tech. J.*, Dec. 1968.

EPILOGUE:

FUTURE POTENTIALS OF SATELLITES

This book has been constrained to discussing *today's* state of the art in communications satellites. It is difficult to escape the conclusion that today's geosynchronous satellites are but an elemental beginning of a technology that is to have a profound effect on the condition of man.

One day readers of this book will look back at WESTAR, the first U.S. domestic satellite, with the amused but admiring sense of history that we have when we look back at Explorer I, America's first grapefruit-sized response to Sputnik. At some time during the 1980s it will become economically reasonable to take hardware weighing many tons into geosynchronous orbit.

The space shuttle represents the beginning of a new era in space. We shall look back at today's throwaway rockets as an incredible waste that was necessary for man to make his first steps into space. Imagine having to throw away a 747 every time it was used. Combined with the space shuttle some means of ferrying hardware into geosynchronous orbit will be needed.

Today we look upon the geosynchronous orbit as a limited resource. At 4/6 GHz the satellites must be kept almost 2000 miles apart to prevent signals transmitted to adjacent satellites from interfering with one another. The limitations can soon be lessened, however. Satellites of different frequencies can be employed without interfering, and higher-frequency satellites can be spaced more closely. Work on high-power lasers is bearing fruit, partly because of massive military expenditure. Some up-links may one day be laser beams, and satellites using these can be very close together without interference.

Lasers operate at more than 10,000 times the frequency of today's uplinks and hence have a much higher potential bandwidth. Lasers may not be used on the down-links because of possible dangers to persons on earth. Highly directional millimeterwave spot beams (10 to 100 GHz) could carry the signals down. With such a technology the main foreseeable limit on satellites is man's own capability to use the available bandwidth.

As we have stressed, the main key to advanced satellite design is weight. It is breathtaking to reflect what could be done with a satellite several times the weight of ATS-6. In the utter stillness of space, with no gravitational force or breath of wind, a large antenna can be a fine gossamer-like structure, impossible on earth. The 100-foot diameter aluminized mylar sphere of the Pageos satellite weighed only 120 pounds. NASA design studies have described antennas for astronomical telescopes 10 *miles* in diameter, deployed in orbit as a fine spinning mesh pulled into position by centrifugal force.

Large solar panels, larger than those on the Canadian CTS satellite, will be used to generate more power. The solar sails of the SKYLAB satellite generated 12,000 watts each—much more than WESTAR's 250 watts or SYMPHONIE's 150 watts.

There is no practical restriction other than cost to the amount of energy that could be generated in geosynchronous orbit. The total solar energy intercepted by a 10-mile-wide strip at the geosynchronous position is hundreds of times greater than the total amount of electricity consumed on earth. Indeed, there have been serious proposals from NASA [1] and Arthur D. Little [2] for building a solar station in geosynchronous orbit generating 10,000 megawatts or so—enough power to supply the whole of New York City. The biggest problem would be getting the power down to earth on a microwave beam sufficiently diffuse not to cause harm. The total cost of power generation in geosynchronous orbit in the 1980s (including the launch) could be as low as $2 per watt if it is done on a large enough scale [2]. Nuclear generators in orbit have also been proposed.

Before the year 2000, mankind could have immensely powerful satellites with large antennas beaming as much information as we are capable of using to our rooftops.

At some time during this century a vast industry will grow up placing massive hardware in orbit in a ring around the earth 22,300 miles above the equator. Eventually, it will become economical to have service vehicles in geosynchronous orbit, repairing, refueling, or assisting in the deployment of the satellite equipment. It has been suggested the geosynchronous orbit will become the right place for certain new manufacturing processes which need the intense vacuum of outer space and perhaps the absence of gravity. The vacuum of geosynchronous orbit is more perfect than in low orbit, and, being stationary above the earth, the production processes can be constantly linked to terrestrial computers and control rooms. Particularly important a geosynchronous satellite is continuously in sunlight so its energy supply is not repeatedly interrupted as with a low-orbit satellite. One can imagine future solid-state logic circuitry or computer memory being fabricated in the utter purity of space, with microscopic components thousands of layers deep being deposited on silicon.

That an immensely powerful communications technology is within our grasp seems beyond doubt. There remains the question of cost. Where will the money come from for the new thrusts that are now technically possible?

The total investment that is being made on the development of terrestrial facilities is huge. AT&T alone is spending $9 to $10 billion per year on capital improvement of the Bell System. AT&T top management has indicated that they intend this level of expenditure to continue. The expenditure on the entire world-wide Intelsat system is only about 1% of this. The annual revenue from telecommunications in the United States is over $40 billion and is growing at about $5 billion per year. Much of the capital expenditure in the telecommunications industry is going into the trunks and trunk switching that satellites and demand-assignment equipment could replace today.

We have indicated that six ATS-6 satellites with appropriate transponders (1974 technology) could carry as much traffic as the peak traffic on the Bell System toll network (i.e., not including *local* telephone calls). If earth stations were associated with the toll offices in the 500 most populous cities, such a satellite network would cost less than $2 billion. If it had a lifetime of 10 years, that would be $200 million per year. (Most telecommunications equipment has a 40-year lifetime.)

The next major thrust in the space segment should capitalize on the economies of scale which today's technology offers. This can be done only by a large organization with large funds. We shall not see the low cost per channel that is now possible, if we continue to launch small satellites such as ANIK, WESTAR, and SYMPHONIE. Government regulation, however, is constraining AT&T from taking full advantage of satellites. NASA, which always launched the most advanced satellites, is being prevented from creating successors to ATS-6. The European administrations, while singing the praises of SYMPHONIE, are unlikely to put up American-launched satellites that would constitute major competition to European telephone trunks.

The barrier to acquiring low-cost satellite channels is thus not technological but political.

It seems likely that in the 1985 to 2000 period satellites will be deployed with much larger antennas and much greater power-generation capability than those of the near future. Powerful microprocessors for control and switching will probably be used on such satellites. NASA and associated corporations have done exploratory designs on such satellites and have carried out detailed studies of their uses.

A detailed design study conducted for NASA by the Aerospace Corporation [4] described a system which could permit wristwatch radio telephones. The satellite would need an antenna 67 meters in diameter. This would give an earth footprint a width of about 80 kilometers. The design used 25 antenna feeds giving 25 spot beams directed at the 25 largest metropolitan areas in the USA. One thousand frequency channels are used in each beam and, with time-division, 100 users could share each beam. In this way a combination of frequency-division and time-division multiple access would give a capacity for one satellite of 100,000 wristphones in each of 25 areas—2.5 million wristphones in total. Larger numbers could be accommodated with more satellites or more beams.

The satellite would have 25 output tubes of 280 watts. It would need a total power generation of 21 kilowatts (less than that used on board Skylab). The total satellite weight is 16,000 pounds. It is thus much smaller than low-orbit satellites already launched for other purposes.

The design used an on-board computer to allocate channels to the calls being made. Each call placed would generate a coded message giving the addressee and addressor. The computer would queue the calls if necessary, allocate a channel to each call, and inform the calling and called parties which channel to use. Emergency calls would have immediate routing, for an extra charge. The computer might enforce discipline to prevent channel hogging. For reliability the on-board computing would be done by a highly redundant configuration of microprocessors.

It is estimated that the wristphones would be about the cost of today's digital watches. They would have a radiated power of 25 milliwatts (about the same as small walkie-talkies which lack the miniaturizing features of microelectronics). The set would have a battery which would have to be recharged every night. The set could be built to weigh less than two ounces.

The designers estimate that the satellite launched with the space shuttle would cost $300 million. It could generate $520 million revenue per year, at a charge per call substantially less than today's telephone rates. If so, it appears like an attractive investment opportunity.

The project is achievable with today's technology and its designers class it as "low risk." One risk they did not assess, however, is the reaction of the FCC.

It is perhaps unlikely that we will see satellites in use for wrist telephones, but a principle of this and other futuristic designs is important. It will be practicable to make the satellites large and complex in the future and when this is done the user equipment can be small, inexpensive, and employed in vast numbers. Satellites can contain many microprocessors for control purposes.

Satellite systems are already in use for mobile military communications and have been demonstrated for use by foot soldiers and car radiotelephones. Search and rescue systems have been designed using inexpensive satellites. Climbers, yachtsmen, or cross-country skiers could carry small portable emergency units which both locate their position and permit voice communication.

Several studies have related to navigation or location-finding systems. By using two satellites far apart which relay pulses with very precise timing devices can be built which compute their exact position on earth to within a hundred meters or less. With small satellites, locating devices for vehicles could be designed. With more powerful or special-purpose satellites, pocket or wristwatch navigation devices could be used. The owner could key in the coordinates of a required destination and the device would tell him how far it is and in which direction.

A particularly valuable use of telephone sets or data devices connected directly to satellites may be in those areas of the earth that are not yet laced with terrestrial communications circuits—remote, rural, coastal, or wilderness areas, remote mines or construction projects, parts of the developing world.

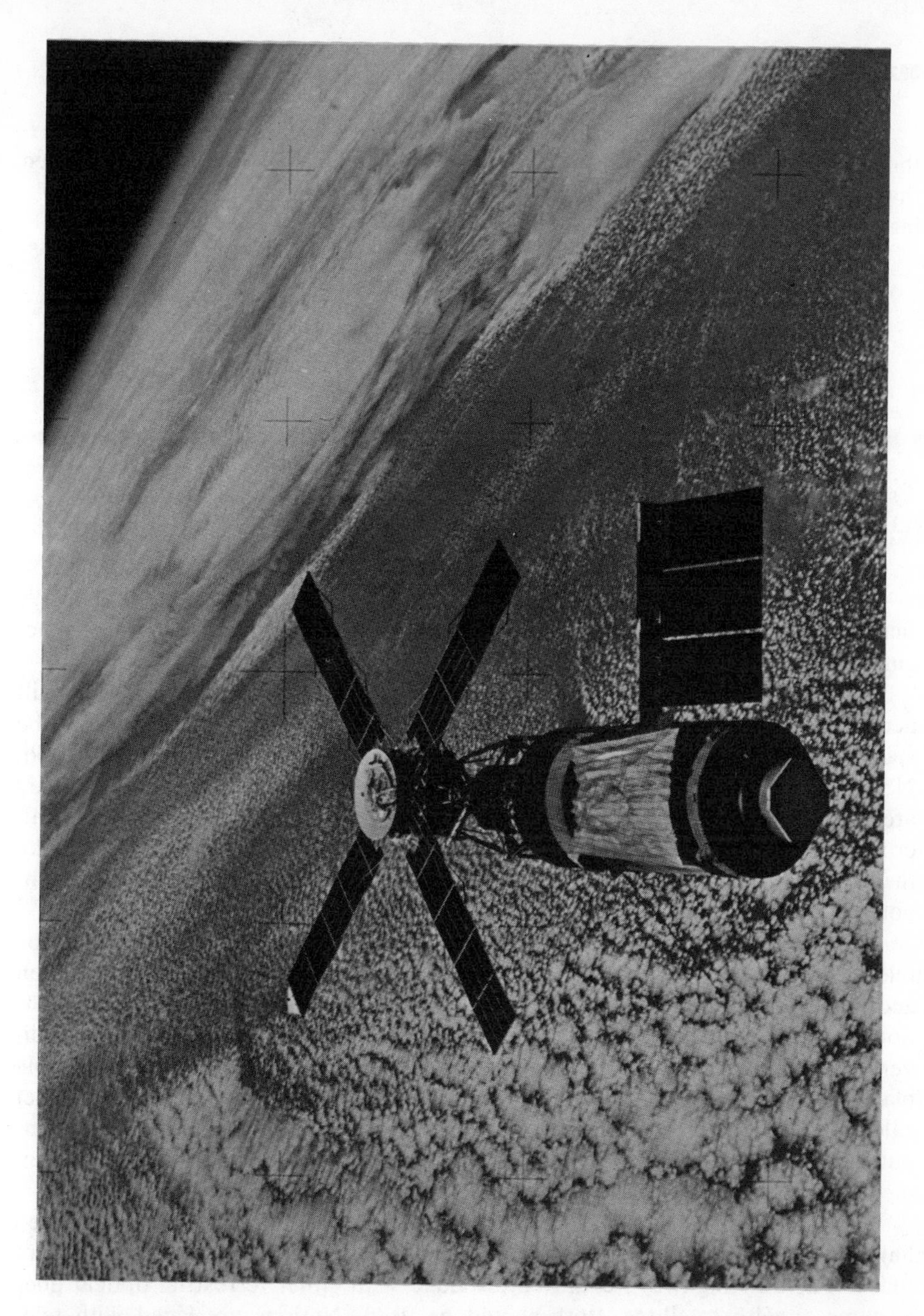

Communications satellites the size and power of Skylab (1973) could have spectacular results on earth. Skylab generated more than a hundred times the power of the SBS satellite. It could have deployed an antenna with ten thousand times the total gain. This could give much lower cost corporate telephone and data networks, cheap electronic mail, videoconferencing, videotraining and education, and varied wideband data links. A redistribution of corporate offices could result that would greatly assist in reducing petroleum consumption [6]. It is unfortunate that NASA gave up communications satellite launches after ATS 6. The world would benefit if they returned to communications satellite construction for the space shuttle era.

It is fascinating to read the original article in *Wireless World* in which Arthur C. Clarke first proposed communications satellites in 1945 [3]. He argued that it would be prohibitively expensive to build a terrestrial network for trunking television or wideband signals:

> The service area of a television station, even on a very good site, is only about a hundred miles across. To cover a small country would require a network of transmitters connected by coaxial lines, wave guides or VHF relay links. A recent theoretical study has shown that such a system would require repeaters at intervals of fifty miles or less. A system of this kind could provide television coverage, at a very considerable cost, over the whole of a small country. It would be out of the question to provide a large continent with such a service, and only the main centers of population could be included in the network.

Clarke suggested that satellites would eventually be far less expensive than terrestrial links, and the details he gave of satellites were surprisingly accurate.

Clarke was wrong in that high-bandwidth terrestrial trunks *were* built across large continents. He greatly underestimated the money that would be spent. The terrestrial networks were in place before satellites could be used. Nevertheless, the simple logic of his argument is valid. Satellites, now that they are practical, are vastly cheaper than continental networks of coaxial cable, microwave, and the new waveguide systems. The problem now is how to make the switch when enormous vested interests are committed to an older technology.

American society does not need many more telephones. Instead, it needs telecommunications to be used for new purposes: to provide better education and medical services, to save gasoline, to develop electronic mail and information services, to lessen the need for commuting into cities, to allow more diverse television and audience response, to make rural communities more viable places to work and live, to make computer terminals as inexpensive as pocket calculators to the mass public, to provide cheap data transmission and computer networks, to provide an infrastructure with which industry and government can operate efficiently with a much higher level of automation.

Many of the most valuable future uses of telecommunications need video links [6]. These need higher bandwidth communications highways than those of today. The new highways could be provided with either terrestrial optical fiber systems or with satellites. Both should be used. Nations are laced with telephone channels today; they should be laced with video channels tomorrow. It is in the building of these new facilities that we have an opportunity to escape from the patterns of the past.

Rather than merely regarding satellites as a substitute for today's telephone systems, it is important to consider new uses for telecommunications.

Telecommunications should be regarded as a future substitute for physical travel. To make it an effective substitute it is necessary to ask this question: What are the best ways for men to communicate with men at a distance? This is a complex question, and there are many different types of answers to it. Most of the answers require transmission at rates higher than that of local telephone loops, often in very brief bursts. Picturephone is one proposed answer, but methods of communication can be found for business and technical discussions which require a much lower total bit rate than Picturephone, and which are more effective because detailed documents, drawings, computer printouts, etc., can be examined, pointed at, and modified.

With rooftop antennas and the flexibility of the various digital demand assignment techniques, a wide variety of forms of communication can be intermixed. Corporate headquarters or plants can have video conference rooms at which meetings can be held with distant members. Conference rooms with many different facilities, thousands of miles apart, can be interconnected. All the types of messages in Box 13.1 could be transmitted as part of normal corporate traffic.

The typical large American corporation spends almost twice as much on air travel as it spends on telecommunications. Some of the largest corporations spend more than $100 million per year on air travel. Some of this travel is unavoidable, but for a large part of it telecommunications could form a substitute if superbly effective video, facsimile, and computer output facilities were available and all capable of being hooked into conference calls. To do so it is necessary to bypass today's telephone loops.

Newsweek [5] estimated that the West needs to spend $500 billion over a period of 10 years on the development of new energy sources if it is to escape the clutches of the OPEC cartel. When new energy sources are listed telecommunications does not normally appear on the list, yet satellite facilities appropriately used could bring a massive reduction in the consumption of petroleum products for human transportation. Roughly $100 billion will be spent on capital improvement of *terrestrial* telecommunications facilities in the U.S. alone over the next 10 years. The cost of superb satellite facilities is small compared with these figures.

As the Club of Rome, and other, models show (however inaccurately), many of the vital resources necessary for running today's industrial society are going to run perilously short. Petroleum is merely one of these resources. At the same time satellites will be spreading television to a vast world population which does not yet see television. In spite of all efforts to control it, the world population will have grown to 6 billion by the year 2000 and by then will be growing by a larger number per year than today. If the impact of satellite broadcasting is to make this vastly growing population demand the way of life of the industrial countries with their cars, plastics, steaks, and heated homes, then the world is in for a traumatic period.

An overly rapid introduction of advanced communications into societies with primitive communications is a formula for chaos. There seems little doubt that mankind is heading into a period of great upheaval. Satellites as much as any technology will contribute to this.

Satellites will give multinational corporations instantaneous global communications. Electronic fund transfer networks will be able to move vast sums of money instantaneously from one country to another. A coup d'etat in Saudi Arabia could be seen immediately on the screens in the Pentagon where military leaders are in satellite communication with the American fleets and have the ability to jam the Arab satellites. Satellite receive-only antennas will be used at the heads of the CATV cables; high-fidelity television will become a new electronics market with the bandwidths available to have wall-sized screens. World shortages of raw materials will become common as developing nations assume that their only way to wealth is to combine in cartels to increase the prices of the raw materials the West does not have. The spread of affluence will be highly selective, with a few nations and groups, such as the Arabs, becoming rich enough to cause economic chaos while much of the burgeoning world population is starving. The bewildered public will watch the effects of economic and political turmoil and world starvation on the high-fidelity wall screens.

All powerful technologies can be used for good or for ill. It is imperative that those persons who influence society should understand the new potentials that are emerging [6]. Western society will change so that the products we consume are different from today—geared to new shortages and to new technological riches. There are limits to growth in many of our existing consumer patterns. But there are no limits near in the consumption of information, the growth of culture, or the development of the mind of man. The new information channels which can access the world's data banks, film libraries, computer-assisted instruction programs, and digitized encyclopedias with built-in film clips, will become available to mankind everywhere.

REFERENCES

1. P. E. Glaser, O. E. Maynard, J. Mackovcisk, Jr., and E. L. Ralph, "Feasibility Study of a Satellite Solar Power Station," *NASA Contractor Report CR-2357,* 1974.

2. "Satellite Solar Power Station: An Option for Power Generation," briefing before the Task Force on Energy of the Committee of Science and Astronautics, U.S. House of Representatives, 92nd Congress, Second Session, Vol. II, U.S. Government Printing Office, Washington, D.C., 1974, 72-902-0.

3. Arthur C. Clarke, *Extraterrestrial Relays,* Wireless World, London, Oct. 1945.

4. Ivan Bekey and Harris Mayer, *1980–2000: Raising Our Sights for Advanced Space Systems,* Astronautics and Aeronautics, July/Aug. 1976.

5. *Newsweek,* Feb. 17, 1975, p. 25.

6. James Martin, *The Wired Society,* Prentice-Hall Inc., Englewood Cliffs, N.J., 1978.

INDEX

A

K

L

M

THE JAMES MARTIN BOOKS ON COMPUTE

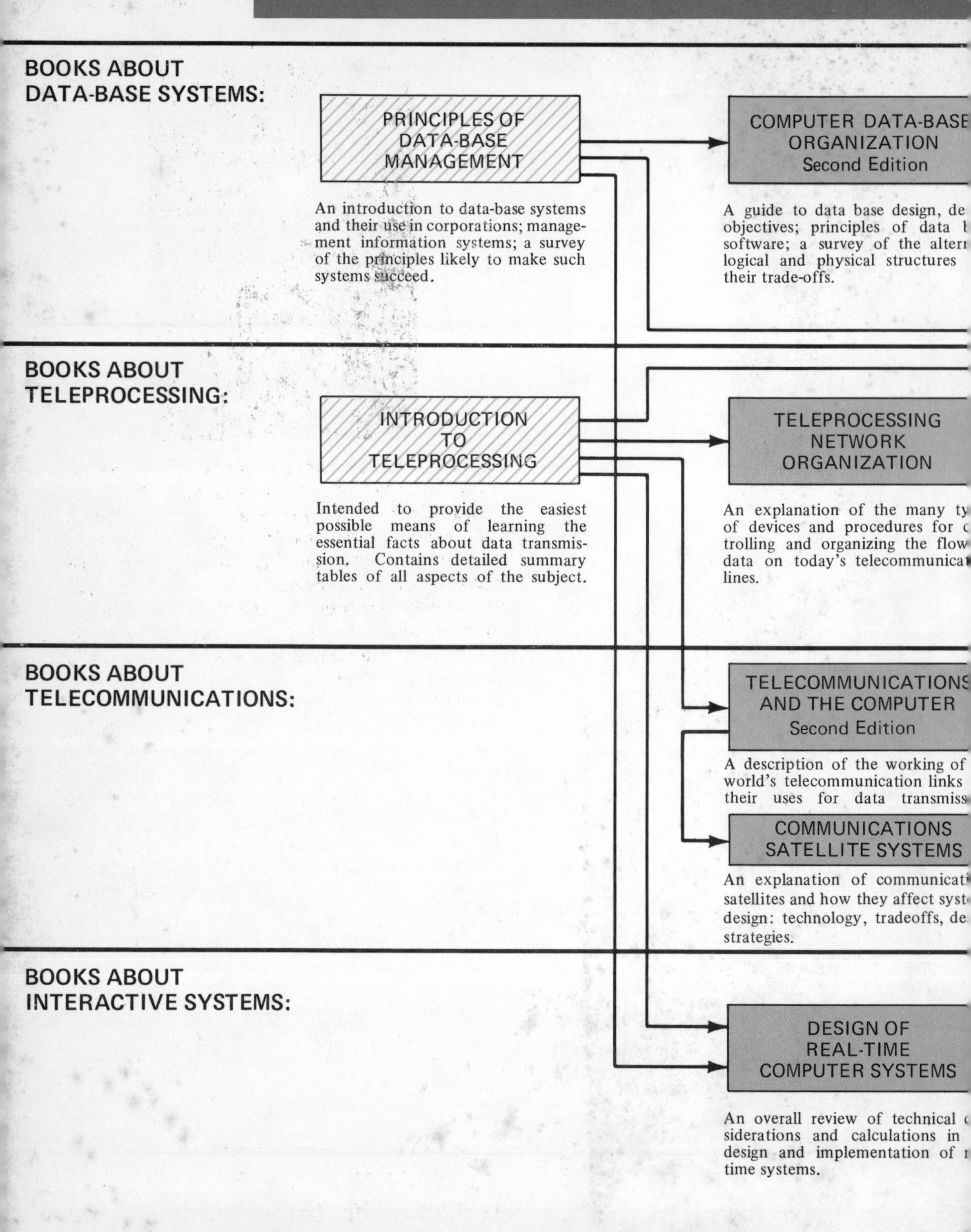
BOOKS ABOUT DATA-BASE SYSTEMS:
PRINCIPLES OF DATA-BASE MANAGEMENT
An introduction to data-base systems and their use in corporations; management information systems; a survey of the principles likely to make such systems succeed.
COMPUTER DATA-BASE ORGANIZATION Second Edition
A guide to data base design, de objectives; principles of data b software; a survey of the altern logical and physical structures their trade-offs.
BOOKS ABOUT TELEPROCESSING:
INTRODUCTION TO TELEPROCESSING
Intended to provide the easiest possible means of learning the essential facts about data transmission. Contains detailed summary tables of all aspects of the subject.
TELEPROCESSING NETWORK ORGANIZATION
An explanation of the many ty of devices and procedures for c trolling and organizing the flow data on today's telecommunica lines.
BOOKS ABOUT TELECOMMUNICATIONS:
TELECOMMUNICATIONS AND THE COMPUTER Second Edition
A description of the working of world's telecommunication links their uses for data transmiss
COMMUNICATIONS SATELLITE SYSTEMS
An explanation of communicati satellites and how they affect syst design: technology, tradeoffs, de strategies.
BOOKS ABOUT INTERACTIVE SYSTEMS:
DESIGN OF REAL-TIME COMPUTER SYSTEMS
An overall review of technical c siderations and calculations in design and implementation of r time systems.

KEY: **INTRODUCTORY BOOKS** These books are an easy-to-read introduction to the subjects. **DETAIL BOOKS**